Ultrasound in Food Processing

Ultrasound in Food Processing

Edited by

MALCOLM J.W. POVEY
Proctor Department of Food Science
University of Leeds
UK

and

TIMOTHY J. MASON
School of Natural and Environment Sciences
Coventry University
UK

BLACKIE ACADEMIC & PROFESSIONAL
An Imprint of Chapman & Hall
London · Weinheim · New York · Tokyo · Melbourne · Madras

Published by Blackie Academic & Professional, an imprint of Thomson Science, 2–6 Boundary Row, London SE1 8HN, UK

Thomson Science, 2–6 Boundary Row, London SE1 8HN, UK

Thomson Science, 115 Fifth Avenue, New York, NY 10003, USA

Thomson Science, Suite 750, 400 Market Street, Philadelphia, PA 19106, USA

Thomson Science, Pappelallee 3, 69469 Weinheim, Germany

First edition 1998

Thomson Science is a division of International Thomson Publishing I(T)P®

Typeset in 10/12 pt Times and produced by Gray Publishing, Tunbridge Wells, Kent

ISBN 978-0-7514-0429-6

A catalogue record for this book is available from the British Library

Library of Congress Catalog Card Number: 97-76697

∞ Printed on acid-free text paper, manufactured in accordance with ANSI/NISO Z39.48-1992 (Permanence of Paper)

Contents

List of contributors

D. Cartwright	Cygnus Instruments Limited, 30 Prince of Wales Road, Dorchester, Dorset DT1 1PW, UK
R.G. Earnshaw	Microbiology Department, Campden and Chorleywood Food Research Association, Chipping Campden, Glos. GL55 6LD, UK
J.A. Gallego-Juárez	Instituto de Acustica, Consejo Superior de Investigaciones Cientificas, Calle Serrano No. 144, Madrid 28006, Spain
B.S. Hoyle	University of Leeds, Department of Electronic and Electrical Engineering, Leeds LS2 9JT, UK
T.G. Leighton	Institute of Sound and Vibration Research, Fluid Dynamics and Acoustics Group, University of Southampton, Highfield, Southampton SO17 1BJ, UK
D.J. McClements	University of Massachusetts at Amherst, Department of Food Science and Nutrition, College of Food and Natural Resources, Amherst, MA 01003, USA
T.J. Mason	Coventry University, School of Natural and Environmental Sciences, Coventry CV1 SFB, UK
M.J.W. Povey	University of Leeds, Procter Department of Food Science, Leeds LS2 9JT, UK

V.J. Pinfield Technical Development Department, Cadbury Ltd, PO Box 12, Bournville, Birmingham B30 2LU, UK

T. Quartly-Watson Production Line Cleaning, High Road, Bressingham, Diss, Norfolk IP22 2AT, UK

M.M. Robins Institute of Food Research, Norwich Research Park, Colney Lane, Norwich NR4 7UA, UK

F.F. Rawson FFR Ultrasonics, Queniborough, Leicestershire, UK

J. Ridgway Faculty of Engineering and Computing, Nottingham Trent University, Burton Street, Nottingham NG1 4BU, UK

E.S. Tarleton Loughborough University, Department of Chemical Engineering, Loughborough, Leicestershire LE11 3TU, UK

Preface

In general terms ultrasound has two main applications in the food industry: at low power and high frequencies it can be used as an analytical tool, and at high power and low frequencies it can assist processing. The use of ultrasound in analysis dates back to 1917 when Langevin suggested a method for the detection of icebergs using a pulse/echo technique which was later adapted for the estimation of depths in water. Many dramatic improvements have occurred since then in both the technical application and theoretical background to these methods. This has led through ASDIC and SONAR to the current use of high-frequency ultrasound in the 1–10 MHz range in the field of non-destructive testing and medical imaging. In other areas there have also been major improvements so that today a significant role for diagnostic ultrasound has been identified in food technology. Paralleling the development of diagnostic ultrasound has been the increasing interest in the uses of power ultrasound for processing (using the lower frequency range of 20–100 kHz). At these lower frequencies the sound energy through a liquid can be made sufficiently large to cause cavitation and hence produce substantial physical and chemical changes in the system. The history of power ultrasound in processing can be traced as far back as 1927 when it was reported that ultrasound was extremely efficient for the production of an oil and water emulsion. Its applications have broadened considerably and many of these are in the chemical and related industries and have been grouped together under the generic term of sonochemistry.

We believe that ultrasound is set to make a considerable impact on the food industry over the next decade. This book covers the uses of ultrasound in both low-power non-destructive measurement techniques and high-power material-altering techniques. This work stems from a conference on Ultrasound in Food Processing, held at Leeds in the late summer of 1996. The Institute of Physics, The Royal Society of Chemistry Sonochemistry Group and the Institute of Food Science and Technology sponsored the conference. The conference contributors were invited to expand their contributions to a length appropriate for a book chapter.

Topics in high-power ultrasound range from sterilization and decontamination including the threat to water safety posed by cryptosporidiosis, through cutting and defoaming to crystallization and drying. In the case of low-power ultrasound many of its applications in remote sensing are addressed. An important aspect of low-power ultrasound is the ease with which it may be integrated with other sensor modalities. This may be important in two ways. First, in enhancing existing process control strategies and second by improving understanding of the process itself. Ultrasound should be viewed in the light of the rapid development of other sensing techniques such as Fourier-transform-infrared and nuclear magnetic resonance. However, these other techniques have their limitations which the strengths of ultrasound, particularly with regard to the ease with which it may be integrated into existing processes, may complement. The lack of appropriate sensor techniques is certainly inhibiting the adoption of advanced process control strategies such as fuzzy logic, model based and neural networks. Perhaps the most remarkable development is the advent of ultrasound particle sizing, with its ability to size particles *in situ* and in-line.

No one concerned with the future of food processing can afford to ignore the rapidly developing field of ultrasound.

MALCOLM J.W. POVEY AND TIMOTHY J. MASON

1 Controlling overfilling in food processing

J. RIDGWAY, K.S. HENTHORN AND J.B. HULL

1.1 Introduction

World-wide interest in the environment has prompted significant commercial interest in the elimination and disposal of waste products and hazardous materials. In addition, despite less publicity in the media, there has been increasing research into the elimination of waste that results from inefficient processing. In food processing, considerable waste produce arises in the form of 'give-aways', in overfilled containers, or through 'throwaways' which are non-conforming underfilled products that end up in rubbish bins. In this work we have investigated the use of ultrasound monitoring, as a means of on-line measurement of fluid levels in food containers.

Ultrasound monitoring can be an accurate measurement technique which is capable of being combined with an appropriate control strategy to provide the most suitable method of reducing or eliminating the waste problem. The objective of the research described in this chapter was to investigate controller design methodologies for a bottling process. The bottling process was modelled using MATLAB® and SIMULINK® packages, and controller designs were implemented to improve the response characteristics of the process. Both digital control and fuzzy logic controller design methods were investigated. The digital controller was designed using the direct digital design method and resulted in an improved system performance. However, when different carousel delays were considered, the performance varied significantly. Fuzzy logic control operates with a predetermined set of rules and has been shown to function very well with a bottling system. Bottle fill times and the resulting stability with respect to changing carousel delays were improved. Available data indicate that fuzzy logic can be used efficiently and effectively, and could provide a valuable asset to the food processing and allied industries.

Food and drink production represents the largest manufacturing sector in the UK economy, and food and drink companies are among the largest UK exporters (£8.2 bn in 1995). The industry makes essential contributions to UK wealth creation and to the quality of life within the UK, and adds value at each of the major stages of the supply chain. Around 45 000 companies are involved in food processing and production in the UK, and 90% of these are classed as small to medium enterprises (SMEs) with less than 250 employees. Most SMEs possess little, if any, research facilities or expertise.

The UK government's Technology Foresight Panel on Food and Drink, which reported in 1996, focused on scientific and technological priorities, and stressed the need for increased investment in R & D to maintain the UK in the first division of food manufacture. Emphasis has been placed on collaborative research with government departments working closely with each other and with industry and academia. There is significant interest in new and innovative research in raw materials, packaging, the supply chain, retailing, safety and processing.

Food processing represents a field of rapid development and implementation of new approaches in process technology, in particular, automation systems, quality procedures, efficiency improvements and sensor systems for in-process monitoring and control. However, until recently, food and drink processing companies have lagged behind in the use of high-technology measurement instrumentation, and most of them employ inaccurate bulk measurement techniques in virtually all areas of process control. Advances in technology and increasing customer awareness have forced manufacturers to rethink their strategies. As a result, they have embarked on the task of streamlining the production processes for improved efficiency as well as quality. Increased competition has also played a major role in this. Technological advancement in sensor design has had an impact on the available measurement techniques for food process engineering, and research has been carried out to develop sensors aimed at establishing systems that can cope with the rapid improvements in processing technology, especially in the area of process control.

UK legislation sets extremely stringent requirements on the volume of liquid in containers destined for the retail market (Trading Standards Service, 1972). In general, food processing companies overfill containers rather than face the dual prospects of legal problems and the damaging adverse publicity that would result from underfilled containers which reach the marketplace. Overfilling of containers results in substantial lost revenue. The UK dairy industry alone currently loses revenue of around £12 million per annum (Hull *et al.*, 1994, 1995a). Within the European food and drink industry, this loss amounts to hundreds of millions of pounds every year.

To overcome this problem, efficient, intelligent monitoring technology and intelligent control systems capable of adapting to a wide variety of industrial environments and process conditions need to be devised and developed. Bottle-filling processes are dependent on the accurate monitoring of level fill and feedback control. Recent research has investigated the application of ultrasound techniques coupled to a PID (proportional–integral–derivative controller), as a method of reducing waste (Zheng *et al.*, 1995; Hull *et al.*, 1995b). The aim of the current research was to extend the investigation of controller design methodologies to include both digital control and fuzzy logic control, in order to establish further improvements to the response characteristics of the processing operations. In order to achieve this aim, a simple bottle-filling process has been modelled using both the MATLAB and SIMULINK software packages, and the various controller designs have been implemented.

1.2 Elements of a bottling process

Bottling machines are designed to automate the filling of glass and plastic containers to produce carbonated drinks, mineral water, beer and wine products. The bottles (transported on a conveyor belt) are spaced out from each other by means of a worm before being taken up by a holder and cleaned. Afterwards, an inlet star that positions them on a carousel takes them up. The bottles are then successively raised towards a filling valve. The filling starts after the bottle has been raised from the filling valve bracket, and the top has touched the opening control. The filling ends when the product in the bottle has reached the end of the spout, therefore setting the level. The filling valve is then set mechanically closed by the 'closing control system' which operates the throttle.

After filling, the bottles are then taken by a star which positions them under the crowner or capper where they are sealed. Another star positioned at the end of the capper picks the bottles and places them on the conveyor belt, where they are again sprayed with water on the outside, then transported to the labeller.

1.3 Ultrasound monitoring

1.3.1 Transducer design

A major requirement for an on-line ultrasonic monitoring system is the design of an effective transducer set-up. A number of design studies have been carried out in order to achieve an optimum combination of transducer and couplant assembly. An approach that allows a flexibility of installation is shown in Figure 1.1. Couplant is pumped through a central vent at the transducer core, and it emerges at the interface between the transducer and the container.

1.3.2 Design implementation

A peristalsis-type pump could be used to circulate the ultrasonic fluid to the transducers (see Figure 1.2). Through the cam action, the fluid within the pipe is pushed forward while the cam's 'rise' (elliptical portion) is in contact

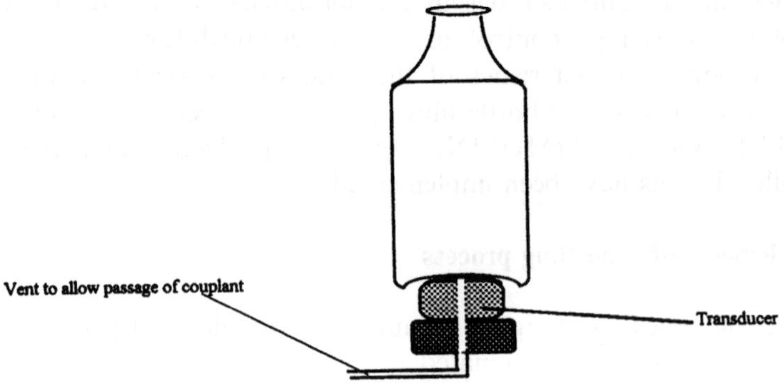

Figure 1.1 Transducer system with couplant pumped though central vent.

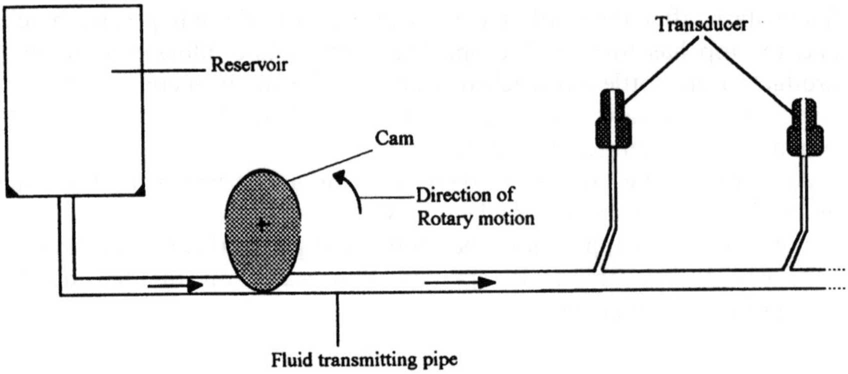

Figure 1.2 Cam action of the peristalsis pump.

with the pipe. During the period when the 'dwell' (circular portion) is in contact with the pipe as a result of the pressure difference between the two sections of the tube, more fluid is allowed into the evacuated section, after which it is pushed further on in the next cycle of the cam.

The ultrasonic fluid could be stored in a reservoir having a constant head to ensure a constant and equal supply of the same to the transducers at any time of their operation.

1.3.3 Ultrasonic couplants

Various fluids and gelatinous solutions may be used that perform to an accepted level of efficiency, the task of transmitting sonic vibrations from the source to the medium under investigation. This may be due to the fact that their acoustic impedance is well matched to the characteristic impedance of the transmitting/receiving transducer. It is more likely due to the ability of the couplant fluid to exclude air from the transducer–sample interface, thereby eliminating the large mismatch due to a solid–air–solid interface.

Acoustic impedance is the characteristic that determines the amount of reflection of ultrasonic waves on the boundary surface of two media, and is a product of density and velocity.

1.3.4 Ultrasound tests

A number of laboratory experiments were carried out to establish the viability of ultrasonic techniques in level measurements. The technique employed was *pulse echo* monitoring of level/level change in soft drink bottles using the experimental set up shown in Figure 1.3.

Despite relatively high attenuation of the ultrasound pulse arising from the large distance traversed in fluid (>400 mm), and the presence of bubbles within the liquid it was possible to attain a measurement accuracy of ≈ 0.1 mm for a nominal fill level in 1-litre bottles. This is well within the requirements for a practical on-line monitoring system.

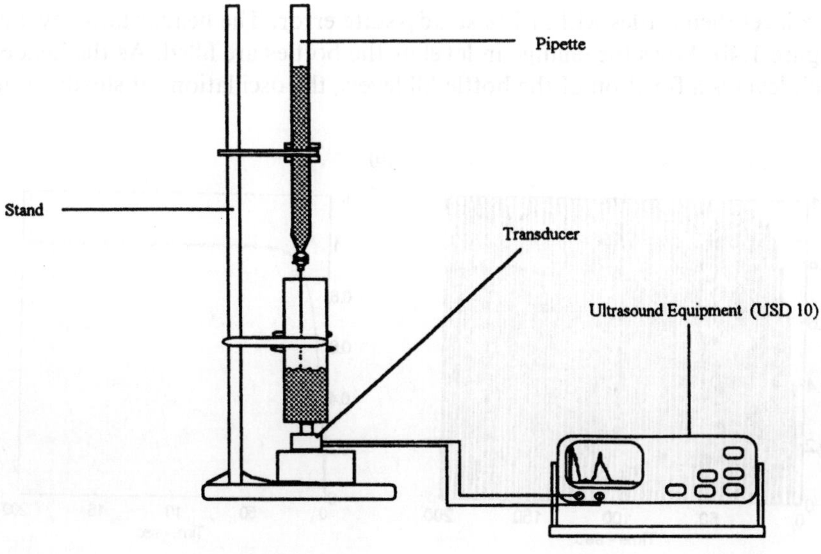

Figure 1.3 Ultrasound experimental set-up.

1.3.5 Ultrasound simulation

Fluid when entering a bottle under filling conditions undergoes turbulence effects that can cause inaccurate results with the ultrasound monitoring equipment. A solution to this problem is to place the ultrasound monitoring equipment further along the filling process. This movement of the sensor produces a system time delay (T), an appropriate name for this is 'the conveyor and carousel delay'.

1.4 Controller design

1.4.1 Three-term controller

Many industrial processes are controlled using a PID controller. Normally a PID controller is used because of its robust performance in a wide range of operating conditions. To implement this type of controller three parameters must be determined: proportional gain, integral gain and derivative gain. However, previous research on the bottling plant investigated only proportional gain (Zheng *et al.*, 1995; Hull *et al.*, 1995b). The integral and derivative gains were set to zero. Another variable, inherent in the process dynamics is, the carousel and conveyor delay time (T), i.e. the time between the filling operation and measurement of the fill level using ultrasound. T was varied to investigate the effects on the system performance. The bottle fill level plots and header tank level plots show the difference that the delay makes.

When $T = 0$ the bottle fill levels reach the desired level in around 50 s. The level then settles within 1% steady-state error. The header tank level in Figure 1.4b shows the change in level as the bottles are filled. As the header tank level is a function of the bottle fill levels, the oscillations at steady-state

(a) (b)

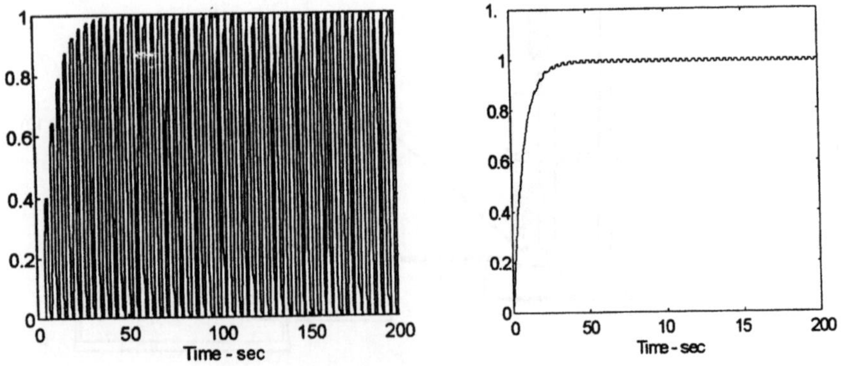

Figure 1.4 (a) Bottle fill level ($T = 0$). (b) Header tank level ($T = 0$).

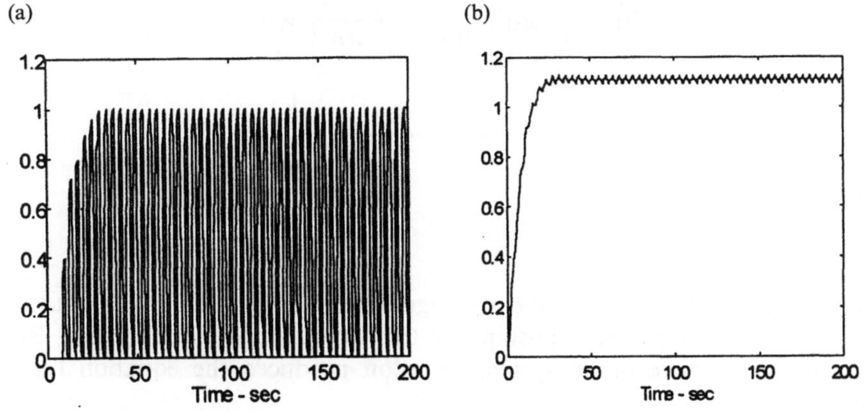

Figure 1.5 (a) Bottle fill level ($T = 4$). (b) Header tank level ($T = 4$).

show where the bottle fills take place at 4-s intervals. When the delay was synchronized with the monitoring equipment delay of 4 s the response was even faster as shown in Figure 1.5a. The time to reach the desired level was reduced to 35 s with a steady-state error within 1%.

When the conveyor and carousel delay was 4 s the header tank level fluctuated more aggressively when steady state was reached, compared to when $T = 0$. As shown in Figure 1.5b the header tank settled slightly higher than the set point. This is a natural effect that is associated with system time delays. The response plots were stable for the PID controller over the simulation period of 200 s.

1.4.2 Digital design

Elements of direct digital controller design:

- determine open loop transfer function of bottling plant
- choose desired closed loop transfer function
- eliminate unstable poles or zeros
- determine controller transfer function.

In order to use a digital controller the process needs to be broken down into discrete phases. This involves changing the continuous *s*-domain terms to a digital *z*-domain. The relationship between the *s*-domain and the *z*-domain is that s maps to z with the relationship, $z = e^{sT}$. In addition, the delay was maintained as a variable within the digital transfer function as T. To analyse the bottling plant model for a digital controller, the plant was expressed as an open-loop transfer function. This enabled the bottling model to be broken down into a simple block diagram as shown in Figure 1.6. Components of the diagram are the bottling process [$G(z)$] and digital

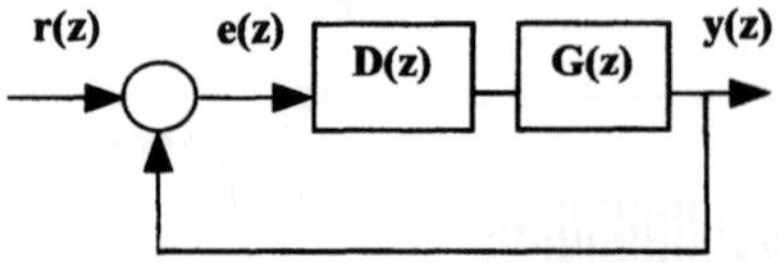

Figure 1.6 Closed-loop system.

controller [$D(z)$]. By using control design techniques the block diagram was reduced to a simple relationship for the closed-loop system [$H(z)$]. Rearranging the closed-loop system equation produced the equation for a digital controller

$$D(z) = \frac{1}{G(z)} \cdot \frac{H(z)}{1 - H(z)} \tag{1.1}$$

1.4.2.1 System model reduction. Variables used to break down the process from the block diagram to a single transfer function are:

- header tank level [$h_1(z)$]
- bottle fill level [$h_2(z)$]
- error signal [$e(z)$].

A relationship was required relating the bottle fill levels to an input signal so that direct digital design can be applied. By reduction techniques a relationship was produced between $h_1(z)$, $e(z)$ and $h_2(z)$. A continuation of the reduction gave a relationship between $h_2(z)$ and $e(z)$ by eliminating $h_1(z)$. The open loop transfer function is shown in Figure 1.7.

1.4.2.2 System stability of discrete systems. Poles are roots of the denominator of a transfer function, zeros are roots of the numerator of a transfer function. A pole or zero that lies outside the z-domain unit circle

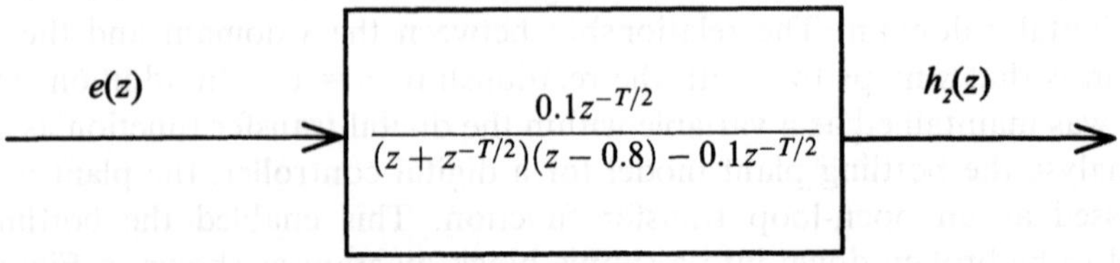

Figure 1.7 Bottle fill level transfer.

will cause unstable results. The only exception to this rule is when a zero lies between 0.5 and unity on the negative real axis. During this period a 'ringing pole' may occur within the digital controller.

For the open loop transfer function to be 'proper', the number of poles must be greater than the number of zeros, and then the system will be causal and physically realizable. An excess of poles over zeros must be maintained with the closed-loop transfer function $H(z)$, also $1-H(z)$ must maintain an equal amount of poles and zeros. This will ensure that any transport delays in the system are maintained.

1.4.2.3 Conveyor and carousel delay time. From the open-loop transfer function values, T was inserted into the relationship, depending on which delay time was being investigated. Time delays used were 0, 2 and 4 s. Each delay gave rise to a different process transfer function [$G(z)$] with varying pole locations. As the denominator of $G(z)$ varied from a second- to a sixth-order polynomial expression, MATLAB was used to determine the pole locations in the z-domain. Stability of $G(z)$ was investigated in combination with the location of the plant poles.

1.4.2.4 Direct digital design. Direct digital design evaluates a controller that cancels the bottling plant dynamics, that is a controller with zeros to cancel plant poles and with poles to cancel plant zeros.

The system dynamic characteristics can be determined by the poles of $H(z)$. The bottling process dynamics are fast compared to the slow sample rates so dead beat control was a realistic aim. With dead beat control the system will settle exactly in a number of samples equal to the order of the system. For a dead-beat response all the closed-loop poles of $H(z)$ are located at $z = 0$.

To satisfy the requirement of dead beat response, the poles of $[1-H(z)]$ were placed at $z = 0$. The number of poles was determined by the number of zeros to satisfy the necessary physical conditions. When the open-loop transfer function had been evaluated with a value for the delay, T, it was clear which poles were unstable. These unstable poles were included as zeros of $[1-H(z)]$. The digital controller transfer function can be expressed by using the equation for $D(z)$ as shown in equation (1.1).

1.4.2.5 Controller calculation and implementation. Determination of the digital controller transfer function can be achieved by manual calculation for low values of T. When T was increased to 2 s and above, the calculations became very lengthy. The Codas software package (Golten and Verner, 1991) was used at this stage in the design. Codas determined the unstable poles of $G(z)$ and requested the operator to enter the desired closed-loop system characteristic equation. Each controller was designed with dead-beat control.

The controller transfer functions were then calculated by the software package. To check the accuracy of Codas, the controller for $T = 0$ s was calculated manually, and compared to the controller designed by Codas. Results of the comparison were that the manual controller design was exactly the same as the software developed controller. Controller designs are entered into the SIMULINK model of the bottling process, replacing the PID controller.

1.4.2.6 Response plots. The response plots that resemble stability when steady state is reached, attain a bottle fill level of 2 when the step input was unity. This is due to a direct current gain that is inherent in the system. For research purposes the step input remained at unity, but if the problem was to be overcome then the step input would be reduced to 0.5 to reduce the bottle fill level.

Controller sample rates can be entered into the digital controller block in the SIMULINK model. Different sample times were investigated and response plots examined. Stable results were only achieved with a sample rate equal to the zero-order hold delay of 2 s.

SIMULINK can use different methods of approximation to find solutions to system models. The Runge–Kutta approximation (Cavallo *et al.*, 1996) was used to evaluate the response plots, as it gave the best results. Runge–Kutta uses a third-order method of evaluating ordinary differential equations with second-order step-size control. Simulation time was 200 s as this represents a fair judgement to the system response.

1.4.2.7 Response when $T = 0$ s. The following open loop pole locations exist on the *z*-domain:

- -1.0539
- $+0.8539$.

Figure 1.8a shows that when the conveyor and carousel delay was 0 s the process settled in approximately 40 s. In comparison to the PID controller with the same delay time, the digital controller had significant improvements. After two bottle fills the output peaked with overshoot, this is due to dead-beat response.

1.4.2.8 Response when $T = 2$ s. The following open loop pole locations exist on the *z*-domain:

- $-0.0288 \pm 1.1024j$
- 0.8576.

The header tank level shown in Figure 1.9b gradually became unstable after 15-s simulation time. This reflects bottle fill levels, since the level drops to unusable levels as time increases.

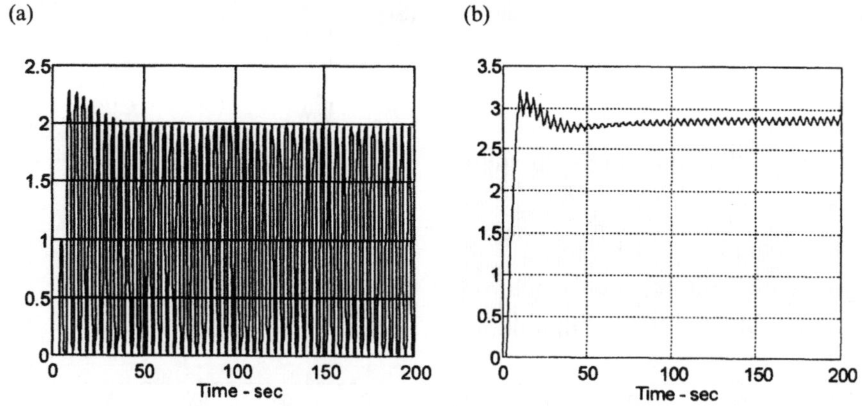

Figure 1.8 (a) Bottle fill levels ($T = 0$). (b) Header tank level ($T = 0$).

1.4.2.9 Response when $T = 4$ s. The following open loop pole locations exist on the *z*-domain:

- $0.4785 \pm 0.8932j$
- -1.0180
- 0.8610.

As with the PID controller, when the conveyor and carousel delay is synchronized with the measuring equipment at 4 s, output response is stable and fast. Settling time is greater than 50 s.

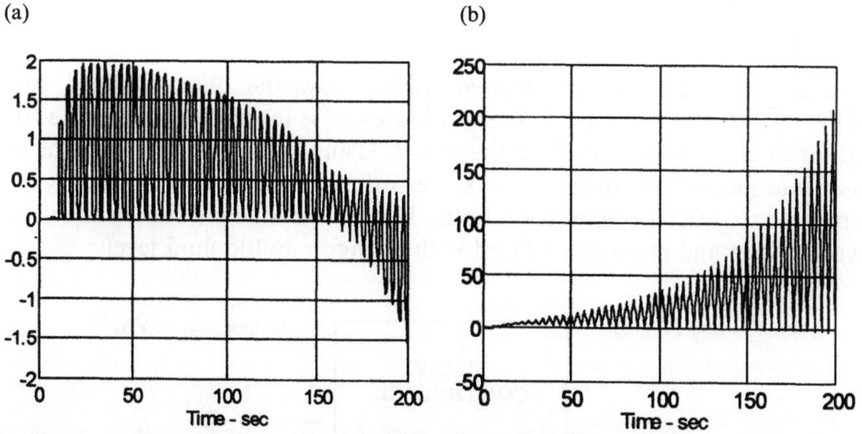

Figure 1.9 (a) Bottle fill levels ($T = 2$). (b) Header tank level ($T = 2$).

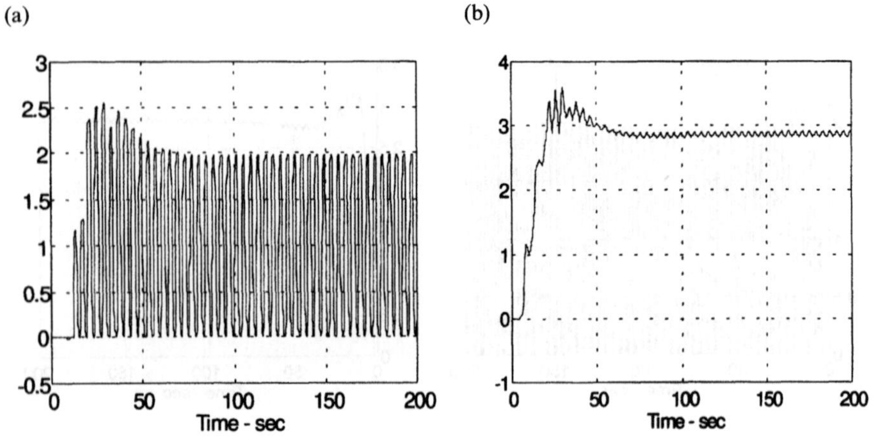

Figure 1.10 (a) Bottle fill levels. (b) Header tank level ($T = 4$).

1.4.3 Fuzzy logic control

1.4.3.1 Fuzzy controller design. To design logic for the controller fuzzy set theory was used. The valve positions and fluid levels are defined in sets and have degrees of membership to that particular set. The input variable being controlled was the bottle level, and the output is the valve position, so the sets were named as LEVEL for the input and VALVE for the output. See Figure 1.11.

After defining the fuzzy rules and both membership functions for the output and input, the controller was then developed. Using a personal computer the fuzzy data were entered into a fuzzy inference system editor (FIS editor). Data were entered in the following order:

- input fuzzy variables
- fuzzy rules
- output fuzzy variables.

The controller was investigated to see how the output varied with different input conditions. A 'fuzzy rule viewer' is incorporated into the FIS editor to view changes made to the inputs. Using a scroll bar on a graphical-user interface (GUI), different x-axis positions indicated how the filler valve operated to different input conditions. This was used to check that the valve would open and close when faced with changing bottle fluid levels.

Figure 1.11 Fuzzy controller.

The FIS editor 'fuzzified' the inputs, then combined the inputs with the rules and the output conditions. After this, the software 'de-fuzzified' the processed data to produce a fuzzy controller in matrix format.

1.4.3.2 Fuzzy controller testing in SIMULINK. The matrix representation of the controller was saved in the MATLAB environment as a variable. SIMULINK included a fuzzy controller block, with the MATLAB variable inserted. After implementing the controller block in to the bottling process SIMULINK model, tests on the system response were carried out.

Results from the SIMULINK output were originally very poor. The fuzzy controller did not help the process settle within a reasonable time span, and the steady state error was too large for practical application. Even with varied conveyor and carousel delay times the control response remained undesirable. Experimentally it was found that to reduce the steady-state error and increase the response time an extra gain was needed to be included in the system. A PID controller with proportional gain was included in-line with the fuzzy logic controller. This provided satisfactory results with a proportional gain of 2.2, with the integral and derivative gains set to zero.

1.4.3.3 Fine tuning of the controller. To achieve the best results, the fuzzy controller needed fine-tuning. This involved changing the position of the input membership functions by minute amounts. Large changes to the membership functions produced erratic results with the system response, but fine changes increased performance. The controller was tuned by trial and error.

1.4.3.4 Delay times. Changes to the delay time *T* gave interesting results. In comparison to the results obtained from the digital controller, the fuzzy controller proved to produce more stable results. Varying the delay time changed the output response of the system. To compensate for the variations different values of proportional gain needed to be included in the PID controller. Along with the gain variations, changes can be made to fine-tune the fuzzy controller to different situations. This is further work to be investigated with the bottle filling system.

1.4.3.5 The fuzzy response plots. The response plots were produced with the proportional gain in the PID controller. By changing the proportional gain and adding integral and differential gains the response plots altered significantly. For the purpose of this research the proportional gain remained constant. When testing the system for different responses, the only variables changed were the conveyor and carousel delay, and fuzzy membership plot positions.

The optimum response plot was for a conveyor and carousel delay time of 4 s. Membership plots were trimmed to suit this performance. Other delay times of 0, 1, 2 and 8 s were tested with the same optimized fuzzy controller that worked well for the 4-s delay. By adjusting the fuzzy membership parameters it would be possible to improve response characteristics for other delay times. All results with the fuzzy logic controller were produced using SIMULINK. The integration method used for modelling the process was 'Linsim'. Simulations were run over a period of 200 s as this gave a fair estimation of the process in steady state.

1.4.3.6 Response when $T = 0\,s$. With no ultrasound monitoring delay, the bottle fill level settled after 80 s, see Figure 1.12a. Fill levels settled with exactly 0.04% steady-state error. In comparison to the PID controlled process with no delay the steady-state error is much lower. The header tank level settled with oscillations that have a small amplitude as shown in Figure 1.12b.

1.4.3.7 Response when $T = 2\,s$. With a delay of 2 s, neither the bottle fill level nor the header tank level settled within the 200-s simulation time (see Figures 13a and b).

1.4.3.8 Response when $T = 4\,s$. When the delay time was synchronized with the monitoring equipment the response with the bottle fill level was very fast. As shown in Figure 1.14a, the process had reached steady-state after five bottle fills and 25 s. The fill level is very accurate with a 0.02% steady-state error. Figure 1.14b shows the header tank level that oscillated with very small amplitude in steady-state. This small oscillation contributes to the fast response of the bottle fills. In comparison to the PID controlled system with the same time delay of 4 s, the fuzzy controller produced a much quicker response saving 10 s on the process start time.

(a) (b)

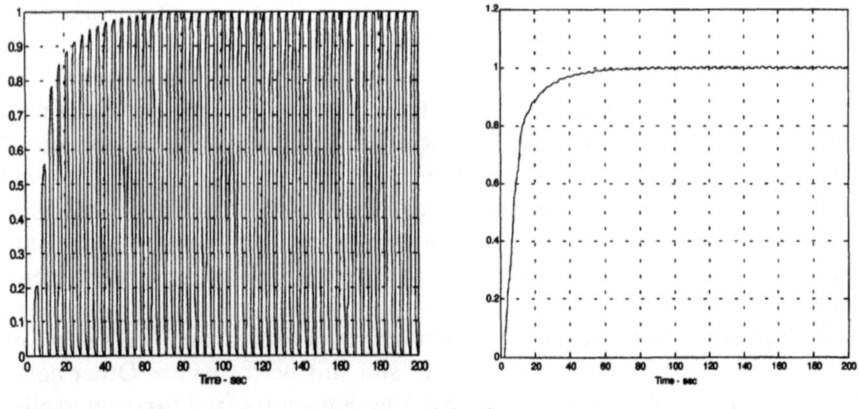

Figure 1.12 (a) Bottle fill level. (b) Header tank level.

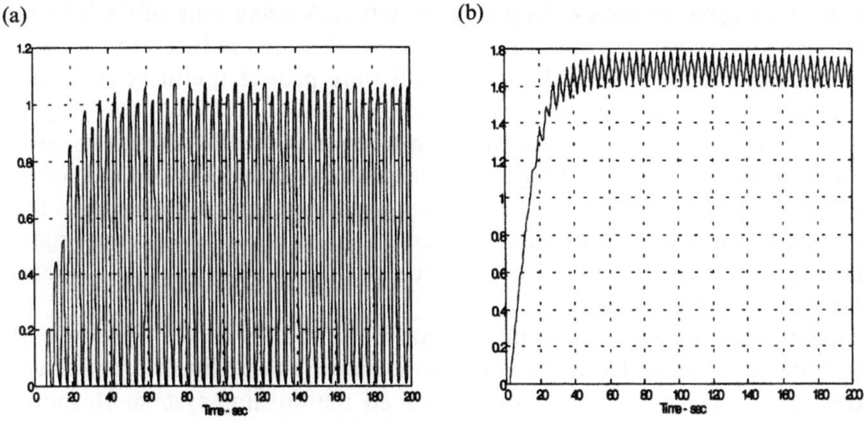

Figure 1.13 (a) Bottle fill level. (b) Header tank level.

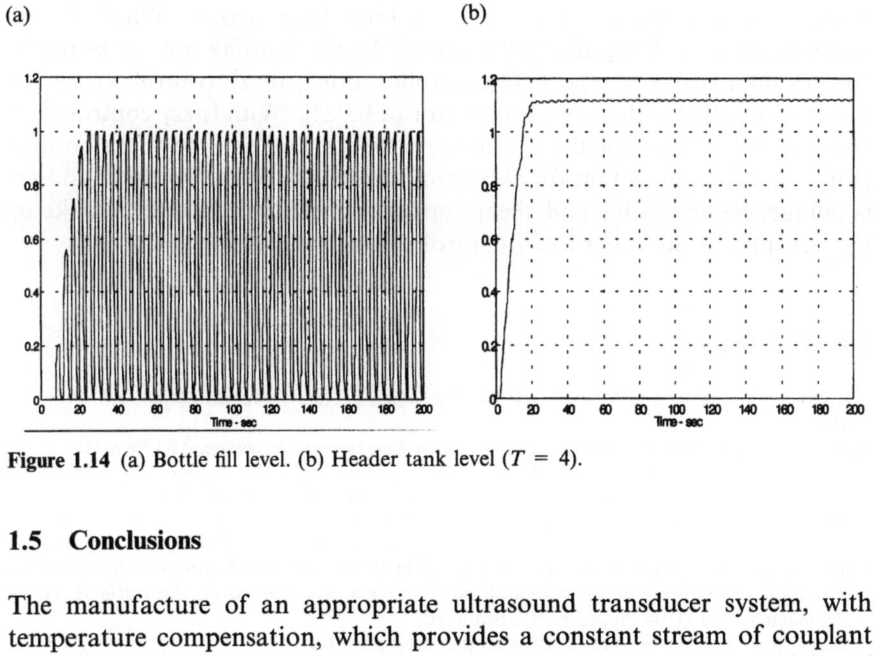

Figure 1.14 (a) Bottle fill level. (b) Header tank level ($T = 4$).

1.5 Conclusions

The manufacture of an appropriate ultrasound transducer system, with temperature compensation, which provides a constant stream of couplant (water) to the interface between the transducer and bottle would be of significant benefit to the success of a practical ultrasound level monitoring device. The versatility of the approach would provide on-line data for monitoring transient changes in density, temperature and possibly composition.

The PID control originally gave very good results with system response. This needed to be improved to remain competitive in a modern market, and has been achieved by use of digital and fuzzy control. However, when conveyor and carousel delay times T were introduced the response either became unstable or took a long time to settle. A delay time of 0 s led to a settling time of 50 s with the PID controller. This was reduced to 40 s with the digital controller. The digital control only proved useful at this stage with no conveyor and carousel delay time.

The bottling process was divided into discrete phases by using two zero-order holds, which meant the system used multisampling. This was the cause of the observed problems with the direct digital design method. Codas had problems calculating controller transfer functions due to multisampling of the plant. Therefore the only controller design that gave a useful response was with a zero time delay T.

Fuzzy logic controller design presents a radically different approach to the bottling process. Design was developed with an intelligent approach to solving the problem, using rules that suited the bottling system. Response times for the fuzzy controller were very impressive. Conveyor and carousel delay time did not make a drastic change to the system response, except for a delay of 2 s, which settled after a long time period. When T was harmonized with the regular pulse rate of 4 s, the bottling process settled in 25 s. In comparison to the PID controller alone, fuzzy control was much faster and settled with a very small error of 0.02%. With fuzzy control there was a 10-s time saving with little error on the starting up time of the bottling plant. The advantage that fuzzy control would give to the bottling industry is unique. Being flexible and cheap, opportunities are available to make up lost revenue by adopting fuzzy control strategies.

1.6 References

Cavallo A., Setola R. and Vasca L. (1996) *Using MATLAB, A Practical Approach*, Prentice-Hall.

Golten, J. and Verner, A., (1991) *Control System Design and Simulation*, McGraw-Hill.

Hull J.B., Muumbo A.M. and Whalley R. (1995a) Controlling waste in food processing. *Proc. Conf. Strategies for Control and Management of Waste*, pp. 35–48, Bradford.

Hull J.B., Henthorn K.S. and Muumbo A.M. (1995b) Controlling waste in the food processing industry. *Advances in Materials and Processing Technologies*, AMPT 95, pp. 31–39, Dublin.

Hull J.B., Peers N., Whalley R. and Zeng Z. (1994) Fluid level monitoring by ultrasound for process control during filling of retail food containers. *Applications of Multivariable System Techniques*, AMST 94, pp. 329–345, Bradford.

Trading Standards Service (1972) *Weights and Measures Act*.

Zheng Z, Whalley R. and Hull J.B. (1995) The monitoring and control of a retail food container filling machine. *Proc. IMechE Part E: J. Process Mech. Eng.*, **209**, 101–115.

2 'Off-the-shelf' ultrasound instrumentation for the food industry

D. CARTWRIGHT

2.1 Introduction

This chapter will only address the use of ultrasound at frequencies in excess of 1 MHz and the so-called longitudinal or compression wave. The level of ultrasonic power 'radiated' will be in the microwatt region and therefore it can be confidently considered to be strictly non-cavitating and non-destructive, we are not addressing uses for sonochemistry reaction and or material alteration.

References to 'the food industry' naturally includes the drinks and beverage industries, it can be assumed that as product systems in the pharmaceuticals, beauty products and household chemicals industries are similar they too can be included without special reference.

2.2 Ultrasound

Low-power ultrasound has been used as a diagnostic medium for many years – the two main 'traditional' application areas being medical diagnostics and industrial inspection, known in industry simply as 'NDT' (non-destructive testing). The use of ultrasound in the food industry is still in its infancy but it is expected to grow in importance.

Whatever the application, ultrasound is generated in the same way (see Figure 2.1). A device known as a transducer (or probe) contains a ceramic crystal, which is excited with a short pulse of electrical voltage (around 100 ns); through the piezoelectric effect this energy is converted to a short (around 1 μS) sound pulse at the fundamental frequency of the transducer. This sound energy is coupled into the material or body under analysis and propagates through that material. Examples of acoustic velocities for different materials are 1500 $m\,s^{-1}$ for pure water and 5900 $m\,s^{-1}$ for mild-steel, at around room temperature. The same piezoelectric element after acting as a transmitter then acts as a receiver, converting any ultrasound

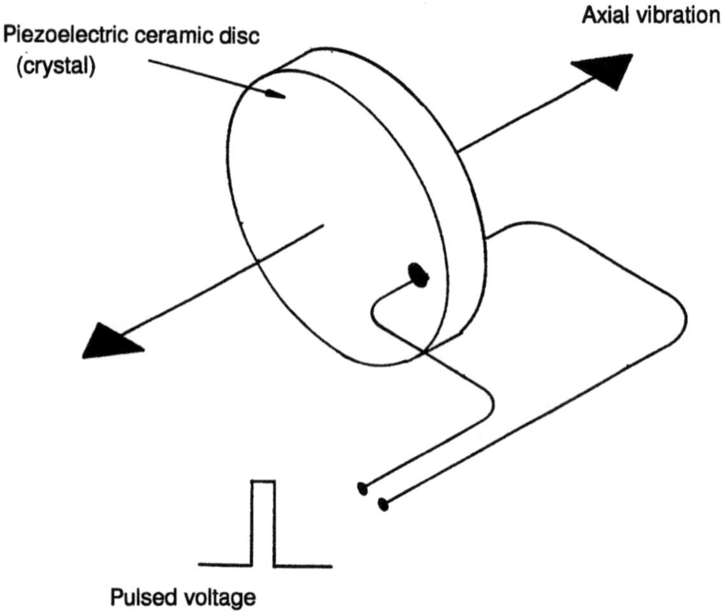

Figure 2.1

impinging on it, back to electrical energy. Medical diagnostics uses probe arrays consisting of many active elements whilst industrial ultrasonics usually uses only one, sometimes two, at any one time.

The use of ultrasound in medicine allows visualization in real-time, of the body's internal organs, things that once required X-rays, such as detection of blockages (stones), gynaecological examination including pre-natal monitoring and cardiovascular examination. These devices are generically referred to as 'ultrasound scanners'.

In industrial inspection ultrasound is used for the analysis of materials (mostly metals) from a quantitative or qualitative point of interest. This may be done, either during manufacture to ensure dimensional conformance and or sectional uniformity for example, or for in service analysis to detect cracking or corrosion in process vessels, pipelines, ships' hulls and structures in general. This equipment falls into two distinct types, the generic analysis tool is usually referred to as a 'flaw-detector' and there are 'thickness gauges' that just give a digital readout of thickness.

To date the only regular and popular use of ultrasound (in the food industry) has been to infer concentration, measure flow velocity and detect liquid content level. These devices are usually application specific and are often set or calibrated at installation for one particular analyte and therefore cannot offer generic analysis. They are designed to be process plant rather than laboratory equipment. Such instrumentation will not be discussed further as it falls outside the present area of interest.

2.2.1 A general overview of ultrasonic equipment

Ultrasound equipment for medical practice has advanced dramatically in recent years and the technology is, of course, being continuously driven by the medical profession – with the manufacturers more than willing to invest in development for the rich and wide market. The world-leaders in electronics include Siemens, Philips and Hewlett Packard. Advances in computer processing power and electronics, particularly in higher speed digital signal processing (DSP), allow these improvements and naturally the electronics vendors and manufacturers follow the demands of this market as they do that of the mobile telephone or audio and video equipment. For example, the US company Analog Devices offer a radio frequency/video amplifier chip-set that has been optimized specifically for use in ultrasonic diagnostic equipment. This device, the AD600/AD602, is a dual channel variable gain amplifier with very precise gain control (linear in dB), low noise and distortion and wide bandwidth. Medical scanners are now capable of showing the undeveloped teeth in a four- to six-month term foetus!

Conversely, the world of industrial ultrasonic inspection is the poor relative. The market is small (in comparison), the inspection industry has little drive and or financial influence on the manufacturers – in some ways the manufacturers lead the users and the market. There are no 'household names' in this sector of the ultrasonics world and there are plenty of 'one-man bands'. U.K. industrial ultrasonic equipment companies who are proper manufacturers, amount to approximately eight with only two or three being solely British. The 'big' names in the whole industry total three, two of whom are US companies with only marketing facilities in the UK. However, some of the largest companies in the world demand ultrasonic inspection. Petrochemical companies which use ultrasound include BP, Shell and ICI, as do aerospace companies like Rolls-Royce, British Aerospace and Boeing. It should be pointed out that the use of ultrasonics for inspection in the (post-production) commercial aerospace industry is not as common or as regular as one would think. It is probable that industrial ultrasound technology is 10 years behind the medical world. Digitization of ultrasonic information, manipulation of the data and DSP techniques are now available as a norm in only the latest inspection equipment. The use of computers as analytical tools or computerization as an inspection aid is only just becoming popular in this industry. Aerospace and chemicals demand completely different levels of diagnostic sophistication and utilize vastly differing levels of electronic complexity, but perhaps surprisingly employ, the same or similar basic ultrasonic techniques.

Ultrasound will propagate through a material until the sound wave impinges on an impedance change, an amount of sound energy is reflected dependent on the size of the interface and the relative impedance differences.

If there are no internal reflectors the wave will continue until it reaches the 'far side' or 'rear wall' of the material, unless the material is so thick that the energy is totally attenuated.

Anything within the ultrasound beam that reflects ultrasound can be detected (see Figure 2.2). It is possible, providing the propagation velocity is known, to determine the distance from the transmitting source to the reflector. If the source were to be moved in real time and all the reflected signals (echoes) displayed in real time, a two-dimensional image of the area

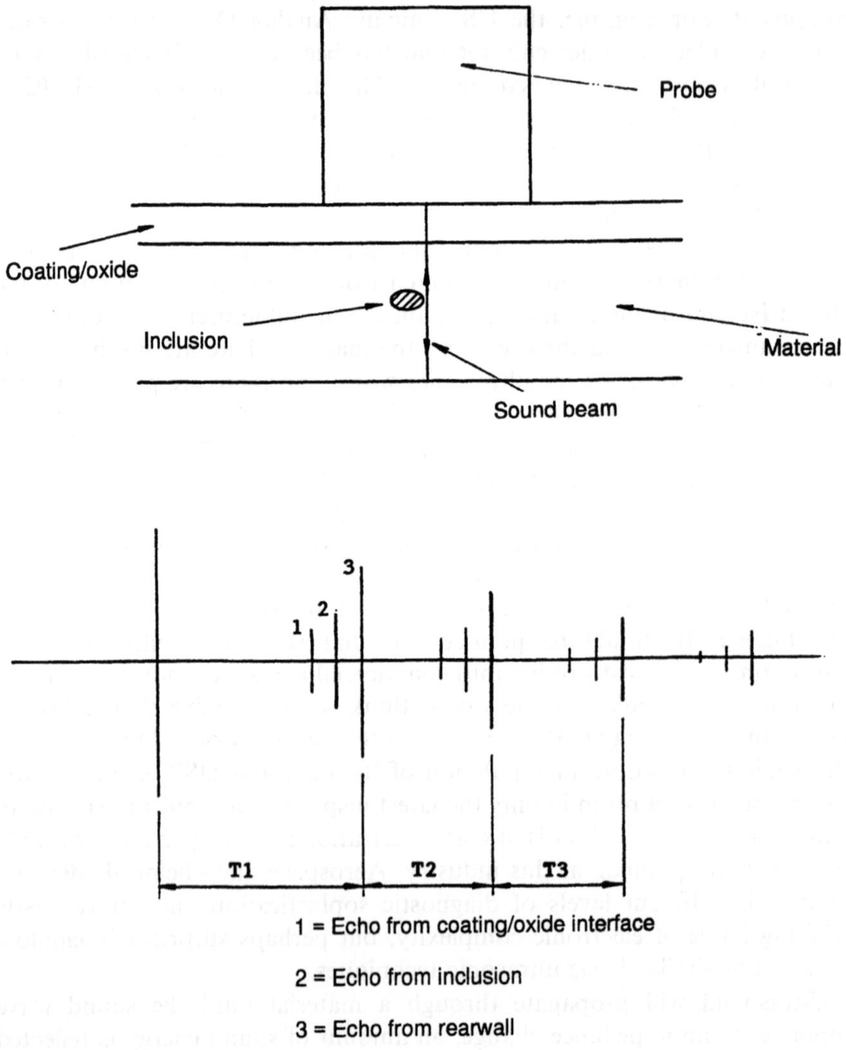

Figure 2.2

beneath the source can be built up. The two main properties exploited are distance and or position from the source. This is the same principle as is used in time-domain radar systems.

Medical diagnostics are viewed on a monitor, the picture is built up by scanning an array of piezoelectric elements which are held stationary by the operator. On the other hand, industrial inspection instrumentation data are presented on a simple cathode-ray tube – an oscilloscope – with the beam representing the path of the ultrasound beam as it travels through the material under test using one probe element alone.

It has proved very difficult, in practice, to adapt existing ultrasound instrumentation for the measurement of foods. The human body is remarkably well suited to ultrasound inspection. On the contrary, foods have very different acoustic properties to human tissue. Thus the amplitude (attenuation) or spectral (frequency) domain information which is used in medical inspection has been difficult to exploit in food applications. It has proved necessary to design *off-the-shelf* ultrasound instrumentation equipment, specifically for use in food applications.

2.3 Amplitude information as a means to detect time-of-flight

The Proctor Department of Food Science at Leeds University approached Cygnus Instruments late in 1988 with a proposal that a joint venture project should be undertaken to develop (in the first instance) an instrument to measure ultrasonic velocities in edible oil-based foods; margarine and similar spreads, mayonnaise, etc. The intention was, initially, that this would be used to determine solid fat content, and at this stage the major food products manufacturer, Unilever, and the Leatherhead Food Research Association were also involved.

It was intended that this should be the beginning of a range of ultrasound products designed specifically for use in the food industry. The project was initially called the 'Adaptable Ultrasonic Sensor'. The idea was that the device would employ a straightforward hardware configuration that could be adapted to a range of applications through software modules, executed in an associated computer to which the hardware was linked. The device developed in this project has since been found to have many analytical applications in the food industry.

2.3.1 Food industry requirements

The initial device specification included time-of-flight measurement accurate to 0.01 μs, and temperature measurement to within 0.1°C. The time-of-flight criteria were set to obtain the highest accuracy in sound velocity measurement obtainable from time-of-flight instrumentation. The temperature specification followed from this because the temperature coefficient of the velocity of sound in water around room temperature is $3\,m\,s^{-1}\,°C^{-1}$. Water is a major component of most foods and therefore temperature control or measurement of the order of 0.1°C was necessary in order to match the accuracy of the sound velocity measurement.

The instrument was to be used in the laboratory so it could be powered from the main electrical supply. This meant power consumption was not an issue, a requirement that could have been difficult to satisfy.

The sample, held in a cell that contains the ultrasonic transducer and the temperature sensor (see Figure 2.3), would be measured continuously in real-time and the data transmitted to an adjacent computer.

The sample-cell should remain inert to the sample and be capable of withstanding samples at temperatures from −15 to 110°C, to operate in a laboratory bath and the sample was to be continuously stirred. The necessity for these provisions will be discussed later.

The sample-cell had to be suitable for cleaning to 'clean-in-place' standards as recognized by the food industry to ensure hygienic conditions. Although this was not really necessary for laboratory work it was used as a target for the specification. The cell had to be large enough to allow the transducer and the temperature sensor to be in contact with the sample and also offer a sensible ranging distance for the velocity measurements, but it should not be so big as to demand an impractical volume of sample. Internal pressures in the cell would be quite high when some of the intended sample

Figure 2.3

materials were crystallizing, the cell would have to withstand this without damage. The cell was also designed to withstand an internal pressure of 1 bar, such as may typically be found in some food processing applications. The instrument was to have no controls (apart from the on/off switch) to minimize the chances of erroneous measurements as a result of incorrect settings. It also had to be ready for use instantaneously from turn on with no 'warm-up' time.

2.3.2 Designing the 'UVM-1'

Our first task at Cygnus Instruments was to investigate the acoustic characteristics of the materials and systems that the UVM-1 was intended to analyse, and how well (or not) our standard transducers were matched to them. Transducers for industrial inspection are made with the output 'faces' matched to metal – improper matching to foodstuffs would result in loss of signal and degradation of instrument performance. It was found, for example, that even 'set' margarines have lower acoustic impedances than metals and that the oils and aqueous-based systems that make up these products do so as well. The standard transducer would match well into these lower impedances.

Sound velocities were in the range of 1000–2000 $m\,s^{-1}$, which meant the wavelengths were shorter than we were used to dealing with in metals testing. This actually made some aspects of the development easier. The relationship between velocity and wavelength is shown below:

$$\lambda = \frac{v}{f} \tag{2.1}$$

where λ is the wavelength in metres; f is the frequency in hertz and v is the velocity in metres per second.

The transducer sound beam has physical 'zones' of differing efficiency (performance) along its length, these are dependent on frequency, the diameter of the active element and the wavelength in the transmission medium. These zones can produce maxima and minima along the beam for any given reflector. We need to work outside the dead zone, the zone adjacent to the crystal face, an area where the sound beam is in great confusion. The next area outward, is the near zone, this lies between the dead zone and the far field. The near zone is the ideal region for time-of-flight measurement because the beam power is at its maximum. Ideally we would want to set our reflecting target at the end of the near field.

In steel the near field of a 2.25 MHz transducer of 13-mm diameter is about 15 mm, in crystallized edible oil it is about 45 mm. These values are given by

$$n = \frac{d^2 f}{4v} \tag{2.2}$$

where n is the near field range (minimum).

It was found that attenuation in the worst case samples resulted in loss of signal at a range of around 80 mm (round trip distance) so the sample cell internal diameter was set at 40 mm m. This meant the whole flight path of the sound pulse would be in the near zone, up to the commencement of the far field.

It was decided to fabricate the cell from high-grade stainless steel with a wall thickness of 13 mm, this being necessary to allow the threading for fitting the transducer and temperature sensors. It was considered that the cell strength would be suitable to withstand the pressures previously mentioned.

The cell height was governed by the space needed to mount the ultrasonic transducer and the temperature sensor and also to allow for a magnetic stirrer 'flea' to rotate without hindrance. A height of 75 mm was sufficient and this gave a sample volume of 95 cm^3. A detachable base was designed for easy cleaning and the forced ejection of viscous materials, an O-ring seals the base when fitted. The base is fabricated from Delrin® (DuPont Plastics, a co-polymer acetal material), this is an inert engineering plastic, unaffected by most chemicals and solvents it would be compatible with the range of sample materials and being non-magnetic would obviously allow the magnetic stirrer to operate freely. Tests showed that a standard Cygnus Instruments transducer at 2.25 MHz with an element diameter of 12.7 mm (0.5 inch), gave the best ultrasound signals for the application. This was mounted radially in the vertical centre of the cell, transmitting across the cell diameter using the opposite wall as a reflector. The transducer was fitted behind an epoxy-resin acoustic lens, which was necessary to preserve segregation from the sample and to protect the transducer face from the sample or the affects of cleaning (see Figure 2.4).

It is possible to transmit the sound through a thin section of the cell wall, this would have been preferred but would have put impractical manufacturing tolerances on the control of the transducer frequency and also on machining of the cell. It would also have been necessary to have 'matched' the transducer to the cell at the time of manufacture but this would have made the manufacturing costs unacceptably high. These complications arise because the thickness of remaining steel would have be exactly a quarter of a wavelength thick for each transducer and cell pair.

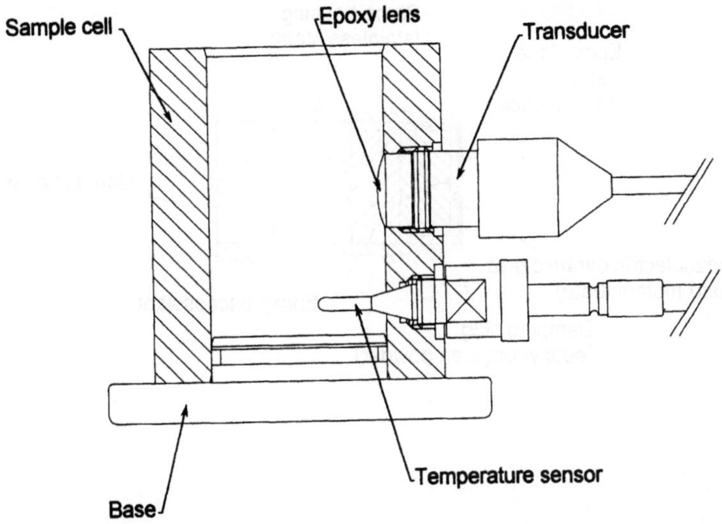

Figure 2.4

Temperature (of the sample) was one of the aspects of the development that caused concern. Generally speaking ultrasound transducers acoustically deteriorate at temperatures in excess of about 85–100°C, although the damage is temporary and reversible. At higher temperatures permanent damage will occur. This is because the epoxy bonding materials that make up the device become unstable and start breaking down. This deterioration is first witnessed by a sharp reduction of sensitivity, then should the temperature exceed the Curie point of the crystal (500°C) it will be irreversibly depolarized, this is permanent and the transducer is rendered useless (see Figure 2.5).

Experiments showed that our transducers suffered no permanent damage at the maximum temperature specified and that the loss of sensitivity at these temperatures was acceptable.

The temperature sensor was a platinum resistance thermometer (PRT), the active element is housed in a stainless steel sheath, and the whole is mounted radially above the transducer such that the tip is in the centre of the sample. The PRT and its associated circuitry and software were manufactured to PRT100 standards, ensuring an overall measurement accuracy of 0.1°C and a precision of 0.01°C.

In use the whole electronic assembly is sealed. This prevents ingress of moisture from the laboratory bath or leakage of sample, thereby ensuring no sample contamination or electrical breakdown can occur. The ultrasound and temperature sensors are connected to the electronics unit via multiway

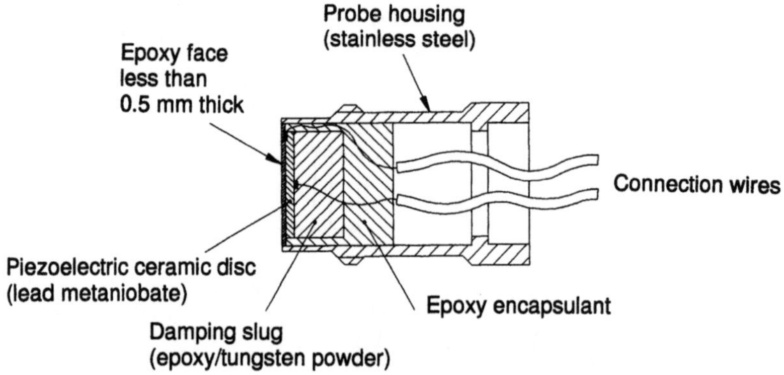

Figure 2.5

and coaxial sockets mounted on its front panel (see Figure 2.6). Because of the likelihood of exposure to high temperatures both the ultrasound and temperature sensor cables are specially insulated.

The electronics unit comprises a power supply, the transducer drive circuitry, the analogue processing and timing circuitry, the temperature measuring circuitry and the communications system that sends the data to the computer. It also offers a local readout of time-of-flight and temperature on a large digital light-emitting diode (LED) display. The front panel also has an LED indicator array showing operational status. The unit fires the transducer at a fixed rate in the kilohertz region. A very accurate crystal controlled timer is started at the instant when the ultrasonic wave exits the acoustic lens (set on manufacture, this negates the propagation period through the lens from the total time-of-flight), the analogue processor then starts searching for an echo. The crystal oscillator-based timing circuits

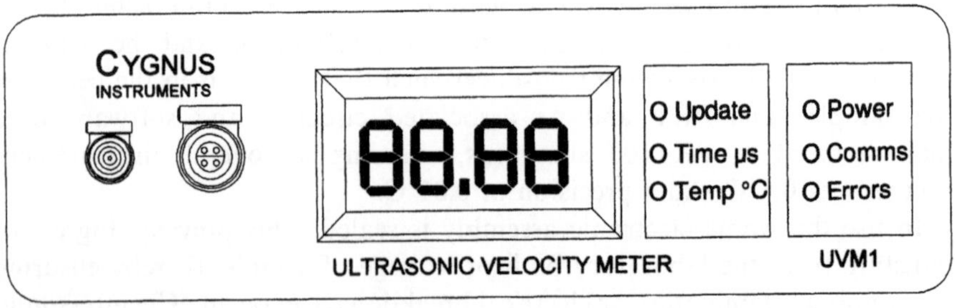

Figure 2.6

provide a high degree of measuring accuracy and also ensure long-term stability for lengthy experiments. When a signal is received the timer is stopped, the time-of-flight is then sent to the computer.

The device measures continually, the timers being reset after every reading. The update rate and the rate at which the readings are transmitted to the computer are wholly dependent on measuring conditions but will always exceed $10\,s^{-1}$. The system also continually measures temperature, the computer is updated every time the time-of-flight is transmitted.

The computer, running special Windows software developed by Dr Malcolm Povey at Leeds University, separates the time-of-flight data and the temperature data, processes them as necessary to obtain the required parameters and then displays them on the computer screen (see Figure 2.7). The temperature is also fed back to the ultrasound device for displaying locally, this also serves as a means of ensuring communications integrity.

The software allows the measured data to be logged in real-time by using Windows dynamic data exchange (DDE). The measurements are sent to a spreadsheet program, which record the data and also allows graphical visualization of the analysis or experiment. The user needs no knowledge of spreadsheet programs to use the instrument, since the instrument controls the spreadsheet program.

The device went through several levels of prototyping and testing to ensure it met the performance criteria defined. The developers allowed external independent assessment by several major food manufacturers to ensure suitability. Intensive use of the instrument by students in a food

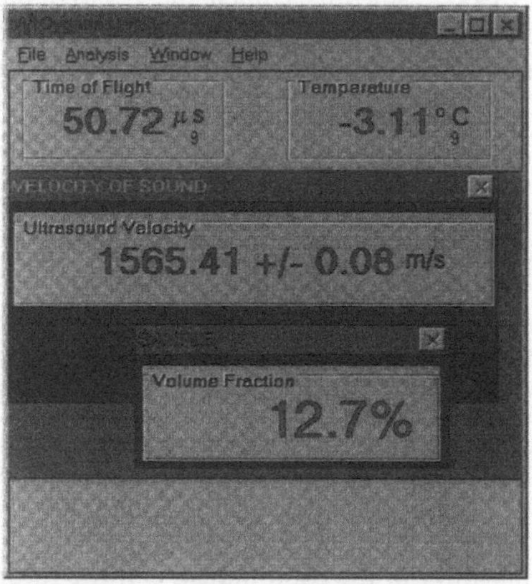

Figure 2.7

science laboratory also gave the designers very useful feedback. The initial design criterion, that the instrument should be easily useable by non-specialists, has been achieved through this intensive testing process.

The UVM-1 was officially launched in early 1993. Since then the device has undergone further development as a result of European Union (EU) legislation. Having satisfied the technical performance specification the UVM-1 had to be subjected to testing for compliance with the EU Electromagnetic Compatibility Directive 89/336/EEC (and amendment by Directive 92/31/EEC). As no application standard (equipment type standard) was available the UVM-1 was tested under the generic emissions and immunity standards EN-50081 (Part-1) and EN-50082 (Part-1), respectively. This directive, which came into force on 1 January 1996, ensures that electrical equipment does not generate electromagnetic interference or suffer in its operation if irradiated by electrical interference from other sources.

2.4 Applications

The UVM-1 was originally designed to measure solid-fat content (Kelly *et al.*, 1990). This was proven successful early on and a number of publications giving entirely new information about crystallization have been obtained using the UVM-1 (Archer *et al.*, 1996; Coupland *et al.*, 1993; McClements *et al.*, 1993). We have shown that the UVM-1 technique is uniquely suited to the study of crystallization and crystallization kinetics *per se*. For example, the UVM-1 is 30 times more accurate than the equivalent nuclear magnetic resonance technique for the measurement of palm oil crystallization in emulsions (Povey, 1997). Since its launch the UVM-1 has been used in a wide variety of applications which include measurement of concentrations of hormone in a solvent carrier, analysis of fixing and finishing chemicals in the paper industry and the measurement of alcohol concentration in brewed and fermented products. It has also been used to determine sugar content in drinks (Contreras *et al.*, 1992) and analyse liquid laundry products and fabric conditioners.

The UVM-1 determines both the ultrasonic velocity and temperature of a given sample. These two quantities are combined in software that is application oriented. The ease with which the user can obtain the data of interest – solid fat content, crystallizing solids, sugar content, etc. It is simple to use, requires no special expertise to calibrate and automatically collects large amounts of data giving a unique view of processes. The data itself are easily accessible since it is stored on Microsoft Excel spreadsheets. Running costs are negligible. The convenience of the instrument and the unique nature of the data it produces suggest that this instrument and others like it will find wide application in the food industry in the future.

2.5 References

Archer G.P., Kennedy C.J. and Povey M.J.W. (1996) Investigations of ice nucleation in water-in-oil emulsions using ultrasound velocity measurements. *Cryoletters*, **17**, 391–6.

Contreras N.I., Fairley P., McClements D.J. and Povey M.J.W. (1992) Analysis of the sugar content of fruit juices and drinks using ultrasonic velocity measurements. *Int. J. Food Sci. Tech.*, **27**, 515–29.

Coupland J., Dickinson E., McClements D.J., Povey M.J.W. and de Rancourt de Mimmerand C. (1993) Crystallisation in simple paraffins and monoacid saturated triacylglycerols dispersed in water. *Food Colloids and Polymers: Stability and Mechanical Properties*, Editors E. Dickinson and P. Walstra, Royal Society of Chemistry, London, pp. 243–9.

Kelly T., McClements D.J. and Povey M.J.W. (1990) Sensor for solid fat determination and use thereof, GB Patent Application No. 9013530.2.

McClements D.J., Dickinson E., Dungan S.R., Kinsella J.E., Ma J.G. and Povey M.J.W. (1993) The effect of emulsifier type on the crystallization kinetics of oil–water emulsions containing a mixture of solid and liquid droplets. *J. Coll. Int. Sci.*, **160**, 293–7.

Povey M.J.W. (1997) *Ultrasonic Techniques for Fluids Characterization*. Academic Press, San Diego, CA.

3 Rapid determination of food material properties

M.J.W. POVEY

3.1 Introduction

The food industry puts great demands on its process-sensing technologies. For hygiene reasons, a sensor should be non-invasive; because of low-profit margins, sensors should be extremely cheap. Food materials tend to be complex and heterogeneous and may be highly variable due to their natural origins. In a processing situation there is a need for a rapid and accurate response. The information required may also be of a highly complex nature, a simple question such as 'has the food been cooked enough?' may be very difficult to answer with the data available from contemporary process sensor technologies. The sensor itself should be inherently safe.

It has been recognized for a number of years that there is a lack of suitable sensors which can provide appropriate information about food attributes during processing (Agricultural Research Council, 1982; Rizvi *et al.*, 1993; Selbome and Hearn, 1987). The nature of ultrasonic radiation makes it a serious contender for the design of sensors for food applications. The pulse nature of the technique means that hundreds of measurements can be made available every second, permitting the measurement of rapidly changing processes. The radiation is non-destructive and safe at the power levels normally used. The technique may be used non-invasively and even non-intrusively, using non-contacting methods of sound generation and detection. Measurements can be made successfully in heterogeneous and multiphase flows in pipes. The sensors are easily integrated into computer-process monitoring and control systems. The measurements are insensitive to noise and vibration. A single sensor may give several distinct items of information.

The question is why have these sensors not been widely adopted outside the level-flow measurement and meat back-fat thickness areas in which they have so far established themselves? The answer lies in the difficulty in interpreting data from ultrasound sensors and the complex nature of the interaction between sound and matter. However, a great deal of progress has been made in recent years in this area and ultrasound should now be re-examined in the light of new developments, to be described here.

As research into ultrasound has progressed, it has become clear that the interaction of sound with matter is a rich source of new information about materials. However, sophisticated data treatment is often necessary to extract the information hidden in the detected sound field. The dramatic fall in the cost of computing power means that it is now feasible to incorporate the computing processing needed for successful ultrasound instrumentation at a relatively low cost.

It is a decade since Povey and McClements wrote a comprehensive review of the applications of ultrasound to food engineering (Povey and McClements, 1988; Povey, 1989) and it is interesting to compare what was said then with the present situation. While our understanding of the interaction of ultrasound with food materials has advanced by leaps and bounds, the available technology has made nothing like the same strides. It is only recently that some of the ideas considered in the past have gained a practical embodiment in new instruments for particle sizing, molecular properties, composition determination and crystallization monitoring.

In the following introduction to the basic principles of ultrasound propagation in food materials an attempt has been made to combine clarity and simplicity with a treatment of a complex subject. That it is necessary to do this may become clear when the following misconceptions and fallacies about ultrasound are explained. Firstly, it is widely believed that the density determines the velocity of sound in materials. While it is true that density affects sound velocity, in practice the major portion of the variation of sound velocity in foods comes from changes in material compressibility. Understanding the ways in which this occurs is crucial to a number of successful applications of the ultrasound technique to the measurement of the properties of foods. Secondly, ultrasound is seen as material altering. While this is true in the case of high-power ultrasound, it is generally not the case for the power levels employed in ultrasonic testing, which are tens of thousands of times less than the levels found in an ultrasound bath.

There is a considerable review literature in the area of ultrasound measurements in food processing. Povey and McClements (1988) have reviewed the theory of ultrasound propagation in processed foods; Self *et al.* (1992a), fruit and vegetables; and Povey (1997a), fluids. McCann (1986), Shirato (1987), Javanaud (1988), Povey (1989), Povey and McClements (1989), Bjorno (1991), Ohashi and Karube (1993), Javanaud and Robins (1993), Samari (1994), McClements (1994a), McClements (1995), Dickinson and McClements (1995a) and McClements (1997) have given general reviews of ultrasound measurements in foods.

3.2 Theory

3.2.1 Pure materials

Sound can propagate through materials in many different ways. Sound may be defined most generally as any form of motion taking place within a material; this definition excludes only motion that involves displacement of the entire body. This motion can take very many forms, the three most important being compression, shear and surface waves. A compression wave (see Figure 3.1) comprises a succession of compression and extension of the material, about an average position, which remains static. The motion of the material is parallel to the direction of motion. A shear wave in contrast involves motion of the material transverse to the direction of motion. Surface waves are akin to sea waves. All three motions interact with the physical properties of the material in different ways. Compression waves tend to move fastest, because the restoring forces are greatest. They are also physically the simplest sort of wave since they are pressure waves whose forces are isotropic. Shear waves generally move more slowly than compression waves and can exhibit polarization, which may be significant if the properties of the material are anisotropic. Shear waves are generally much more attenuated than compression waves. Surface waves are even more complex and require special detection means. So compression waves are most widely used in ultrasound measurement because they are most easily detected over the greatest distances. Shear waves will normally only be used if there is special interest in the shear modulus.

It is generally accepted that sound frequencies above about 16 kHz, the average threshold for human hearing, are called ultrasound frequencies.

Ultrasound measurement is normally non-destructive and involves power levels well below that at which the physical or chemical properties of the food might be altered (Povey and McClements, 1988).

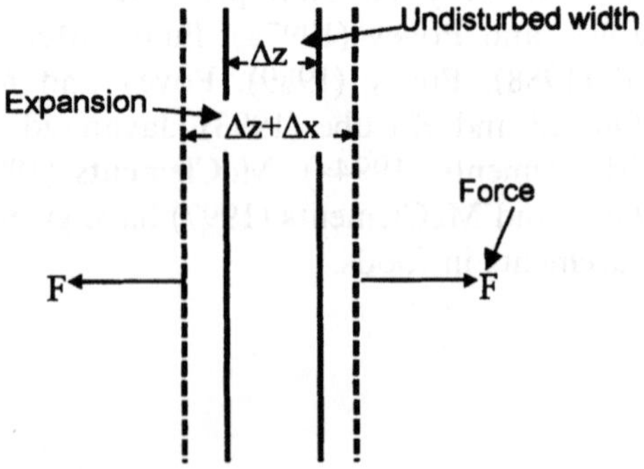

Figure 3.1 The expansion phase of the passage of a sound wave (from Povey, 1997a).

3.2.1.1 Directly measured variables. A detailed discussion of ultrasound propagation in fluids can be found in Povey (1997a), from which much of the following discussion (generalized to cover propagation in solids) has been borrowed. The interaction between ultrasound and food is usually characterized by the following parameters: frequency ($f = \omega/2\pi$), attenuation (α), velocity (v), wavelength ($\lambda = v/f$) and impedance ($Z = \omega\rho/kZ$). Attenuation is a measure of the decrease in signal amplitude of a compression wave (ξ) with distance (z) and the characteristic impedance is the product of density (ρ) and velocity (v). These quantities are derived from measurements of frequency, signal phase ($i\alpha$, where $i = \sqrt{-1}$) and signal amplitude, which quantities may be direction and distance dependent. These quantities are related as follows:

$$\xi = \xi_0 e^{i\omega t} e^{ikz} \tag{3.1}$$

where x_0 is the amplitude of the signal at the point where it is launched into the sample and

$$k = k' + ik''$$

Here

$$k' = \frac{\omega}{v} \tag{3.2}$$

$$k'' = i\alpha \tag{3.3}$$

In general, the frequency (ω) and wave vector (k) are related through an equation called a dispersion relation

$$\frac{\omega}{k} = \sqrt{\frac{E}{\rho}} \tag{3.4}$$

where E is a generalized complex elastic modulus. This equation contains a great deal of detail, since both k and E are complex quantities, which are generally frequency dependent.

In most ultrasound measurements $k' \gg k''$ because it is difficult to make measurements when $k'' \geq k'$ and the following discussions assume that this is always the case.

3.2.1.2 Physical variables measureable with ultrasound. It is shown below that the velocity of sound depends on an elastic constant and on the density. So it is often worth combining ultrasound velocity measurement with a density determination because the elastic constant can be determined directly. This is the standard method of determining the adiabatic compressibility of materials (Povey, 1997a). It is important to stress that static methods of determining elastic constants, such as the use of an Instron, determine the isothermal elastic constants that can differ quite considerably from the adiabatic constants determined by ultrasound.

The velocity of sound is often very temperature dependent For example, in water at 20°C the temperature coefficient of velocity is $3\,\mathrm{m\,s^{-1}\,°C^{-1}}$. The temperature coefficient of velocity in water is also the opposite sign to that of most other materials. It is therefore usual to combine ultrasound measurement with temperature measurement and the quality of the data is greatly enhanced by doing this (Povey, 1997a).

When the velocity of sound is combined with density, frequency and other thermophysical data (ratio of specific heats, thermal conductivity and specific heat) information about the many physical parameters may be obtained. These include adiabatic compressibility, bulk modulus, shear modulus and shear viscosity, the real and imaginary parts of the elastic modulus and Young's modulus. It should be emphasized that these parameters are those appropriate at the measurement frequency.

If phase and amplitude are measured as a function of frequency then a much wider range of physical information is potentially measurable. This includes particle size, shear and bulk viscosity of the fluid in a small region surrounding the particle and qualitative information about the state of aggregation of particles.

3.2.1.2.1 Ultrasound data. Collections of data on the ultrasound properties of foods may be found in Povey and McClements (1988), Javanaud (1988) and McClements (1997). More general collections of data are in Weast (1988), Kaye and Laby (1986) and Schaafs (1967).

3.2.1.3 The interpretation of attenuation data. In contrast to light detectors, ultrasound transducers are phase sensitive. This can result in a considerable loss in detected signal, which is not due to absorption of the ultrasound signal but is due to scattering. For this and other reasons, measurement of ultrasound attenuation tends to be much less reliable than measurement of velocity. The received signal normally requires correction for diffraction (Khimunin, 1972), losses at interfaces and relaxation effects (Frindi *et al.*, 1994). The reader is referred to Povey (1997a) for a detailed discussion of the corrections needed for attenuation measurements.

The measured attenuation coefficient is obtained by fitting an exponentially decaying curve to the signal amplitude measured as a function of distance

$$\xi = \xi_0 e^{-\alpha z} \tag{3.5}$$

3.2.1.4 Propagation in pure materials. The velocity of compression sound (v) in a fluid is simply described by the Wood equation (Wood, 1964)

$$v = \sqrt{1/(\kappa\rho)} \tag{3.6}$$

The equivalent equation for solids and viscoplastic materials is

$$v = \sqrt{[B + (4/3)G]/\rho} \tag{3.7}$$

where $B = 1/\kappa$, κ is adiabatic compressibility, B is bulk modulus, G is rigidity modulus and ρ is density. The velocity of a shear wave in a solid (v_s) is given by

$$v_s = \sqrt{G/\rho} \tag{3.8}$$

These equations may be derived from Hooke's law and Newton's second law as described below.

A sound wave causes local compression and extension in the material through which it passes. Suppose that a material segment of width Δz, area A and bulk modulus B is stretched by a force F to a width of $\Delta z + \Delta x$ (Figure 3.1).

The strain, defined as change in length per unit length, in this segment is $\Delta x/\Delta z$. The fractional increase in volume associated with this strain is called the dilation which is negative for hydrostatic pressure

$$\frac{(A + \Delta A)(\Delta x + \Delta z) - A\Delta z}{A\Delta z} \cong \frac{\Delta x}{\Delta z} = D \tag{3.9}$$

here the products of finite differences are assumed to be negligible, this is called the linear approximation. So the strain approximately equals the dilation. The adiabatic compressibility is defined in terms of the dilation D as follows:

$$\kappa = -\frac{1}{V}\frac{\partial V}{\partial p} \cong -\frac{D}{\Delta p} \tag{3.10}$$

This gives the stress (Δp), in terms of the strain and is a statement of Hooke's law if the adiabatic compressibility is independent of strain.

The above statement of Hooke's law can be restated in terms of the force and the bulk modulus ($B = 1/\kappa$), bearing in mind that the force is opposite in sign to the hydrostatic pressure

$$\frac{F}{A} = B\frac{\Delta x}{\Delta z} \tag{3.11}$$

The excess force ΔF that causes an infinite thin segment, Δz of the material to accelerate is given by

$$\Delta F = A \cdot B \cdot \left(\frac{\partial^2 x}{\partial z^2}\right) \cdot \Delta z \tag{3.12}$$

This force is given, from Newton's second law, by the product of mass ($\rho A \Delta z$) and acceleration ($\partial^2 x/\partial t^2$) so that the acceleration can be written as:

$$\frac{\partial^2 x}{\partial t^2} = \frac{B}{\rho} \cdot \frac{\partial^2 x}{\partial z^2} \tag{3.13}$$

which is Newton's second law.

The velocity of sound v, for a liquid is then given by

$$v = \frac{\partial z}{\partial t} = \sqrt{\frac{B}{\rho}} = \sqrt{\frac{1}{\kappa\rho}} \tag{3.14}$$

This is called the Wood equation or the Laplace equation. This is the velocity of a compressional wave that is the only propagational acoustic mode possible in liquids.

It may also be shown that the attenuation of the sound wave is given by

$$\alpha = \frac{\omega}{2v^3}\left\{\frac{(\eta_v + \frac{4}{3}\eta_s)}{\rho} + \sigma(\gamma - 1)\right\} \tag{3.15}$$

where η_v and η_s are the volume and shear viscosity, respectively, σ is the thermal diffusivity and γ, the ratio of the specific heats, ω is the radial frequency. This is correct only for a homogeneous fluid in which relaxation and scattering effects are absent, apart from relaxation effects attributable directly to viscosity.

The corresponding result for an isotropic elastic solid may be obtained in the same way as above but replacing the bulk modulus with the appropriate elastic modulus, namely

$$v = \sqrt{\frac{[B + \frac{4}{3}G]}{\rho}} \tag{3.16}$$

where G is the rigidity, which has units of pascal, which is shown in equation (3.7).

The results for viscoplastic materials are complicated by memory effects such as polymer entanglement, but relatively simple results can be obtained by allowing the elastic moduli B and G to be complex and frequency dependent (Povey, 1997a), so that

$$M(\omega) = B(\omega) + \frac{4}{3}G(\omega) + M' + iM'' \tag{3.17}$$

defining a complex velocity of sound as

$$v(\omega) = \sqrt{\frac{M(\omega)}{\rho_1}} \tag{3.18}$$

then the phase velocity will be

$$v = \sqrt{\frac{2\sqrt{M'^2 + M''^2}}{\rho(1 + M'^2/\sqrt{M'^2 + M''^2})}} \tag{3.19}$$

and the attenuation

$$v = \omega\sqrt{\rho\frac{1 + M'/\sqrt{M'^2 + M''^2}}{\sqrt{M'^2 + M''^2}}} \tag{3.20}$$

Similar equations are quoted without proof in Pethrick (1983).

If the attenuation is small enough that the real part of the wavenumber $k' \gg \alpha$ then

$$v = \sqrt{\frac{M}{\rho_1}} = \sqrt{\frac{B + 4/3G}{\rho_1}} \cong \sqrt{\frac{B}{\rho_1}} \tag{3.21}$$

$$\alpha = \frac{\omega M''}{2M}\sqrt{\frac{\rho_1}{M}} = \frac{B'' + 4/3G''}{2\rho_1 v^3}\omega \tag{3.22}$$

3.2.2 *Multiphase materials*

The above equations may be used for pure materials to relate physical properties of interest to the food manufacturer, such as elasticity, stiffness, compliance, loss and storage moduli and compressibility, to the measured velocity of sound. Unfortunately, very few food materials are 'pure' in the chemical sense and in general the velocity of sound is a complicated function of a large number of physical, chemical and biochemical properties.

The simplest treatment for multiphase materials was suggested by Wood (1964) and was stated explicitly by Urick (1947) in an equation which is now named after him. It is a restatement of the Wood equation in which the compressibility and density are volume averages of the components, which make up the mixture

$$v = \sqrt{\frac{1}{\kappa\rho}} \tag{3.23}$$

where

$$\kappa = \sum \phi_i \kappa_i$$

$$\rho = \sum \phi_i \rho_i$$

and ϕ_i is the volume fraction of the ith component in the mixture. This is called a homogeneous description because the properties of each phase contribute independently to the properties of the system.

3.2.3 Scattering

In general, scattering complicates propagation in mixtures and the reader is referred to Povey (1997a) for a comprehensive discussion of acoustic scattering theory in the long wavelength limit.

While the formalism of scattering theory is difficult, the underlying physics is interesting and informative. The passage of a sound wave will affect particles suspended in the medium through which it passes. We consider the case where the wavelength (Figure 3.3) of the sound wave is much greater than the size of the particle (Figure 3.2).

The compression and expansion due to the sound wave heats and cools the material. This heating and cooling is adiabatic, this means that the heat flow between the compression and expansion phases of the sound wave is insignificant. At the interface between a particle and its surroundings, the temperature rise and fall undergoes a change, due to the different volume expansivity and heat capacity of the materials. Heat will then flow in order to maintain continuity of temperature and the boundary region is heated and cooled differently to the bulk of the material. This causes a differential expansion and contraction of the boundary region that acts as a source of

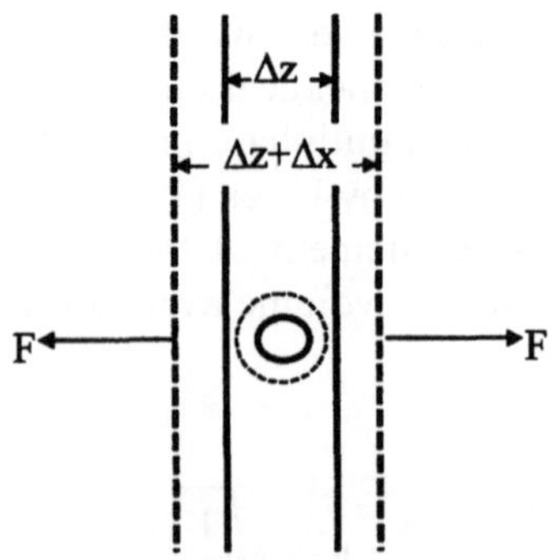

Figure 3.2 Expansion of an element containing a small particle (from Povey, 1997a).

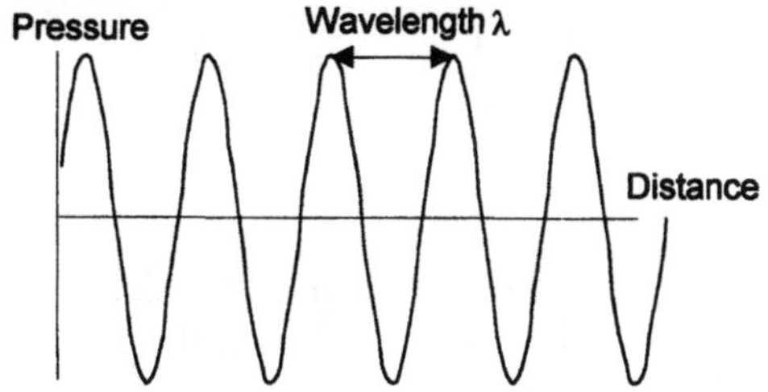

Figure 3.3 Pressure, wavelength and distance in an acoustic wave (from Povey, 1997a).

sound. This process is called thermal scattering and its effects can be surprisingly large. Pinfield and Povey (1997) show that the relative temperature difference between an alcohol molecule and the water surrounding it in solution can vary between −25.6 at 0°C through 0 at 4°C to 17 at 10°C. The magnitude of thermal scattering is related to the relative temperature fluctuation arising from the sound wave, which is given by the equation for R in equation (3.24) below. Thermal scattering can affect the velocity of sound, the attenuation and the apparent compressibility of the dispersed material, in a frequency dependent manner.

The difference in density between particles and their surroundings can also scatter sound. This occurs because the particles may move relative to their surroundings, due to their different inertia. The viscosity of the surrounding material may affect this relative motion and hence the effect has been called visco-inertial scattering. If the viscosity is high the particle will move with its surroundings and visco inertial scattering will be small, regardless of the density difference. If the viscosity is small, the reaction between the particle and the displaced fluid volume will be small and once again visco-inertial scattering will be small. Therefore, visco-inertial scattering will depend on frequency, density and viscosity in a complicated way. However, in the long wavelength limit, visco-inertial scattering becomes insignificant.

The effects of scattering are considered by Julian McClements in his chapter on particle sizing (Chapter 5) and form the basis of ultrasound particle sizing. In this chapter the effects of scattering will be considered from the point of view of their effect on the determination of the apparent compressibility of particles and molecules. The determination of apparent compressibility is very important in the physical chemistry of proteins and solutes in solutions (Pinfield and Povey, 1997). Applications of compressibility measurements to food materials are summarized in Table 3.1.

The conventional use of the Wood equation (equation (3.6)) for the determination of compressibility is flawed by the effects of thermal scattering Pinfield and Povey, 1997). In its place, the following form of the Urick equation that is appropriate for the determination of adiabatic compressibility in binary solutions and mixtures should be used.

$$\frac{1}{v^2} = \frac{1}{v_1^2}(1 + \alpha\phi + \delta\phi^2)$$

where

$$\alpha = \left[\frac{\kappa_{a2} - \kappa_{a1}}{\kappa_{a1}} + \theta + \frac{\rho_2 - \rho_1}{\rho_1}\right]$$

and

$$\delta = \left(\frac{\kappa_{a2} - \kappa_{a1}}{\kappa_{a1}} + \theta\right)\left(\frac{\rho_2 - \rho_1}{\rho_1}\right) + \frac{2(\rho_2 - \rho_1)^2}{3\rho_1^2}$$

$$\theta = (\gamma - 1)\frac{\rho_2 C_{p2}}{\rho_1 C_{p1}} R^2$$

where

$$R = \left[\frac{\dfrac{\beta_2}{\rho_2 C_{p2}} - \dfrac{\beta_1}{\rho_1 C_{p1}}}{\dfrac{\beta_1}{\rho_1 C_{p1}}}\right] \tag{3.24}$$

Here v_1, velocity in pure solvent; ϕ, volume fraction of the solution occupied by solute; α and δ are coefficients defined above, which can be related to the monopole and dipole scattering coefficients a_0 and a_1. Quantities referring to the solvent are subscripted 1, those referring to the solute are subscripted 2. β is the volume thermal expansivity (K^{-1}). This is called the 'modified Urick equation' (Pinfield *et al.*, 1996; Pinfield and Povey, 1997).

This equation can be used in a straightforward way to analyse the dependence of the velocity of sound on the concentration of a single dispersed component. If the velocity is measured against concentration on a volume basis, then the inverse square of the velocity may be plotted against a quadratic function of concentration and the scattering coefficients determined. The compressibility can then be calculated from the first scattering coefficient α. It is particularly useful for the determination of the volume fraction of dispersed material from the velocity of sound and is used in ultrasound profiling applications (Povey, 1997b; Pinfield *et al.*, 1995). It is necessary here to make density measurements in order to obtain the volume fraction occupied by the solute molecules (Povey, 1997a).

3.2.3.1 Determination of molecular properties. The above technique provides a very powerful method for determining the molecular properties of food materials. In particular, it is no longer necessary to extrapolate to infinite dilution in order to obtain the compressibility of a solute molecule. Thus, lower precision, cheaper equipment may be used, bringing these measurements within the budget of the process engineer and the factory laboratory. A further advantage of the technique is that it gives actual compressibilities, not apparent compressibilities and conclusions may be drawn about the overall state of a molecule, for example, whether it is hydrophobic or hydrophilic. The denaturation of protein could be studied this way for example.

3.2.4 *Solid content measurement*

In the case of systems containing crystallizing material, somewhat surprisingly, an even simpler relationship (Pinfield *et al.*, 1995; McClements, 1988) allows a very accurate determination of the amount of crystallized material

$$\phi = \frac{\frac{1}{v^2} - \frac{1}{v_l^2}}{\frac{1}{v_s^2} - \frac{1}{v_l^2}} \Phi \tag{3.25}$$

Here ϕ is the calculated *volume fraction* of solid present in the entire system; Φ_s, the volume fraction of the dispersed phase, v the measured velocity of sound; v_s, the extrapolated velocity of sound for an emulsion containing wholly solid droplets and v_l the extrapolated velocity of sound for an emulsion comprising wholly liquid droplets.

In the following sections a number of practical applications of ultrasound to the measurement of food properties are reviewed.

3.2.5 *Sound attenuation in the presence of relaxation and scattering*

It is very important to separate out the different contributions to the measured attenuation of sound (Povey, 1997a). In an emulsion for instance contributions may come from relaxation effects associated with exchange between surfactant micelles and the solution (Frindi *et al.*, 1994). Contributions can also be expected from scattering from individual particles of the dispersed phase, from multiple scattering effects and from the 'classical' absorption due to the viscosity and thermal conductivity inherent in the continuous phase. The process of crystallization may also contribute towards relaxation processes (see Section 3.5.3)

$$\alpha = \alpha_{\text{classical}} + \alpha_{\text{scattering}} + \alpha_{\text{relaxation}} \tag{3.26}$$

The different contributions may be separated out by a calibration procedure in which measurements are first made on the continuous phase and then on the emulsion.

3.3 Techniques

3.3.1 Pitch and catch technique

Perhaps the simplest approach to the measurement of ultrasound velocity is the 'pitch and catch' technique. Here one transducer generates a sound pulse and the other detects it (Figure 3.4). A variation on this uses one transducer that acts both as generator and receiver (Figure 3.5). In its simplest form, a timer is started when the transducer is first stimulated and the timer is stopped when the resultant pulse is detected. The time measured, divided into the distance travelled by the pulse gives the velocity. The distance can often be calibrated using distilled water and an accuracy of the order of $1\,m\,s^{-1}$ is easily achievable with very simple and cheap circuitry. David Cartwright in Chapter 2 describes an instrument based on this technique.

A review of ultrasound measurement techniques can be found in Povey (1997a).

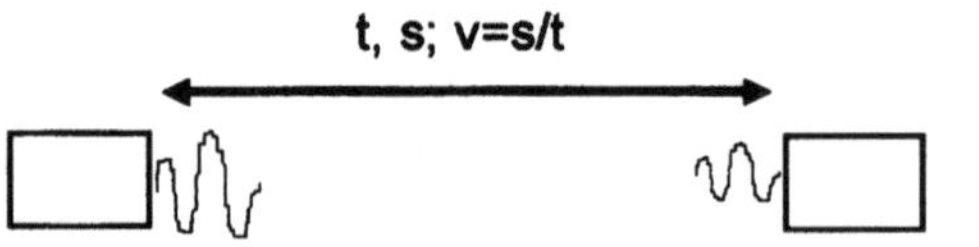

Figure 3.4 Principle of velocity measurement by 'pitch and catch' technique, also known as pulse echo time-of-flight measurement (Povey, 1997a).

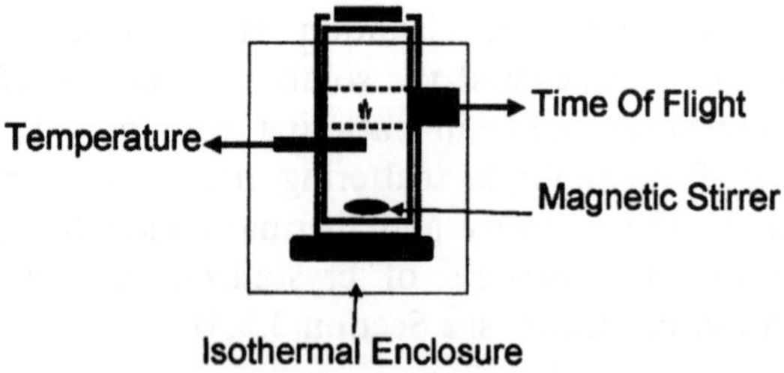

Figure 3.5 Diagram of a commercially available ultrasound velocity meter (Cygnus Limited, Dorchester, UK).

3.3.2 *Ultrasound spectrometry*

By analysing the frequency components of the pulse, ultrasound spectrometry may extend the information obtainable from a pulse echo system and it is possible to obtain both the frequency dependent velocity and the frequency dependent attenuation. Such data may be used to obtain particle size information. This approach is described by Julian McClements in Chapter 5. Secondly, interferometry may be used. There is a wide range of interferometric techniques available (Eggers and Kaatze, 1996; Sakka and Mitaku, 1987). They use either continuous or semi-continuous sources of ultrasound and depend on the analysis of the interference of one part of the ultrasound wave with another to give very accurate data on signal attenuation and wavelength (Figure 3.6).

Techniques such as these have been widely used for the study of chemical and surfactant relaxation but have not been widely used on food materials.

3.3.3 *Ultrasound profiling*

Ultrasound profiling consists of the measurement of the velocity of sound along the height of a tube containing dispersion. As the concentration of the dispersed phase changes with time, under the influence of gravity and other destabilizing forces, the sound velocity changes also. It is possible to compute from equation (3.24) the volume fraction of dispersed phase at any height in the tube and from this build up a picture of the processes by which the dispersion is destabilized. This is a very powerful technique for characterizing the stability of emulsions and dispersions which Margaret Robins describes in Chapter 12.

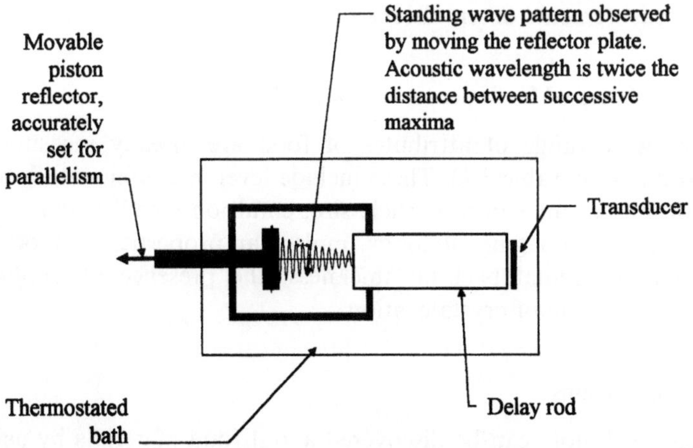

Figure 3.6 Diagram of an ultrasound spectrometer (Povey, 1997a).

3.3.4 Process tomography

Medical ultrasound shows that very detailed pictures of the interior of human beings may be obtained. It is tempting to believe that these techniques may be applied directly to foods. However, food in general presents a far greater variation in acoustical contrast than does the human body and it has proved difficult to transfer medical tomography techniques directly to food processing. Brian Hoyle, in Chapter 4, describes recent progress made in the application of ultrasound tomography to processing.

3.3.5 Particle sizing

A number of ultrasound instruments have recently appeared on the market, claiming the ability to size particles in undiluted samples and, in some cases, in-line. Examples include the Opus instrument, manufactured by Sympatech, Germany, the Ultrasizer from Malvem Instruments, UK, the Matec Acoustosizer (USA) and Penkem (USA). Julian McClements reviews ultrasound particle sizing in Chapter 5.

3.3.6 Process control

For many manufacturers ultrasound will be of interest because of the information it may supply to enhance process control. Jonathon Ridgway and his co-authors, in Chapter 1, describe how ultrasound can be integrated into a complete process control system.

3.3.7 Computer modelling

The quality of the information available from an ultrasonic sensor may be greatly enhanced by comparison with a computer model of the data provided by the sensor. Valerie Pinfield in Chapter 13 describes how computer modelling techniques may be used to refine the data obtained from measurements on emulsions.

3.4 Attributes of food measurable with ultrasound

A very wide range of attributes of food are already measureable with ultrasound (see Table 3.1). These include level measurement, flow rate and flow profile, composition, particle size, emulsion stability, characterization of dispersions, imaging of foods, molecular properties of foods, state of aggregation, animal back-fat thickness, the presence of bubbles, phase changes and state of crystallization.

3.5 Applications

A great deal more can be discovered about food materials by using sound than may at first be apparent. A more general derivation of equations for velocity and attenuation may be obtained from scattering theory, via a generalized dispersion relation and the stress–strain rate relationship (Povey, 1997a; Kittel, 1971). These scattering theory results may be used to interpret frequency dependent velocity and attenuation results in terms of particle size. For example, there is the possibility of sizing particles in the range 10 nm to 1 mm, in undiluted suspensions (see Chapter 5).

Table 3.1 Ultrasound measurements on food materials

Aerated foods	Bugman and Von Stockar (1989); Fairley (1992); Fairley *et al.* (1991a, b, 1992); Manetou (1990)
Alcohols	Antosiewicz and Shugar (1984); D'Angelo *et al.* (1994); Kumar (1991); Rajendran (1993); Sakurai *et al.* (1994); Yasunaga *et al.* (1964)
Alkyl halides and other salts	Conway and Verrall (1966); Freyer (1931); Garnsey *et al.* (1969); Jha and Jha (1986); Kleis and Sanchez (1991); Miecznik (1993); Mikhailov and Shutilov (1965); Millero *et al.* (1977, 1978); Onori (1988, 1990); Owen and Simons (1957); Pillai *et al.* (1983); Rowe and Chou (1970); Sakurai *et al.* (1975)
Amino acids and nucleosides	Buckin *et al.* (1989); Ivnitskii *et al.* (1987); Millero *et al.* (1978); Ogawa *et al.* (1984a, b); Tamura *et al.* (1993)
Bread and starch	Hae *et al.* (1992); Lee and Luan (1992); Povey and Rosenthal (1983)
Carbohydrates	Berchiesi *et al.* (1987); Contreras Montes de Oca *et al.* (1992); Fedotkin *et al.* (1980); Galema and Høiland (1991); Juszkiewicz and Antosiewicz (1986); Kaulgud and Dhondge (1988); Maxwell *et al.* (1984); Nomoto and Kishimoto (1957); Pryor and Roscoe (1954); Shiio (1958); Smith and Winder (1983)
Concentration determination (see also specific food components)	Alig *et al.* (1992)
Crystallization	Akulichev and Bulanov (1982); Archer (1996); Coupland *et al.* (1993); Dickinson and McClements (1995b); Dickinson *et al.* (1990, 1991a, b, 1993a, 1994a, 1996); Dobromyslov and Koshkin (1970); Hiki and Tamura (1981); Hussin and Povey (1984); Kelly *et al.* (1990); McClements and Povey (1988a, b, 1990a); McClements *et al.* (1991, 1993a, b); Miles and Cutting (1974); Pinfield (1996); Povey (1984, 1995); Tamura *et al.* (1986)
Dairy products – eggs	Bae (1996); Choi *et al.* (1987); Javanaud *et al.* (1984); Gould (1972); Mayer and Hiedemann (1958); Povey and Wilkinson (1980); Voissey and Hamilton (1976)
Dairy products – cheese	Bachman *et al.* (1978); Hae *et al.* (1992); O'Donnal *et al.* (1996); Orlandini and Annibaldi (1983)
Dairy products – milk	Ahvenainen *et al.* (1989a, b, 1991); Ay and Gunasekaran (1994a, b); Bachman *et al.* (1978); Benguigui *et al.* (1994); Bhatti *et al.* (1986); Delwiche *et al.* (1980); Dion and Champagne (1986); Fitzgerald *et al.* (1961); Gestrelius (1994); Gestrelius *et al.* (1993, 1994); Griffin and Griffin (1990); Gunasekaran and Ay (1994); Hae *et al.* (1992); Lee and Luan (1992); Mattila (1986); Miles *et al.* (1990); Mog and Winder (1971); Povey (1997b); Saraf *et al.* (1982); Saraf and Samal (1984); Winder *et al.* (1961)

Table 3.1 Continued

Distance and level detection	Brusewitz and Venable (1987); Harrenstein and Brusewitz (1987); Lynnworth (1989); Povey (1989)
Emulsion stability	Alig *et al.* (1992); Cao *et al.* (1988, 1990); Dickinson *et al.* (1993a, b, 1994b, 1996, 1997); Fillery-Travis *et al.* (1992, 1993); Gunning *et al.* (1988); Hibberd *et al.* (1986, 1997); Howe *et al.* (1985, 1986); McClements (1988, 1991, 1994b); McClements and Povey (1987b); McClements *et al.* (1990b); Pinfield *et al.* (1994, 1995, 1996, 1997); Povey (1988, 1993, 1995, 1997b); Povey and McClements (1989)
Flow rate measurement	Brusewitz and Venable (1987); Delwiche *et al.* (1980); Harrenstein and Brusewitz (1987); Lenn *et al.* (1989); Lynnworth (1989); Peek Measurement Ltd (1994); Tucker (1993)
Foreign body detection (includes bacteria)	Ahvenainen *et al.* (1989a, b, 1991); Gestrelius (1994); Gestrelius *et al.* (1993, 1994); Wirtanen *et al.* (1992)
Fouling	Withers (1992, 1994a, b)
Freezing	Archer *et al.* (1996); Hiki and Tamura (1981); Miles and Cutting (1974); Raschi *et al.* (1989)
Fruit and vegetables	Cheng and Haugh (1994); Dull (1986); Garrett and Furry (1972); Hayes and Chingon (1982); Liljedahl and Abbott (1994); Mizrach *et al.* (1989a, b, 1991, 1994a, b); Nielsen (1997); Self (1991); Self *et al.* (1990, 1991, 1992a, b, 1993, 1994); Yamamoto *et al.* (1981)
Gelation	Audebrand *et al.* (1995); Acton *et al.* (1986); Bae (1996); Benguigui *et al.* (1994); Choi *et al.* (1987); Povey (1997b); Pryor *et al.* (1958); Reddy and Suryanarayana (1981); Tanaka *et al.* (1993)
Hydrocarbons	Rossini *et al.* (1953); Wang and Nur (1991)
Meat and fish	Freese and Makow (1968); Fisher *et al.* (1989); Javanaud (1992); Kolb and Nilter (1993); McCauley *et al.* (1994); Miles and Cutting (1974); Miles *et al.* (1991); Ophir *et al.* (1994); Selborne and Hearn (1987); Shore and Miles (1988); Swatland (1995)
Oils and fats	Gladwell *et al.* (1985); Hustad *et al.* (1970); Grigor'ev *et al.* (1976); Hussin and Povey (1984); Javanaud *et al.* (1991); Javanaud and Rahalker (1988); Kelly *et al.* (1990); Kuo (1971); Kuo and Weng (1975); McClements (1988); McClements and Povey (1987a, 1988a–c); Mog and Winder (1971); Murthy *et al.* (1993); Povey (1984)
Packaging	Ahvenainen *et al.* (1989a); Rizvi *et al.* (1993)
Proteins	Acton *et al.* (1986); Almagor *et al.*(1990); Bae (1996); Chalikian *et al.* (1995); Choi *et al.* (1987); Gekko and Hasegawa (1989); Gekko and Noguchi (1979); Gekko and Yamagami (1991); Kharakoz (1991); Kharakoz and Sarvazyan (1993); Mitaku *et al.* (1985); Nomura *et al.* (1987); Pavlovskaya *et al.* (1992a–c); Sarvazyan (1991); Sarvazyan and Hemmes (1979); Tamura *et al.* (1993)
Protein aggregation (includes milk proteins)	Ay and Gunasekaran (1994a, b); Bachman *et al.* (1978); Benguigui *et al.* (1994); Griffin and Griffin (1990); Gunasekaran and Ay (1994); Povey (1997b)
Solids content (see also crystallization)	Fitzgerald *et al.* (1961); Kelly *et al.*(1990); McClements and Povey (1987a); Mog and Winder (1971); Winder *et al.* (1961); Zacharias and Parnell (1972)
Texture measurement	Hayes and Chingon (1982); Juodeikiene *et al.* (1990, 1993, 1994); Lee *et al.* (1988); McCauley *et al.* (1994); Nielsen (1997); Povey and Harden (1981); Tesch *et al.* (1996); Vickers (1984, 1985); Vickers and Wasserman (1979)
Temperature measurement	Cooper (1994); Richardson and Povey (1990)
Wine (see also alcohol determination)	Winder *et al.* (1970)

3.5.1 Water

Water is a component of practically all food materials. However, its properties are very unusual and its ultrasonic properties are no exception. Below 70°C the temperature coefficient of the velocity of sound is positive whilst that of nearly all other liquids is negative. The temperature dependence of water mixtures and solutions is therefore, in general, a complicated function of temperature and concentration. This is illustrated below for the case of a dispersion of sunflower oil-in-water (see Figure 3.7).

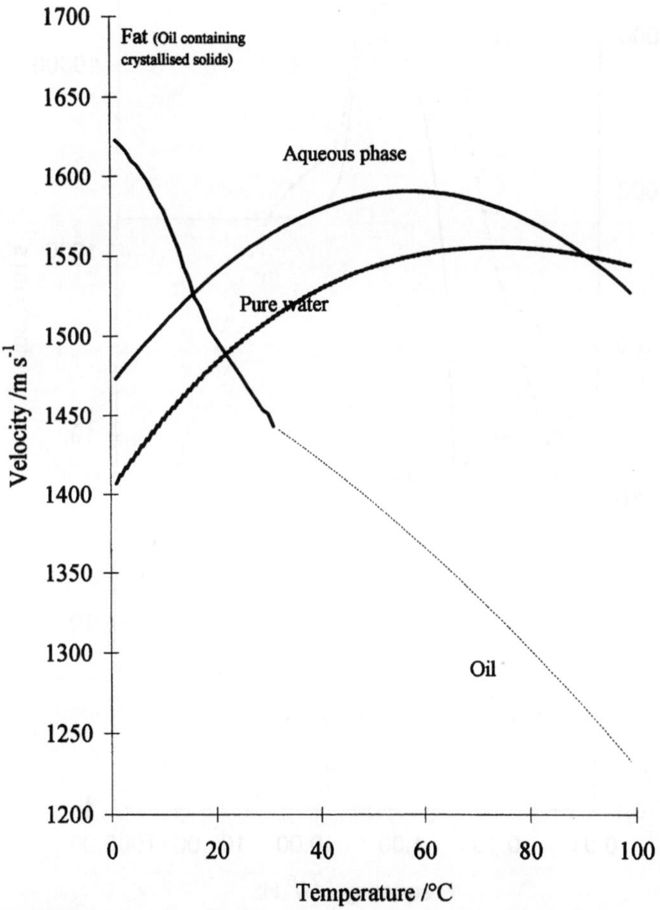

Figure 3.7 Plot of the velocity of sound in pure water and in the components of a sunflower oil-in-water margarine.

3.5.2 Air bubbles

Air bubbles may have a very dramatic effect on ultrasound propagation (Fairley, 1992, Fairley *et al.*, 1991a, b). Normally, the presence of bubbles will attenuate the signal so much that it is no longer detectable. If it were detectable, the bubbles may change the velocity substantially (Figure 3.8). Ultrasound is therefore a very effective detector of bubbles. On the other hand, it is difficult to make any measurements when bubbles are present.

Methods for removing air bubbles in preparation for ultrasound measurements are discussed by Povey (1997a).

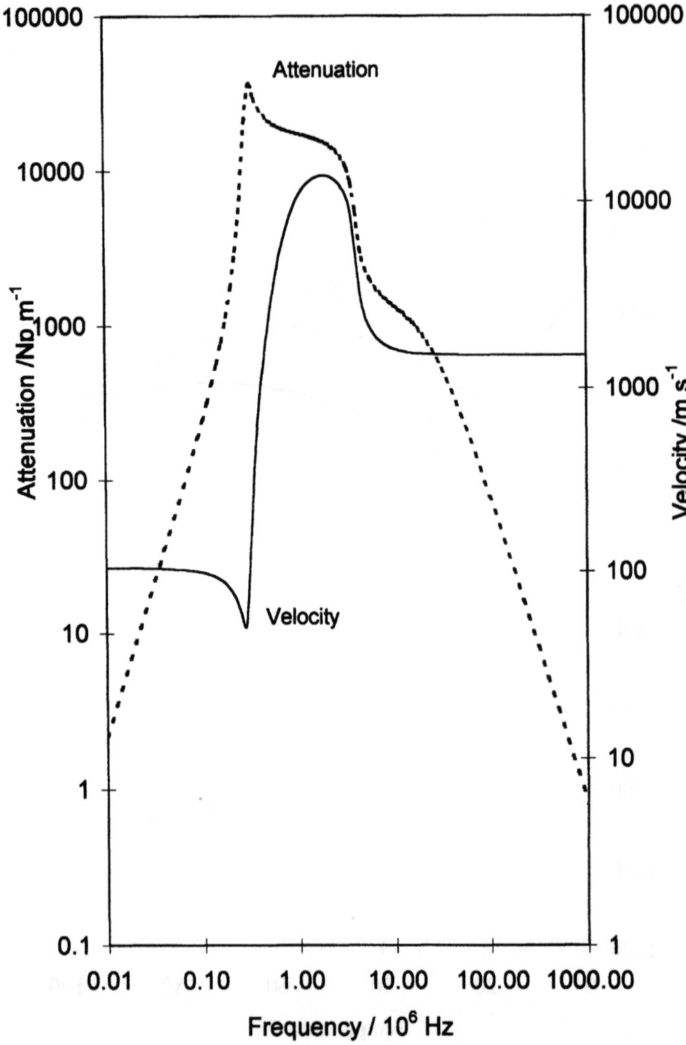

Figure 3.8 Velocity and attenuation plotted against frequency for bubbles of 10-μm diameter and a concentration of 2% v/v (Fairley, 1992; Fairley *et al.*, 1991).

3.5.3 Crystallization

Ultrasound is a very accurate sensor of crystallization and solid content (McClements and Povey, 1987a; Povey, 1997a). An example of the use of ultrasound to measure solid fat content in cocoa butter is given in Figure 3.9. The ultrasound velocity depends on the solid content of fat and of the lecithin, which has been added to the chocolate. The contribution of the lecithin can be clearly seen at temperatures above about 33°C. It has a melting point of just over 80°C. The lecithin is all-solid at the point where the fat begins to crystallize, so the contributions of the solid fat and the lecithin can be studied separately.

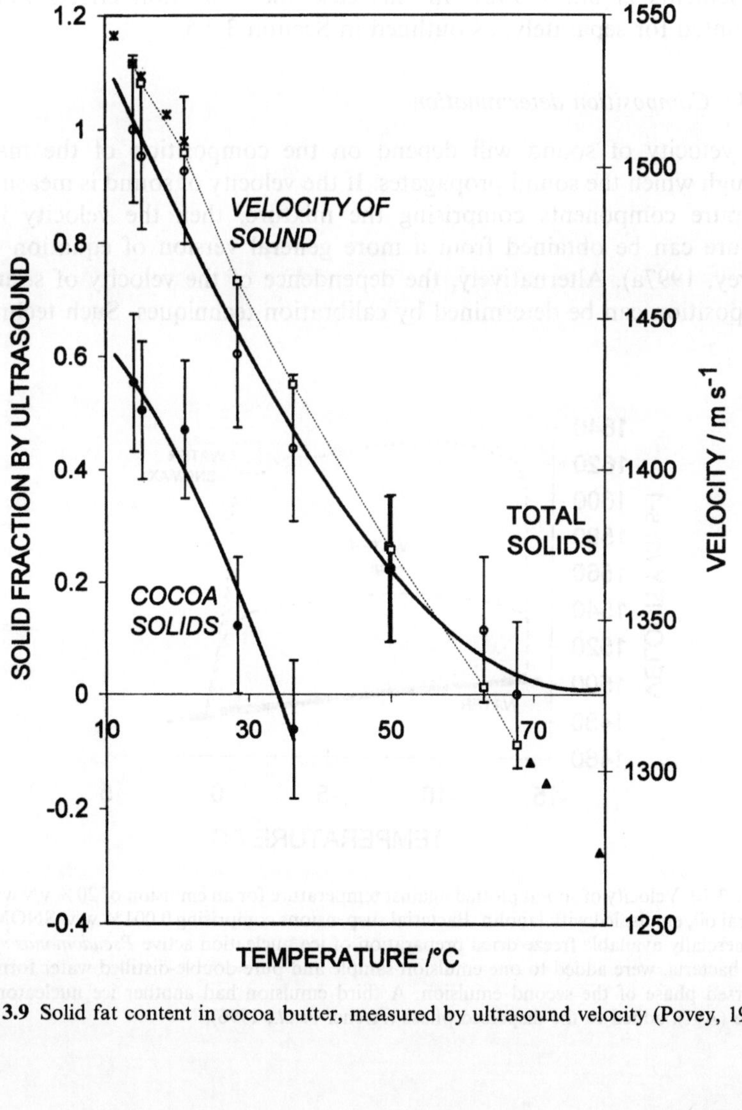

Figure 3.9 Solid fat content in cocoa butter, measured by ultrasound velocity (Povey, 1997).

A different example of ultrasound's use to study crystallization is given in Figure 3.10. Here the effects of a biological ice nucleator are studied using ultrasound velocity measurements of a dispersion of water droplets in oil (Archer *et al.*, 1996). Dispersing the water as fine droplets permits the elimination of the effects of dust and other particles in ice nucleation, allowing the effects of the nucleators to be studied with precision. Here it is seen that the biological ice nucleator SNOMAX, is a much more effective nucleator of ice than silver iodide.

Under some circumstances, crystallization can affect sound propagation through relaxation effects that occur in addition to the scattering effects summarized by equation (3.25). (See, for example, Akulichev and Bulanov, 1982; Dobromyslov and Koshkin, 1970; Hiki and Tamura, 1981; McClements *et al.*, 1993b). In this case the relaxation effects must be accounted for separately, as outlined in Section 3.2.5.

3.5.4 Composition determination

The velocity of sound will depend on the composition of the material through which the sound propagates. If the velocity of sound is measured in the pure components comprising the mixture, then the velocity in the mixture can be obtained from a more general version of equation (3.24) (Povey, 1997a). Alternatively, the dependence of the velocity of sound on composition can be determined by calibration techniques. Such techniques

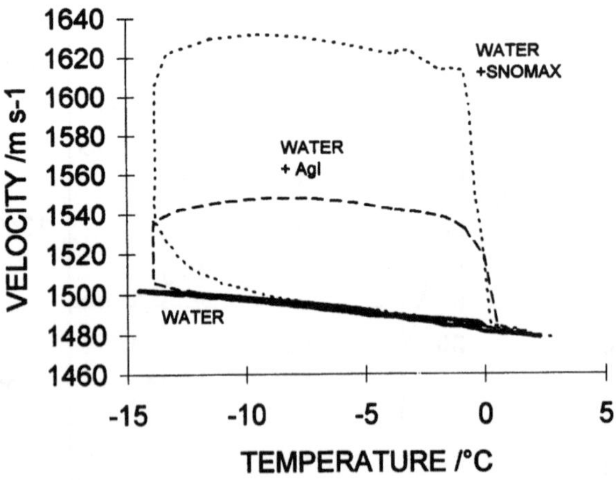

Figure 3.10 Velocity of sound plotted against temperature for an emulsion of 20% v/v water-in-mineral oil, emulsified with lanolin. Bacterial suspensions comprising 0.001% w/w SNOMAX, a commercially available freeze-dried preparation of ice nucleation active *Pseudomonas syringae 31 A* bacteria, were added to one emulsion sample and pure double-distilled water formed the dispersed phase of the second emulsion. A third emulsion had another ice nucleator, silver iodide (AgI), added to the dispersed phase (Archer *et al.*, 1996).

have been used to determine the concentration of sugar in drinks (Contreras Montes de Oca *et al.*, 1992). The technique has also been applied to a very wide range of other solutes (Table 3.1) such as alcohols, salts, amino acids and nucleosides, carbohydrates, hydrocarbons and proteins.

One of the most widespread uses for ultrasound testing is in the determination of animal backfat thickness (Fisher *et al.*, 1989; Freese and Makow, 1968; Javanaud, 1992; Kolb and Nilter, 1993; McCauley *et al.*, 1994; Miles and Cutting, 1974; Miles *et al.*, 1991; Ophir *et al.*, 1994; Selborne and Hearn, 1987; Shore and Miles, 1988; Swatland, 1995). This depends on a measurement of the time-of-flight of an ultrasound pulse, which is reflected at the boundary between the fat and muscle layers on the animal's back. There is a range of commercially available instruments which are portable and robust and give rapid, objective and reproducible results (Swatland, 1995).

3.5.5 Viscoplastic materials

The term 'viscoplastic' applies to the class of materials that possess properties intermediate between solid and liquid. Examples include yoghurt, soft cheese and thick sauces. Ultrasound attenuation measurement is capable of detecting the aggregation in these materials (McClements, 1994b; Povey, 1997a, b). Ultrasound profiling also provides this sort of information, as Margaret Robins points out in Chapter 12.

3.5.6 Dairy products

A considerable amount of research has been carried out into the application of ultrasound to the monitoring of dairy products (Table 3.1). There have been four main areas of effort. Firstly the characterization of milk, pioneered by Winder and his group. Secondly the evaluation of eggs, pioneered by Mayer and Hiedemann (1958). Thirdly the study of the properties of cheese, pioneered by Orlandini and Annibaldi (1983). Fourthly monitoring the aggregation of milk proteins (O'Donnal, Shilton, Cosgrove *et al.*, 1996, Povey, 1997b).

Mayer and Hiedemann (1958), Gould (1972) and Voissey and Hamilton (1976) showed that ultrasound is a viable technique for monitoring egg shell thickness, using a pulse-echo time-of-flight technique. However, changes in egg production and distribution mean that it is not necessary to measure eggshell thickness. Povey and Wilkinson (1980) demonstrated that the attenuation in egg white was related to the quality of the white. The functionality of egg white depends on the proportion of 'thick' to 'thin' white. As the egg ages, so the proportion of thick white falls. Since 'thick' white is essential for meringue production and for a high-quality egg, a method for detecting high levels of 'thick' white might be expected to be attractive to egg producers.

The work on cheese has been hampered by the complicated nature of the product. Some authors have attempted to use ultrasound to characterize the 'texture' (Orlandini and Annibaldi, 1983), others to control cheese manufacture through the identification of the coagulation point of curd (O'Donnal *et al.*, 1996). The evolution of gas bubbles during cheese manufacture is a further obstacle to the adoption of ultrasound monitoring.

The monitoring of aggregation of proteins may yet prove to be a very useful application of ultrasound, particularly when applied the development of networks in proteins such as casein. It is likely that ultrasound spectrometric techniques will come into their own in providing unique insights into the aggregation of casein micelles.

Despite all the effort, I know of no present industrial application of ultrasound in the processing of dairy products. Although there is no particular reason to think that this situation will continue, it is worth examining the difficulties associated with making the sort of measurements referred to here.

Dairy products tend to be very complex systems, which undergo even more complex changes during processing. Many of the difficulties faced by the developer of ultrasound techniques for the monitoring of dairy products are faced by the developers of all sensor modalities (O'Donnal *et al.*, 1996). Small air bubbles in dairy products frequently interfere with ultrasound measurements.

Many authors have attempted to use the changes that undoubtedly occur in the attenuation of sound whilst dairy products are processed to obtain quantitative information. For example texture (Orlandini and Annibaldi, 1983), cheese processing (O'Donnal *et al.*, 1996), egg white quality (Povey and Wilkinson, 1980) and milk clotting (Ay and Gunasekaran, 1994a, b; Bachman *et al.*, 1978). The difficulty is that accurate measurements of sound attenuation are hard to make. On the other hand, it is easy to make accurate measurements of sound velocity and techniques that measure sound velocity are beginning to show some promise. The future for sound attenuation measurement probably lies with the development of low cost acoustic spectrometric techniques, such as those discussed by Julian McClements in Chapter 5. These are likely to be allied to sound velocity measurements.

A considerable effort has been expended on the study of ultrasound for the characterization of gels and gelling systems. This has failed so far because the shear modulus of most food gels is far smaller than their bulk modulus (Povey, 1997b). The result is that the velocity of sound changes very little during gelation and the attenuation changes are difficult to monitor, as explained above. However, the situation may change with the development of ultrasound spectrometry, as pointed out above (McClements, 1997).

3.5.7 *Level measurement*

The ultrasound measurement of level employs the pulse-echo technique to give a time-of-flight measurement (Figure 3.11).

The most sophisticated versions of level detection employ three transducers. A pair is bolted to the sides of the vessel containing the liquid whose level is to be measured. These measure the velocity of sound in the liquid in the tank, by virtue of their fixed acoustic path-length. A time-of-flight measurement of the pulse determines the velocity of sound from 'velocity is distance divided by time'. The probe placed on the bottom of the tank measures the time-of-flight for the pulse to travel to the top of the liquid and back. The distance travelled can then be calculated from the known velocity of sound. This technique gives results that are independent of temperature and are relatively insensitive to foam. However, it can suffer from bubbles contained in the tank liquid.

3.5.8 *Flow measurement*

There are a variety of techniques for flow measurement and the reader is referred to Lynnworth (1989). The two fundamental approaches to the question of flow rate measurement by ultrasound are ultrasound Doppler and ultrasound time-of-flight.

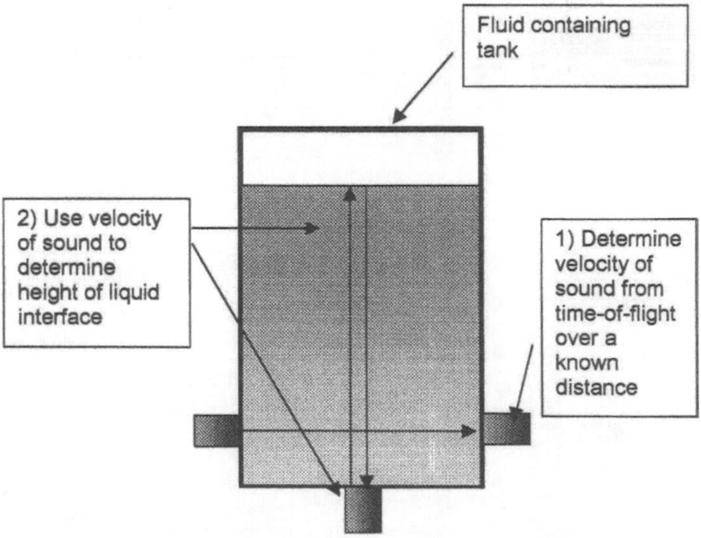

Figure 3.11 Level detection system employing three transducers. The horizontal pair measures the velocity of sound in the tank liquid and the vertical transducer measures the liquid height in the tank.

In the Doppler system, ultrasound transmitted at one frequency is scattered back from acoustical discontinuities suspended in the flowing liquid. The frequency of the backscattered signal is compared to the original frequency and the difference is related to the velocity of the scatterer, relative to the ultrasound probe (Figure 3.12). This approach has the advantage that probes may be bolted onto the surface of a pipe, permitting non-invasive measurement of flow. This is a very attractive prospect since most flow measuring devices are invasive and cause problems from the point of view of hygiene. However, various problems are associated with this sort of device. Firstly, the scattering objects need not be travelling at the same speed or even in the same direction as the liquid conveying them. The scatterers may be bubbles, variations in liquid density arising from turbulence or suspended particles and they can behave in different ways at different parts of a pipeline. Secondly, the scattering objects on which the device depends may disappear from the flow. Bubbles may dissolve and disappear for example. The result is that Doppler flow meters can be unreliable. As the technology develops, Doppler flow metering may be

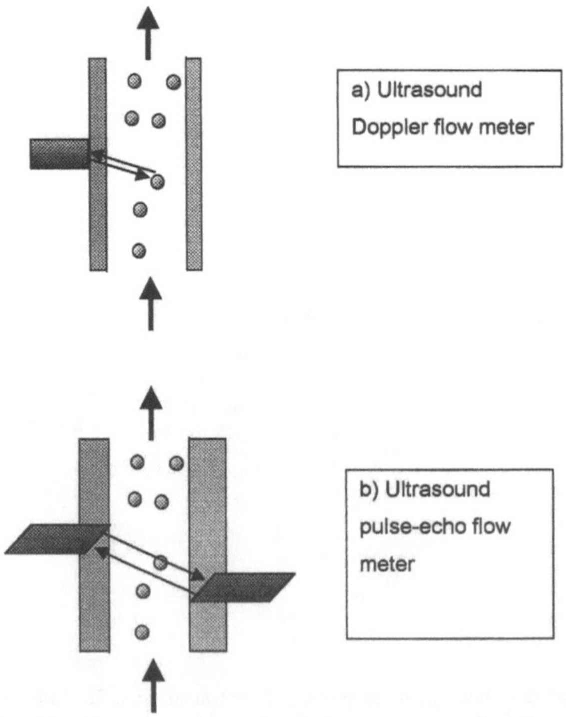

Figure 3.12 Diagrams of (a) a Doppler flow meter and (b) a time-of-flight flow meter.

expected to become much more reliable and it has great potential. For example, there are commercial devices that claim to give the flow profile for a liquid.

A far more reliable approach than ultrasound Doppler is to embed a pair of ultrasound transducers into the pipeline, at an angle to the flow (see Figure 3.12). The time-of-flight measured upstream and downstream of the flow are compared to determine the flow rate.

However, this obviously involves more interference with the pipeline than does the Doppler approach.

3.5.9 Temperature measurement

Ultrasound temperature measurement depends on the measurement of the speed of sound in a calibrated sensor. This is potentially a very accurate method of measuring temperature and Richardson and Povey (1990) suggested that an accuracy of 0.1°C was obtainable from a practical sensor for use in microwave ovens. The absence of a commercial ultrasound temperature sensor is probably due to the widespread availability of other temperature sensing techniques with which people are more familiar. Nevertheless, ultrasound is worth considering where other techniques fail, for example at high temperature, at high pressure and in microwave fields.

3.5.10 Current applications of ultrasound measurement in foods

Applications of ultrasound measurement to foods are summarized in Table 3.1. A cursory examination of the table will indicate that a wide variety of applications of ultrasound have been examined over the years. The number of references is not necessarily a good guide to the success of a particular application. In fact, where an application has matured and in wide use, there may be very few references. Probably the most widespread application of ultrasound is in level and flow measurement. The relative lack of references in these areas is because Table 3.1 is restricted to references to foods. In fact a large literature exists on level and flow measurement that is not restricted to any one type of material; details can be found in Lynnworth (1989). Measurement of meat quality is another commercialized and popular use of ultrasound. The characterization of emulsions is a rapidly developing area in which commercial instrumentation is now appearing. Ultrasound is now being adopted in industry as a method for monitoring crystallization in processes.

3.6 The future – integration with other sensors

Ultrasound techniques are likely to be most effective when integrated with other technologies such as densitometry and electrical tomography. This is because the properties of interest to the food manufacturer are usually not those measured directly by ultrasound. A food manufacturer would only be interested in the compressibility of a protein, for instance, if it told her or him something about the state of denaturation of the molecule. In turn, this information will require a temperature measurement and possibly also a density measurement. Whilst this might appear an obstacle to the development of ultrasound, it is in fact a general feature of all sensor modalities. Where the direct measurement of a single process parameter is possible with a straightforward application of a single sensor modality, this has already been achieved.

We can expect to see significant developments over the next decade in particle sizing, the study of crystallization, particle aggregation, protein denaturation, the molecular properties of foods and the study of phase changes in foods.

Ultrasound non-destructive evaluation has a very exciting future. The possibility already exists of making on-line particle-size measurements in processes. This could lead to the effective removal of serious public health threats such as cryptosporidium in our water supplies. The monitoring of phase transitions such as crystallization is already a reality. Ultrasound is likely to take its place in tomographic inspection systems of the future offering 100% inspection of food materials.

It is time for the food manufacturer to take ultrasound measurement seriously.

3.7 References

Acton J., Clay D., Robinson K.E., Dick R.L. and Acton J. (1986) Structural characteristics of protein gels as determined by an ultrasound imaging technique. *J. Food Sci.* **51**, 524–5.

Agricultural Research Council (1982) Sensors in Agriculture. *Report of Agricultural Research Council Working Party on Transducers* 1, Agricultural Research Council, HMSO, London.

Ahvenainen R., Mattila-Sandholm T. and Wirtanen G. (1989a) Ultrasound penetration through different packaging materials-a non-destructive method for quality control of packed UHT milk. *Lebensm. Wiss. Technol.* **22**, 268–72.

Ahvenainen R., Wirtanen G. and Manninen M. (1989b) Ultrasound imaging – a nondestructive method for monitoring the microbiological quality of aseptically-packed milk products. *Lebensm. Wiss. Technol.* **22**, 382–6.

Ahvenainen R., Wirtanen G. and Mattila-Sandholm T. (1991) Ultrasound imaging – a non-destructive method for monitoring changes caused by microbial enzymes in aseptically-packed milk and soft ice-cream base material. *Lebensm. Wiss. Technol.* **24**, 397–403.

Akulichev V.A. and Bulanov V.N. (1982) Sound propagation in a crystallizing liquid. *Sov. Phys. Acoust.* **27**, 377–81.

Alig I., Haeussler G., Dinger F., Walther L., Blume T. and Stieber F. (1992) Process for monitoring and control of the concentration of suspensions, emulsions and solutions. German Patent De 4023 977 A1.

Almagor A., Yedgar S. and Gavish B. (1990) Solvent viscosity effects on protein dynamics studied by ultrasonic-absorption. *Biotechnology*, **27**, 605–10.

Antosiewicz J. and Shugar D. (1984) Hydration of alcohols by ultrasonic velocity measurements in ternary systems. *J. Solution Chem.*, **13**, 493–503.

Archer G.P., Kennedy C.J. and Povey M.J.W. (1996) Investigations of ice nucleation in water-in-oil emulsions using ultrasound velocity measurements. *Cryoletters*, **17**, 391–6.

Audebrand M., Doublier J.L., Durand D. and Emery J.R. (1995) Investigation of gelation phenomena of some polysaccharides by ultrasonic spectroscopy. *Food Hydrocoll.*, **9**, 195–203.

Ay C. and Gunasekaran S. (1994a) Evaluating milk coagulation time with ultrasonics. *Food Tech.*, **48**, 74.

Ay C. and Gunasekaran S. (1994b) Ultrasonic attenuation measurements for estimating milk coagulation time. *Trans. Am. Soc. Agric. Engng*, **37**, 857.

Bachman S., Klimaczak B. and Gasnya Z. (1978) Non-destructive viscometric studies of enymic milk coagulation. I An ultrasonic viscometric method for the measurement of milk clotting activity of proteolytic enzymes. *Acta Aliment. Pol.*, **4**, 55–62.

Bae J.-R. (1996) Ultrasonic study of gelation process in egg-white proteins. *Jpn. J. Appl. Phys.*, **35**, 2934–8.

Benguigui L., Emery J.R., Durand D. and Busnel J.P. (1994) Ultrasonic study of milk-clotting. *Lait*, **74**, 197–206.

Berchiesi G., Amico A., Vitali G., Amici L. and Litargini P. (1987) Ultrasonic investigation in aqueous solutions of sucrose. *J. Mol. Liq.*, **33**, 157–81.

Bhatti S. S., Bhatti R. and Singh S. (1986) Ultrasonic testing of milk. *Acustica*, **62**, 96–9.

Bjomo L. (1991) Ultrasound in the food industry. *Proceedings of Ultrasonics International '91*, 23–9.

Brusewitz G.H. and Venable P.B. (1987) Technical notes: sound level measurements of flowing grain. *Trans. Am. Soc. Agric. Engng*, **30**, 863.

Buckin V.A., Kankiya, B.I. and Kazaryan, R.L. (1989) Hydration of nucleosides in dilute aqueous-solutions ultrasonic: velocity and density-measurements. *Biophysical Chem.*, **34**, 211–23.

Bugman G. and Von Stockar U. (1989) Characterizing bubbles in bioreactors by ultrasound. *Trends in Biotechnology*, **7**, 166–9.

Cao Y., Dickinson E. and Wedlock D.J. (1988) Influence of polysaccharides on the creaming of casein-stabilized emulsions. *Food Hydrocoll.*, **2**, 443–54.

Cao Y., Dickinson E. and Wedlock D.J. (1990) Creaming and flocculation in emulsions containing polysaccharide. *Food Hydrocoll.*, **4**, 185–95.

Chalikian T.V., Gindikin V.S. and Breslauer K.J. (1995) volumetric characterizations of the native, molten globule and unfolded states of cytochrome-C at acidic Ph. *J. Mol. Biol.*, **250**, 291–306.

Cheng Y. and Haugh C.G. (1994) Detecting hollow heart in potatoes using ultrasound. *Trans. Am. Soc. Agric. Engng*, **37**, 217–24.

Choi P.K., Bae J.-R. and Takagi K. (1987) Gelation and ultrasonic hysteresis in egg-white. *Jpn. J. Appl. Phys.*, **26**, 32–4.

Contreras Montes de Oca N.I., Fairley P., McClements D.J. and Povey M.J.W. (1992) Analysis of the sugar content of fruit juices and drinks using ultrasonic velocity measurements. *Int. J. Food Sci. Tech.*, **27**, 515–29.

Conway B.E. and Verral R.E. (1966) Partial molar volumes and adiabatic compressibilities of tetraalkyllammonium and aminium salts in water. I. Compressibility behaviour. *J. Phys. Chem.*, **70**, 3952–61.

Cooper T.J.R. (1994) Developments in temperature measurement and control. In *Food Technology International Europe*, Editor A. Turner, SPG, London.

Coupland J., Dickinson E., McClements D.J., Povey M.J.W. and de Rancourt de Mimmerand C. (1993) Crystallization in simple paraffins and monoacid saturated triacylglycerols dispersed in water. *Food Colloids and Polymers: Stability and Mechanical Properties*, Editors E. Dickinson and P. Walstra, Royal Society of Chemistry, Cambridge.

D'Angelo M., Onori G. and Santucci A. (1994) Self-association of monohydric alcohols in water: compressibility and infrared absorption measurements. *J. Chem. Phys.*, **100**, 3107–12.

Delwiche M.J., Scott M.J. and Drost C.J. (1980) Ultrasonic measurement of teat milk flow. *Trans. Am. Soc. Agric. Engng*, **23**, 746.

Dickinson E. and McClements D.J. (1995a) Ultrasonic characterization of food colloids. In *Advances in Food Colloids*, Editors E. Dickinson and D.J. McClements, Blackie, Glasgow.

Dickinson E. and McClements D.J. (1995b) Crystallization in oil-in-water emulsions. In *Advances in Food Colloids*, Editors E. Dickinson and D.J. McClements, Blackie, Glasgow.

Dickinson E., Goller M.I., McClements D.J., Peasgood S. and Povey M.J.W. (1990) Ultrasonic monitoring of crystallisation in an oil-in-water emulsion. *J. Chem. Soc. Faraday Trans.*, **86**, 1147–8.

Dickinson E., McClements D.J. and Povey M.J.W. (1991a) Ultrasonic investigation of the particle size dependence of crystallisation in *n*-hexadecane–water emulsions. *J. Coll. Interface Sci.*, **142**, 103–10.

Dickinson E., Goller M.I., McClements D.J. and Povey M.J.W. (1991b) Monitoring crystallisation in simple and mixed oil-in-water emulsions using ultrasonic velocity measurement. *Food Polymers, Gels and Colloids*, 171–9.

Dickinson E., Kruizenga F., Povey M.J.W. and van der Molen M. (1993a) Crystallization in oil-in-water emulsions containing liquid and solid droplets. *Coll. Surf. A*, **81**, 273–9.

Dickinson E., Goller M.I. and Wedlock D.J. (1993b) Creaming and rheology of emulsions containing polysaccharide and non-ionic or anionic surfactants. *Coll. Surf. A*, **75**, 195–201.

Dickinson E., Ma J. and Povey M.J.W. (1994a) Creaming of concentrated oil-in-water emulsions containing Xanthan. *Food Hydrocoll.* **8**, 481–97.

Dickinson E., Ma J., Pinfield V.J. and Povey M.J.W. (1994b) Ultrasonic studies of the creaming of concentrated emulsions. In *Food Macromolecules and Colloids*, Editors E. Dickinson and D. Lorient, Royal Society of Chemistry, Cambridge.

Dickinson E., Ma J. and Povey M.J.W. (1996) Crystallization kinetics in oil-in-water emulsions containing a mixture of solid and liquid droplets. *J. Chem. Soc. Faraday Trans.*, **192**, 1213–15.

Dickinson E., Golding M. and Povey M.J.W. (1997) Creaming and flocculation of oil-in-water emulsions containing sodium caseinate. *J. Coll. Interface Sci.*, **185**, 515–29.

Dion J.L. and Champagne C. (1986) Detection of particles in milk by ultrasonics. *Technique Laitiere & Marketing* , **1013**, 27–8.

Dobromyslov N.A. and Koshkin N.I. (1970) Velocity of sound in molecular crystals near the melting point. *Sov. Phys. Acoust.*, **15**, 386–7.

Dull G.G. (1986) Nondestructive evaluation of quality of stored fruits and vegetables. *Food Tech.*, **40**, 106–10.

Eggers F.G. and Kaatze U. (1996) Broad-band ultrasonic measurement techniques for liquids. *Meas. Sci. Technol.*, **7**, 1–19.

Fairley P. (1992) Characterisation of bubbly liquids using ultrasound. PhD thesis, Leeds.

Fairley P., McClements D.J. and Povey M.J.W. (1991a) Ultrasonic characterisation of aerated foodstuffs. *Ultrasonic International '91 Conference Proceedings*, **13**, 79–82.

Fairley P., McClements D.J. and Povey M.J.W. (1991b) Ultrasonic characterisation of some aerated foodstuffs. *Proc. Inst. Acoust.*, **13**, 63–70.

Fairley P., McClements D.J. and Povey M.J.W. (1992) An ultrasonic investigation of aerated cream. In *Developments in Acoustics and Ultrasonics*, Editors M.J.W. Povey and D.J. McClements, Institute of Physics, Bristol, pp. 239–42.

Fedotkin I.M., Sukauskas V.A., Chepurnoi M., Shnaider V.E. and Klimenko M.I. (1980) Propagation of ultrasonic waves in sugar solutions and sugar juices. *Pishchevaya Promyshlennost', Respublikanskii Mezhvedomstvennyi Nauchno Tekhnicheskii*, **26**, 33–7.

Fillery-Travis A.J., Gunning P.A., Hibberd D.J. and Robins M.M. (1992) Mechanisms of flocculation in oil-in-water emulsions. Differentiation by creaming behaviour. In *Gums and Stabilizers for the Food Industry 6*, Editors G.D. Phillips, P.A. Williams and D.J. Wedlock, IRL, Oxford, pp. 363–70.

Fillery-Travis A.J., Gunning P.A., Hibberd D.J. and Robins M.M. (1993) Coexistent phases in concentrated polydisperse emulsions flocculated by non-adsorbing polymer. *J. Coll. Interface Sci.*, **159**, 189–97.

Fisher J.C., Miles C.A. and Newman R. (1989) Evaluation of velocity of ultrasound, video image analysis and light reflectance probes in carcass classification schemes. *Proceedings of the 35th International Congress of Meat Science and Technology*, 20–25 August 1989, pp. 257–62.

Fitzgerald J.W., Ringo G.R. and Winder W.C. (1961) An ultrasonic method for measurement of solids-not-fat and milk fat in fluid milk. I Acoustic properties. *J. Dairy Sci.*, **44**, 1165.

Freese M. and Makow D. (1968) High-frequency ultrasonic properties of freshwater fish tissue. *J. Acoust. Soc. Am.*, **44**, 1282–9.

Freyer E.B. (1931) Sonic studies of the physical properties of liquids. II The velocity of sound in solutions of certain alkali halides and their compressibilities. *J. Am. Chem. Soc.*, **53**, 1313–20.

Frindi M., Michels B. and Zana R. (1994) Ultrasonic-absorption studies of surfactant exchange between micelles and the bulk phase in aqueous micellar solutions of amphoteric surfactants. *J. Phys. Chem.*, **98**, 6607–11.

Galema S.A. and Høiland H. (1991) Stereochemical aspects of hydration of carbohydrates in aqueous-solutions 3. Density and ultrasound measurements. *J. Phys. Chem.*, **95**, 5321–6.

Gamsey R., Boe R.J., Mahoney R. and Litovitz T.A. (1969) Determination of electrolyte apparent molal compressibilities at infinite dilution using a high-precision ultrasonic velocimeter. *J. Chem. Phys.*, **50**, 5222–8.

Garrett P.R. and Furry R.B. (1972) Velocity of sonic pulses in apples. *Trans. Am. Soc. Agric. Engng*, **15**, 770.

Gekko K. and Hasegawa Y. (1989) Effect of temperature on the compressibility of native globular proteins. *J. Phys. Chem.*, **93**, 426–9.

Gekko K. and Noguchi H. (1979) Compressibility of globular proteins in water at 25°C. *J. Phys. Chem.*, **83**, 2706–14.

Gekko K. and Yamagami K. (1991) Flexibility of food proteins as revealed by compressibility. *J. Ag. Food Chem.*, **39**, 57–62.

Gestrelius H. (1994) Ultrasonic doppler – a possible method for non-invasive sterility control. *Food Control*, **5**, 103–5.

Gestrelius H., Hertz H., Nuamu T.G., Persson H.W. and Lindstrom K. (1993) A nondestructive method for microbial quality control of aseptically packaged milk. *Lebensm. Wiss. Technol.*, **26**, 334.

Gestrelius H., Mattila-Sandholm T. and Anvenainen T. (1994) Methods for non-invasive sterility control in aseptically packaged foods. *Tr. Food Sci. Technol.*, **5**, 379–3.

Gladwell N.R., Javanaud C., Peers K.E. and Rahalker R.R. (1985) Ultrasonic behavior of edible oils – correlation with rheology. *J. Am. Oil. Chem. Soc.*, **62**, 1231–6.

Gould R.W. (1972) Non-destructive egg shell thickness measurements using ultrasonic energy. *Poult. Sci.*, **51**, 1460–1.

Griffin W.G. and Griffin M.C.A. (1990) The attenuation of ultrasound in aqueous suspensions of casein micelles from bovine-milk. *J. Acoust. Soc. Am.*, **87**, 2541–50.

Grigor'ev S.B., Manucharov Y.S., Milhailov I.G. and Khakimov O. (1976) Acoustic relaxation in vegetable oils. *Sov. Phys. Acoust.*, **21**, 331–4.

Gunasekaran S. and Ay C. (1994) Evaluating milk coagulation with ultrasonics. *Food Tech.*, **48**, 74–8.

Gunning P.A., Hibberd D.J., Howe A.M. and Robins M.M. (1988) Gravitational destabilization of emulsions flocculated by non-adsorbed xanthan. *Food Hydrocoll.*, **2**, 119–29.

Hae, O.K., Lee, H. and Luan, H. (1992) Use of an ultrasonic technique to evaluate the rheological properties of cheese and dough. *J. Food Engng*, **16**, 127–50.

Harrenstein A. and Brusewitz G.H. (1987) Sound level measurements of flowing wheat. *Trans. Am. Soc. Agric. Engng*, **29**, 1114.

Hayes C.F. and Chingon H.T.G. (1982) Acoustic properties of papaya. *Text. Stud.*, **13**, 397.

Hibberd D.J., Holmes A.K., Garrood M., Fillery-Travis A.J., Robins M.M. and Challis R.E. (1997) Ultrasonic characterization of flocculation in oil-in-water emulsions. In *Food Colloids – Proteins, Lipids and Polysaccharides*, Editor E. Dickinson, Royal Society of Chemistry, Cambridge.

Hibberd D.J., Howe A.M., Mackie A.R., Purdy P.W. and Robins M.M. (1986) Measurement of creaming profiles in oil-in-water emulsions. *Food Emulsions and Foams*, Editor E. Dickinson, Royal Society of Chemistry, Cambridge.

Hiki Y. and Tamura J. (1981) Ultrasonic-attenuation in ice crystals near the melting temperature. *Journal de Physique*, **42**, 547–52.

Howe A.M., Mackie A.R. and Robins M.M. (1986) Technique to measure emulsion creaming by velocity of ultrasound. *J. Disp. Sci. Tech.*, **7**, 231–43.

Howe A.M., Mackie A.R., Richmond P. and Robins M.M. (1986) Creaming of oil-in-water emulsions containing polysaccharides. In *Gums and Stabilizers for the Food Industry 3*,

Editors G.O. Phillips, D.J. Wedlock and P.A. Williams. Elsevier Applied Science Publishers, Barking, England, pp. 295–310.

Hussin A.B.B. and Povey M.J.W. (1984) A study of dilation and acoustic propagation in solifying fats and oils. II – Experiment. *J. Am. Oil. Chem. Soc.*, **61**, 560–4.

Hustad G.O., Richardson T., Winder W.C. and Dean M.P. (1970) Factors affecting sound velocity in fats and oils. *J. Dairy Sci.*, **53**, 1525–31.

Ivnitskii D.M., Priev A.I. and Shilnikov G.V. (1987) Ultrasonic velocimetry of immunoglobulins. *Sov. Phys. Acoust.*, **33**, 390–2.

Javanaud C. (1988) Applications of ultrasound to food systems. *Ultrasonics*, **26**, 117–23.

Javanaud C. (1992) Two aspects of the use of acoustics in the quantitative assessment of fish stocks. In *Developments in Acoustics and Ultrasonics*, Editors M.J.W. Povey and D.J. McClements, Institute of Physics, Bristol, pp. 49–68.

Javanaud C., Gladwell N.R., Gouldby S.J., Hibberd D.J., Thomas A. and Robins M. M. (1991) Experimental and theoretical values of the ultrasonic properties of dispersions: effect of particle state and size distribution. *Ultrasonics*, **29**, 331–7.

Javanaud C. and Rahalker R.R. (1988) Velocity of sound in vegetable oils. *Fett Wissenschaft Technologie*, **90**, 73–5.

Javanaud C. and Robins M.M. (1993) Ultrasonic methods. In *Food Process Monitoring Systems*, Editors A.C. Pinder and G. Godfrey, Blackie, Glasgow.

Javanaud C., Rahalker R.R. and Richmond, P. (1984) Measurement of speed and attenuation of ultrasound in egg white and egg yolk. *J. Acoust. Soc. Am.*, **76**, 670–5.

Jha D.K. and Jha B.L. (1986) Free volumes of some I and IIB group salt solutions in water from ultrasonic velocity. *Colloids and Surfaces*, **14**, 44–9.

Juodeikiene G., Petrauskas A., Riedel O. and Mohr E. (1990) Possibilities of application of ultrasonics to determination of the texture of wafer sheets. *Ernaebrung*, **14**, 507–11.

Juodeikiene G., Basinskiene L., Kunigelis V., Adomatis V., Riedel O. and Mohr E. (1993) Modelling and experimental evaluation of the correlation between acoustically-determined texture characteristics and quality of wafer products. *Ernaebrung*, **17**, 92–4.

Juodeikiene G., Schleining G., Kunigelis V. and Adomaitis V. (1994) Theoretical and experimental investigated correlation between acoustic and structural properties of crackers. *Ernaebrung* **18**, 235.

Juszkiewicz A. and Antosiewicz J. (1986) Ultrasonic velocity studies on carbohydrates in aqueous ethanolic: solutions. *Zeitschrift fur Physikalische Chemie Neue Folge*, **148**, 185–95.

Kaulgud M.V. and Dhondge S.S. (1988) Apparent molal volumes and apparent molal compressibilities of some carbohydrates in dilute aqueous solutions at different temperatures. *Ind. J. Chem. A*, **27**, 6–11.

Kaye G.W.C. and Laby T.H. (1986) *Tables of Physical and Chemical Constants*, 15th edition, Longman Science, London, p. 29.

Kelly T., McClements D.J. and Povey M.J.W. (1990) Sensor for solid fat determination and use thereof. GB Patent Application No. 9013530.2.

Kharakoz D.P. (1991) Volumetric properties of proteins and their analogs in diluted water solutions. 2. Partial adiabatic compressibilities of amino-acids at 15–70°C. *J. Phys. Chem.*, **95**, 5634–42.

Kharakoz D.P. and Sarvazyan A.P. (1993) Hydrational and intrinsic compressibilities of globular-proteins. *Biopoylmers*, **33**, 11–26.

Khimunin A.S. (1972) Numerical calculation of the diffraction corrections for the precise measurement of ultrasound absorption. *Acustica*, **27**, 173–81.

Kittel C. (1971) Phonons and lattice vibrations. In *Introduction to Solid State Physics*, John Wiley, New York.

Kleis S.J. and Sanchez L.A. (1991) Dependence of sound-velocity on salinity and temperature in saline solutions. *Solar Energy*, **46**, 371–5.

Kolb R. and Nilter G. (1993) Digitized ultrasonic images on live pigs for prediction of meat proportion in the belly. *Zuchtungskunde*, **65**, 297.

Kumar A. (1991) Thermodynamic properties of pentanol deduced from ultrasonic velocity: data of its aqueous-solutions. *Phys.Chem. Liq.*, **23**, 87–92.

Kuo H.L. (1971) Variation of ultrasonic velocity and absorption with temperature and frequency in high viscosity vegetable oils. *Jpn. J. Appl. Phys.*, **10**, 167–70.

Kuo H.L. and Weng J.S. (1975) Temperature and frequency dependence of ultrasonic velocity and absorption in sperm and seal oils. *J. Am. Oil. Chem. Soc.*, **52**, 167–8.

Lee H.O. and Luan H. (1992) Use of and ultasonic technique to evaluate the rheological properties of cheese and dough. *J. Food Engng*, **16**, 127.

Lee G.H., Deibel A.E., Glembin C.T. and Munday E.G. (1988) Analysis of food crushing sounds during mastication: frequency–time studies. *J. Text. Stud.*, **19**, 27.

Lenn C.P., Sanderson M.L., Spence L. and Richardson P. (1989) Feasibility of ultrasonic techniques for the measurement of food particle flow in aseptic process plant. Technical Memorandum, Campden Food & Drink Research Association, Campden.

Liljedahl L.A. and Abbott J.A. (1994) Changes in sonic resonance of delicious and golden delicious apples undergoing accelerated ripening. *Trans. Am. Soc. Agric. Engng*, **37**, 907–12.

Lynnworth L.C. (1989) Ultrasonic measurements for process control: theory, techniques, applications. In *Ultrasonic Measurements For Process Control*, Academic Press, San Diego, CA.

Manetou A. (1990) Development of an ultrasonic sensor for monitoring air content. Report No. 667, Leatherhead Food Research Association, Leatherhead.

Mattila T. (1986) Ultrasound echography – a non-destructive method for quality control of packed UHT milk. *Proceedings of Symposium on Mastitis Control and Hygienic Production* of *Milk*, **1**, 293.

Maxwell J.L., Kurtz F.A. and Strelka B.J. (1984) Specific volume (density) of saccharide solutions (corn syrups and blends) and partial specific volumes of saccharide–water mixtures. *J. Ag. Food Chem.*, **32**, 974–9.

Mayer W.G. and Hiedemann E.A. (1958) On the feasibility of ultrasonic grading of shell eggs. *Food Research*, **24**, 97–103.

McCann J.D. (1986) A review of ultrasonic techniques and applications, with particular reference to low frequency systems and their use in the food processing industry. *AERE–R 11271*, Editor J.D. McCann, UK Atomic Energy Authority, Harwell.

McCauley J.D., Thane B.R. and Whittaker A.D. (1994) Fat estimation in beef by ultrasound images using texture and adaptive logic networks. *Trans. Am. Soc. Agric. Engng*, **37**, 997.

McClements D.J. (1988) *The use of ultrasonics for characterising fats and emulsions*. PhD thesis, Leeds.

McClements D.J. (1991) Ultrasonic characterisation of emulsions and suspensions. *Adv. Coll. Int. Sci.*, **37**, 33–72.

McClements D.J. (1994a) Ultrasonic NDT of foods and drinks. *International Advances in Nondestructive Testing*, **17**, Editor W.J. McGonnagle, Gordon & Breach, Basel, p. 63.

McClements D.J. (1994b) Ultrasonic determination of depletion flocculation in oil-in-water emulsions containing a non-ionic surfactant. *Coll. Surf. A*, **90**, 25–35.

McClements D.J. (1995) Ultrasonics in food processing. In *Food Processing: Recent Developments*, Editor A.G. Gaonkar, Elsevier, London.

McClements D.J. (1997) Ultrasonic characterization of foods and drinks: principles, methods and applications. *Critical Reviews in Food Science and Nutrition*, **37**, Editor G. Clydesdale, CRC Press, Boca Raton, FL, pp. 1–46.

McClements D.J. and Povey M.J.W. (1987a) Solid fat content determination using ultrasonic velocity measurements. *Int J. Food Sci. Tech.*, **22**, 419–28.

McClements D.J. and Povey M.J.W. (1987b) Ultrasonic velocity as a probe of emulsions and suspensions. *Adv. Coll. Int. Sci.*, **27**, 285–316.

McClements D.J. and Povey M.J.W. (1988a) Ultrasonic velocity measurements in some triglycerides and vegetable oils. *J. Am. Oil. Chem. Soc.*, **65**, 1787–9.

McClements D.J. and Povey M.J.W. (1988b) Comparison of pulsed NMR and ultrasonic velocity techniques for determining solid fat contents. *Int J. Food Sci. Tech.*, **23**, 159–70.

McClements D.J. and Povey M.J.W. (1988c) Investigation of phase transitions in glyceride/paraffin oil mixtures using ultrasonic velocity measurements, *J. Am. Oil. Chem. Soc.*, **65**, 1791–6.

McClements D.J., Dickinson E. and Povey M.J.W. (1990a) Crystallisation in hydrocarbon-in-water emulsions containing a mixture of solid and liquid droplets. *Chem. Phys. Letts.*, **172**, 449–52.

McClements D.J., Povey M.J.W., Jury M. and Betsanis E. (1990b) Ultrasonic characterization of a food emulsion. *Ultrasonics*, **28**, 266–72.

McClements D.J., Povey M.J.W. and Dickinson E. (1991) Ultrasonic monitoring of melting/crystallization in emulsions. *Ultrasonic International '91 Conference Proceedings*, pp. 107–10.

McClements D.J., Dickinson E., Dungan, S.R., Kinsella, J.E., Ma J. and Povey, M.J.W. (1993a) Effect of emulsifier type on the crystallization kinetics of oil-in-water emulsions containing a mixture of solid and liquid droplets, *J. Coll. Int. Sci.*, **160**, 293–7.

McClements D.J., Povey M.J.W. and Dickinson E. (1993b) Absorption and velocity dispersion due to crystallization and melting of emulsion droplets. *Ultrasonics*, **31**, 433–7.

Miecznik P. (1993) Investigations of complexing in aqueous-solutions of dimethylulfoxide and zinc-chloride by an ultrasonic spectroscopy method. *Acustica*, **78**, 36–45.

Mikhailov I.G. and Shutilov V.A. (1965) Nonlinear acoustical properties of aqueous electrolytic solutions. *Sov. Phys. Acoust.*, **10**, 385–9.

Miles C.A. and Cutting C.L. (1974) Technical note: changes in the velocity of ultrasound in meat during freezing. *J. Food Tech.*, **9**, 119–22.

Miles C.A., Fursey G.A.J., Fisher J.C. and Page S.J. (1991) Estimation of lamb carcass composition from measurements of the speed of ultrasound in the soft tissues of live animals and carcasses. *Meat Science*, **30**, 245–56.

Miles C.A., Shore D. and Langley K.R. (1990) Attenuation of ultrasound in milks and creams. *Ultrasonics*, **28**, 394–400.

Millero F.J., Ward G.K. and Chetirkin P.V. (1977) Relative sound velocities of sea salts at 25°C. *J. Acoust. Soc. Am.*, **61**, 1492–8.

Millero F.J., Lo Surdo A. and Shin C. (1978) The apparent molal volumes and adiabatic compressibilities of aqueous amino acids at 25°C. *J. Phys. Chem.*, **82**, 784–92.

Mitaku S., Kagayama T. and Kataoka R. (1985) Ultrasonic properties of proteins. *Jpn. J. Appl. Phys.*, **24**, 43–45.

Mizrach A., Galili N. and Rosenhouse G. (1989) Determination of fruit and vegetable properties by ultrasonic excitation. *Trans. Am. Soc. Agric. Engng*, **32**.

Mizrach A., Galili N., Rosenhouse G. and Teitel D.C. (1991) Acoustical, mechanical and quality parameters of winter grown melon tissue. *Trans. Am. Soc. Agric. Engng*, **34**, 2135.

Mizrach A., Galili N. and Rosenhouse G. (1994a) Determining quality of fresh products by ultrasonic excitation. *Food Tech.*, **48**, 68–71.

Mizrach A., Galili N., Teitel D.C. and Rosenhouse G. (1994b) Ultrasonic evaluation of some ripening parameters of autumn and winter-grown Galia melons. *Scientia Horticulturae*, **56**, 291–7.

Mog C.F. and Winder W.C. (1971) Development of an ultrasonic method for continuously monitoring the fat and solids-non-fat content of fluid milk. *J. Dairy Sci.*, **54**, 756.

Murthy V.R., Siva Kumar K.V., Ranga Reddy R.N.V. and Narasimha Murthy Y. (1993) Ultrasonic studies of natural oils and fats of plant origin. *Cryst. Res. Technol.*, **28**, 723–7.

Nielsen M. (1997) *Texture measurement of fruits and vegetables by the use of ultrasonics*. PhD thesis, The Royal Veterinary and Agricultural University, Copenhagen.

Nomoto O. and Kishimoto T. (1957) Molecular sound velocity of aqueous sugar solutions. *J. Phys. Soc. Jpn.*, **12**, 311.

Nomura H., Kawaizumi F. and Iida T. (1987) Partial specific compressibility and its relation to solvation of solute based on measurents of suspension system. *Bull. Chem. Soc. Jpn.*, **60**, 25–30.

O'Donnal C., Shilton N., Cosgrove N. and O'Callaghan (1996) On-line sensors in cheese making. *Food Technolology International Europe*, Editor A. Turner, SPG, London, pp. 103–5.

Ogawa, T., Mizutani, K. and Yasuda, M. (1984a) The volume, adiabatic compressibility and viscosity of amino acids in aqueous alkali–chloride solutions. *Bull. Chem. Soc. Jpn.*, **57**, 2064–8.

Ogawa T., Yasuda M. and Mizutani K. (1984b) Volume and adiabatic compressibility of amino acids in urea–water mixtures. *Bull. Chem. Soc. Jpn.*, **57**, 662–6.

Ohashi E. and Karube I. (1993) Sensors for the food industry. *Food Control 4*, 183–8.

Onori G. (1988) Ionic hydration in sodium chloride solutions. *J. Chem. Phys.*, **89**, 510–16.

Onori G. (1990) Reply to the comment on: 'Ionic hydration in sodium chloride solutions'. *J. Chem. Phys.*, **92**, 7720–1.

Ophir J., Miller J., Ponnekauti R.K., Cespedes I. and Whittaker A.D. (1994) Elastography of beef muscle. *Meat Science*, **36**, 239.

Orlandini I. and Annibaldi S. (1983) Investigation about new techniques for the valuation of parmigiano-reggiano cheese structure. Ultrasounds and X-rays. *Scienza e Tecnica Lattiero-Casearia*, **34**, 20–30.

Owen B.B. and Simons H.L. (1957) Standard partial molal compressibilities by ultrasonics. I Sodium chloride and potassium chloride at 25°C. *J. Phys. Chem.*, **61**, 479–82.

Pavlovskaya G.E., McClements D.J. and Povey M.J.W. (1992a) A preliminary study of the influence of dextran on the precipitation of legumin from aqueous salt solutions. *Int. J. Food Sci. Tech.*, **27**, 629–35.

Pavlovskaya G.E., McClements D.J. and Povey M.J.W. (1992b) Ultrasonic investigation of aqueous solutions of a globular protein. *Food Hydrocoll.*, **6**, 253–62.

Pavlovskaya G.E., McClements D.J. and Povey M.J.W. (1992c) Ultrasonic studies of aqueous salt solutions of a globular protein. *Developments in Acoustics and Ultrasonics*, Editors M.J.W. Povey and D.J. McClements, Institute of Physics, Bristol, pp. 243–6.

Peek Measurement Ltd (1994) Peek Measurement launches new Istt-P flowmeter. *The European Food and Drink Review*, January, p. 15.

Pethrick R.A. (1983) Ultrasonic studies of macromolecules. *Prog. Polym. Sci.*, **9**, 197–295.

Pillai S.O., Natarajan S. and Palanisarny P.K. (1983) Free-volume of sodium-chloride solutions in water dimethyl formamide: mixtures from ultrasonic velocity. *J. Acoust. Soc. Am.*, **73**, 1616–18.

Pinfield V.J. (1996) *Studies of creaming, flocculation and crystallisation in emulsions: computer modelling and analysis of ultrasound propagation.* PhD thesis, University of Leeds, Leeds.

Pinfield V.J. and Povey M.J.W. (1997) Thermal scattering must be accounted for in the determination of adiabatic compressibility. *J. Phys. Chem. B*, **101**, 1110–12.

Pinfield V.J., Dickinson E. and Povey M.J.W. (1994) Modelling of concentration profiles and ultrasound velocity profiles in a creaming emulsion: importance of scattering effects. *J. Coll. Interface Sci.*, **166**, 363–74.

Pinfield V.J., Povey M.J.W. and Dickinson E. (1995) The application of modified forms of the Urick equation to the interpretation of ultrasound velocity in scattering systems. *Ultrasonics*, **33**, 243–251.

Pinfield V.J., Povey M.J.W. and Dickinson E. (1996) Interpretation of ultrasound velocity creaming profiles. *Ultrasonics*, **34**, 695–8.

Pinfield V.J., Dickinson E. and Povey M.J.W. (1997) Modelling of combined creaming and flocculation in emulsions. *J. Coll. Interface Sci.*, **186**, 80–9.

Povey M.J.W. (1984) A study of dilatation and acoustic propagation in solidifying fats and oils. I – Theory. *J. Am. Oil. Chem. Soc.*, **61**, 558–9.

Povey M.J.W. (1988) Ultrasonics as a probe of food emulsions and dispersions. In *Advances in Food Emulsions and Foams*, Editors E. Dickinson and G. Stainsby, Elsevier, London, pp. 285–327.

Povey M.J.W. (1989) Ultrasonics in food engineering II. *J. Food Engng*, **9**, 1–20.

Povey M.J.W. (1993) Ultrasonic emulsion analyser. *Sensor Review*, **13**, 18–19.

Povey M.J.W. (1995) Ultrasound studies of shelf-life and crystallization. In *New Physico-chemical Techniques for the Characterization of Complex Food Systems*, Editor E. Dickinson, Chapman & Hall, London.

Povey M.J.W. (1997a) *Ultrasonic Techniques for Fluids Characterization*, Academic Press, San Diego, CA.

Povey M.J.W. (1997b) Measuring aggregation in colloids using ultrasound velocity and attenuation. In *Food Colloids – Proteins, Lipids and Polysaccharides* Editors E. Dickinson and B. Bergenstahl, Royal Society of Chemistry, Cambridge.

Povey M.J.W. and Harden, C. A. (1981) An application of the ultrasonic pulse echo technique to the measurement of the crispness of biscuits. *J. Food Sci.*, **16**, 167–75.

Povey M.J.W. and McClements, D. J. (1988) Ultrasonics in food engineering: Part I. Introduction and experimental methods. *J. Food Engng*, **8**, 217–45.

Povey M.J.W. and McClements, D. J. (1989) Ultrasonic measurements in food emulsions and dispersions. In *Food Colloids*, Royal Society of Chemistry Special Publication No. 75, Editors R. D. Bee, P. Richmond and J.Mingins, Royal Society of Chemistry, Cambridge.

Povey M.J.W. and Rosenthal, A.J. (1983) Ultrasonic detection of the degradation of starch by α-amylase. *J. Food Tech.*, **19**, 115–19.

Povey M.J.W. and Wilkinson, J.M. (1980) Application of ultrasonic pulse-echo techniques to albumen quality testing – a preliminary report. *Br. Poult. Sci.*, **21**, 489–95.

Pryor A.W. and Roscoe R. (1954) The velocity and absorption of sound in aqueous sugar solutions. *Proc. Phys. Soc.*, **67**, 70–81.

Pryor A.W., Reed R.D.C. and Richardson E.G. (1958) The propagation of ultrasonic waves in sols and gels. In *Recent Advances in Gelatin and Glue Research*, Editor G. Stainsby, Pergamon Press, Oxford.
Rajendran V. (1993) Internal-pressure studies of binary-mixtures of higher alcohols in tea at 303.15 K. *Ind. J. Pure & Appl. Phys.*, **31**, 812–14.
Raschi A., Mugnozza G.S., Surace R., Valentini R. and Vazzana C. (1989) The use of ultrasound technique to monitor freezing and thawing of water in plants. *Agriculture Ecosystems & Environment*, **27**, 411–18.
Reddy M.A. and Suryanarayana C.V. (1981) Ultrasonic-absorption in aqueous-solutions of starch and gelatin. *Acustica*, **48**, 19–22.
Richardson, P. and Povey M.J.W. (1990) Ultrasonic temperature measurement and its potential for food processing systems. *Food Control*, 54–7.
Rizvi S.S.H., Singh R.K., Hotchkiss J.H., Heldman D.R. and Leung H.K. (1993) Research needs in food engineering, processing and packaging. *Food Tech.*, **47**, 26S.
Rossini F.D., Pitzer K.S., Arnet R.L., Braun R.M. and Pimentel G.C. (1953) *Selected Values of Physical and Thermodynamic Properties of Hydrocarbons and Related Compounds*, Carnegie Press, Pittsburgh.
Rowe Jr, A.M. and Chou J.C.S. (1970) Pressure–volume–temperature-concentration relation of aqueous NaCl solutions. *J. Chem. Engng Data*, **15**, 61–6.
Sakka T. and Mitaku S. (1987) Ultrasonic spectrometer for solutions of biological-material. *Jpn. J. Appl. Phys.*, **26**, 64–6.
Sakurai M., Nakajima T., Komatsu T. and Nakagawa T. (1975) Apparent molar compressibility of sodium chloride in water. *Chem. Lett.*, 971–6.
Sakurai M., Nakamura K. and Takenaka N. (1994) Apparent molar volumes and apparent molar adiabatic compressions of water in some alcohols. *Bull. Chem. Soc. Jpn.*, **67**, 352–9.
Sakurai, M., Nakamura, K., Nitta, K. and Takenaka, N., (1995) *J. Chem. Eng. Data*, **40**, 301–310.
Samari S. (1994) Ultrasonic inspection methods for food products. *Lebens.-Wiss. U.-Technol.*, **27**, 210–13.
Saraf B. and Samal K. (1984) Ultrasonic velocity and absorption in reconstituted powdered milk, Part II. *Acustica*, **56**, 61–6.
Saraf B., Mishra S.C. and Samal K. (1982) Ultrasonic velocity and absorption in reconstituted powdered milk. *Acustica*, **52**, 40–2.
Sarvazyan A.P. (1991) Ultrasonic velocimetry of biological compounds. *Ann. Rev. Biophys. Biophys. Chem.*, **20**, 321–42.
Sarvazyan A.P. and Hemmes P. (1979) Relaxational contributions to protein compressibility from ultrasonic data. *Biopolymers*, **18**, 3015–24.
Schaafs W. (1967) Sound velocity in binary mixtures and solutions. In *Landolt-Börnstein New Series*, Editors K.-H. Hellwege and A.M. Heliwege, Springer, New York.
Selborne L. and Hearn J. (1987) *Report of the Agricultural and Food Research Council for the Year 1986–1984*. AFRC, London.
Self G. (1991) Ultrasonic determination of harvest maturity in banana. *Proceedings of the 10th ACORBAT Reunion Conference*.
Self G., Chan G., Povey M.J.W. and Wainwright H. (1991) Ultrasonic non-destructive evaluation of the maturity and quality of fresh fruit and vegetables. *Ultrasonics International '91*, pp. 83–6.
Self G., Povey M.J.W. and Wainwright H. (1990) Non-destructive methods of evaluating maturity ripening and quality of tropical fruits. *Proc. XXIII Int. Hort. Cong., Florence.*
Self G., Povey M.J.W. and Wainwright H. (1992a) What do ultrasound measurements in fruit and vegetables tell you? In *Developments in Acoustics and Ultrasonics*, Editors M.J.W. Povey and D. J. McClements, Institute of Physics, Bristol, pp. 129–64.
Self G., Povey M.J.W. and Wainwright H. (1992b) Ultrasonic evaluation of post-harvest attributes of fruit and vegetables. *Abstracts of the Association of Applled Biologists 1992*, pp. 82–5.
Self G., Wainwright H. and Povey M.J.W. (1993) Non-destructive quality determination for fruit and vegetables. *Post Harvest News and Information* **4**, 18–19.
Self G., Ordozgoiti E., Povey M.J.W. and Wainwright H. (1994) Ultrasonic evaluation of ripening avocado flesh. *Postharvest Biology and Technology*, **4**, 111–16.

Shiio H. (1958) Ultrasonic interferometer measurements of the amount of bound water. Saccharides. *J. Am. Chem. Soc.*, **80**, 70–3.

Shirato M. (1987) Applications of supersonic waves to the food industry (e.g. ultrasonic cleaning, measurements using ultrasonic probe). *Up to Date Food Proc. [Shokuhin Kaihatsu]* **22**, 32–7.

Shore D. and Miles C.A. (1988) Attenuation of ultrasound in meat. *Ultrasonics*, **26**, 31.

Smith D.E. and Winder W.C. (1983) Effects of temperature, concentration and solute structure on the acoustic properties of monosaccharide solutions. *J. Food Sci.*, **48**, 1822–5.

Swatland H.J. (1995) Objective assessment of meat yield and quality. *Tr. Food Sci. Technol.*, **6**, 117.

Tamura J., Kogure Y. and Hiki Y. (1986) Ultrasonic-attenuation and dislocation damping in crystals of ice. *J. Phys. Soc. Japan*, **55**, 3445–61.

Tamura J., Suzuki K. and Mihashi K. (1993) Adiabatic compressibility of myosin subfragment-1 and heavy-meromyosin with or without nucleotide. *Biophys Jnl*, **65**, 1899–905.

Tanaka Y., Sakurai M. and Nakamura K. (1993) Ultrasonic velocities in aqueous gellan solutions. *Food Hydrocoll.*, **7**, 407–15.

Tesch R., Normand M.D. and Peleg M. (1996) Comparison of the acoustic and mechanical signatures of 2 cellular crunchy cereal foods at various water activity levels. *J. Sci. Food Agric.*, **70**, 347–54.

Tucker G.S. (1993) Novel techniques for characterizing the flow of solids and liquids. *Trans. Am. Soc. Agric. Engng*, **4**, 243–6.

Urick R.J. (1947) A sound velocity method for determining the compressibility of finely divided substances. *J. Appl. Phys.*, **18**, 983–7.

Vickers Z.M. (1984) Crispness and crunchiness: a difference in pitch? *J. Text. Stud.*, **15**, 157–63.

Vickers, Z.M. (1985) The relationships of pitch, loudness and eating technique to judgements of the crispness and crunchiness of food sounds. *J. Text. Stud.*, **16**, 85–95.

Vickers Z.M. and Wasserman S. S. (1979) Sensory qualities of food sounds based on individual perceptions. *J. Text. Stud.*, **10**, 319–32.

Voissey P. and Hamilton R.M.G. (1976) Ultrasonic measurement of egg shell thickness. *Poult. Sci.*, **55**, 1319–24.

Wang Z. and Nur A. (1991) Ultrasonic velocities in pure hydrocarbons and mixtures. *J. Acoust. Soc. Am.*, **89**, 2725–30.

Weast R C. (1988) *Handbook of Chemistry and Physics*, 68th edition, CRC Press, Boca Raton, FL.

Winder, W.C. Aulick D.J. and Rice A.C. (1970) An ultrasonic method for direct and simultaneous determination of alcohol and extract content of wines. *Am. J. Enol. Vitic.*, **21**, 1.

Winder W.C., Consigny N.P. and Rodriguez-Lopez B. (1961) An ultrasonic method for measurement of solids-non-fat and milk fat in fluid milk. II. An evaluation of method. *J. Dairy Sci.*, **44**, 1165.

Wirtanen G., Anvenainen R. and Mattila-Sandholm T. (1992) Non-destructive detection of spoilage of aseptically-packed milk products – effect of frequency and imaging parameters on the sensitivity of ultrasound imaging. *Lebensm. Wiss. Technol.*, **25**, 126–32.

Withers P.M. (1992) Sanitation of food processing plants. An ultrasonic hygiene sensor, 1st year report. Technical Memorandum No. 656, CFDRA, Campden.

Withers P.M. (1994a) Ultrasonics in the detection of fouling. *Food Technology International Europe*, Editor A. Turner, SPG, London, pp. 96–9.

Withers P.M. (1994b) Ultrasonic sensor for the detection of fouling in UHT processing plants. *Food Control*. **5**, 67–72.

Wood A. B. (1964) *A Textbook of Sound*, 3rd edition, Bell, London.

Yamamoto H., Iwamoto M. and Haginuma S. (1981) Acoustic impulse response method for measuring natural frequency of intact fruits and preliminary applications to internal quality evaluation of apples and watermelons. *J. Text. Stud.*, **11**, 117–36.

Yasunaga T., Usui I., Iwata K. and Miura M. (1964) Ultrasonic studies of the hydration of various compounds in an ethanol–water mixed solvent. II. The hydration of organic compounds. *J. Chem. Soc. Am.*, **37**, 1658–60.

Zacharias E. M. and Parnell R. A. (1972) Measuring the solids content of foods by sound velocimetry. *Food Tech.*, **26**, 160–6.

4 Ultrasound process tomography – the prospects in food processing and inspection

B.S. HOYLE

4.1 Introduction

Ultrasound has found many applications in the wider food industry through its provision of a range of one-dimensional measurements. The use of new process tomography techniques using ultrasound may provide new opportunities for process monitoring and product quality inspection. The use of ultrasonic transducers for process tomography is in its infancy relative to many other techniques, in particular considering the widespread use of ultrasonic transducers in a variety of non-tomographic sensing applications.

Multidimensional tomographic measurements are provided through the use of multiple sensors whose data are combined through a reconstruction algorithm resulting in a pseudo-image of the process state. Modern information technology systems, possibly including parallel computation, enable the reconstruction algorithm to be computed at high speeds, delivering acceptable real-time performance for many industrial processes.

This chapter briefly reviews the background science of process tomography using ultrasound, followed by examples of generic systems, and their associated advantages and difficulties, including the critical time required for data collection. Suggestions are made for the exploitation of this technology in the food industries and for its potential quality and cost benefits. Finally a review is included of research investigations into the use of ultrasonic-based process tomography systems.

4.2 Overview of process tomography

Process tomography is a novel technique that seeks to apply the well-established principles of medical tomography to industrial processes, such as mixing, pipeline transport and inspection and monitoring. It is concerned with the derivation of information relating to two or three dimensions. Such derivations require information, or *projections*, from a number of different viewpoints, preferably symmetrically surrounding the object or field of interest. The time taken to collect such multiple projections may not be critical in many applications, where the process of interest has slow dynamics, e.g. the mixing of an injected material, where the requisite number of projections could be obtained sequentially. In other applications the dimension of time may be critical, e.g. the on-line measurement of a flowing mixture will require short projection intervals in which flow evolution is insignificant.

It is an important fact that measurement epochs in ultrasonic-sensed systems are intrinsically limited by the velocity of sound, which is very low compared with electromagnetic waves. A typical sonic velocity in water is 0.0015 m/μs (or 1.5 mm/μs), compared with 300 m/μs for electromagnetic waves, a ratio of 200 000. This places an upper limit on the intrinsic dynamics of processes for which a process tomography system is likely to be usable. The use of fan-shaped beams, in contrast to the narrow beams used in non-tomographic applications, provides the potential to insonify larger regions in a given time interval.

4.3 Ultrasonic sensors for tomography

Ultrasonic sensors have a long history of success in a variety of non-tomographic applications: in process measurement (Asher, 1983; Lynnworth, 1989; Plaskowski *et al.*, 1992; Hoyle and Luke; 1994); and in non-destructive testing (McMaster, 1963; Silk, 1984); widely in single viewpoint medical imaging (Havelice and Taenzer, 1979) and in food (Javanaud, 1988; McCann, 1986; Pinder and Godfrey, 1993; Povey, 1989, 1995). In general, the object or field will interact with an ultrasound beam through some form of acoustic scattering; the interaction may then be sensed to yield information about the object or field. This is viable only where a significant interaction occurs, generally related closely to modulus of elasticity and density variations, e.g. a liquid containing gas bubbles.

Process systems in which ultrasonic tomography has been investigated have been confined to those involving only liquids and gases. In such media an ultrasonic beam advances as a longitudinal wavefront (and also as a shear wavefront), in common with all sound waves. For a single medium, if the influence of boundaries can be ignored, the medium appears unbounded and a *bulk wave* can be assumed. Under these conditions, for propagation within a single (gas or liquid) medium, an approximate model can be used in which the speed of the longitudinal wave is simply related to the bulk properties of the medium: the modulus of elasticity and density. Such simple models are particularly crude for ultrasound in gases and totally inappropriate for solids. Furthermore, it is also noted that longitudinal waves propagate over long distances only in liquids (ignoring viscoelastic liquids), and gases. (A rigorous treatment of the general propagation of ultrasound may be found in many textbooks, e.g. Povey, 1997. A basic introduction is given in Chapter 1.)

The density of a material is clearly related to volume and temperature. For simplicity we assume uniform temperature, although systems have been developed for the tomographic estimation of temperature distribution (as noted in Section 4.4). The wavelength of the ultrasound wave is related simply to its velocity and frequency. For the approximate model discussed above, speed is constant for a particular material under the same ambient conditions; and the frequency will be determined by the transducer. Thus, the wavelength may be considered to arise from this frequency and the material through which the beam passes. In principle wavelength is directly linked to the resolving power of the sensor system; in a given material increasing the frequency may be expected to increase the resolution of the system. Thus the longitudinal wavelength is linked to the corresponding *axial* resolution. The *lateral* resolution will also be of interest and demands consideration of a two-dimensional field model which encompasses the profile of the beam. Objects smaller than the wavelength of the ultrasonic beam will not block or reflect the acoustic pressure wave, but will diffract and scatter it.

Most process tomography systems which have been investigated depend on blocking and reflective interactions and with transmission attenuation, since these are the easiest to detect and interpret. A lower limit on the band of frequencies which are likely to be useful will generally be simple to determine. Food processes where process tomography is likely to be viable will be based on liquids; for example, water has a sonic velocities of approximately 1 mm/μs. Thus an ultrasonic beam having a frequency greater than 1 MHz could theoretically resolve objects less than 1 mm in diameter. Other limitations, concerned with the detailed design of the transducer, make this limit difficult to approach in practical systems. Although a reduced wavelength may be advantageous to a point, and can be achieved by increasing the frequency of the transducer, this also will incur fundamental costs in terms of attenuation of the ultrasonic beam with distance. This is due to losses which to a crude approximation, increase linearly with frequency for a given material under the same ambient conditions. Physical data giving pressure wave attenuation in terms of frequency for materials of general interest in process tomography can be used to optimize parameters (e.g. McMaster, 1963).

Process tomography using ultrasonic sensing will rely on detectable interactions both in a homogeneous transmission medium and from interfaces, for example at gas bubbles in a liquid. A variety of interactions may arise. The attenuation of incident waves due to the object or field of interest may be detected between a transmitter and receiver. Variation of the speed of sound in an inhomogeneous material may be measured. Backscattering at interfaces may also be sensed. Thus, the sensed interaction data may be used as source data for tomography. Ultrasonic tomography for imaging of the human body is a well-known example, however, typical medical systems are designed to deal primarily with weakly inhomogeneous object fields (the body consists mainly of water!). In contrast, most process applications which contain distinct particles are likely to present strongly inhomogeneous object fields.

A useful descriptor of the interaction of ultrasound with a material is its *acoustic impedance* (Wells, 1969), analogous to electrical impedance (here the complex ratio of sound pressure to particle velocity). It is clear that the greater the difference in impedance at the interface, the greater will be the amount of energy reflected, conversely where impedances are similar, most will be transmitted. Thus reflections provide a basis for tomography where materials offer distinct impedances, for example water and air. The complete interaction behaviour is complex and depends not simply on the differences in acoustic impedance, but on the physical size and shape of the interface.

Interfaces due to objects which are much smaller than the acoustic wavelength interact according to Rayleigh scattering; basically related to the fourth power of frequency. Conversely, where the wavelength is small relative to the smallest object particles, a straight-line *ray* behaviour can be assumed. An ultrasound beam may be expected to be incident at interfaces in a non-normal fashion and will split into reflected and transmitted components. As in the analogous optical case, for the reflected wave, the angle of reflection will equal the angle of incidence. The transmitted wave will obey Snell's law relating the angles of incidence and refraction to the speed of sound in the two materials. The straight-line ray model is clearly violated when the ray transits through part of the object field which exhibits strong inhomogeneity, e.g. a gas bubble in a liquid. However, if the system primarily makes use of reflection and of blocking in transmission interactions, such refractions may be secondary in importance.

Tomography systems are concerned with derivation of information relating to two- or three-dimensions, however, basic transduction principles may be appreciated from a simple one-dimensional case. Here interaction would be sensed either: (i) from the same viewpoint by the partial reflection of the ultrasonic acoustic energy; or, (ii) from an opposing viewpoint by transmission, where the energy is partly attenuated.

From either viewpoint the time of flight to an object may be estimated from both transmission and reflection sensor data. This may be considered in terms of a section through a liquid containing gas bubbles: $f(x, y)$ may be used to describe the attenuation of the object field to the ultrasonic energy; a complementary function, $g(x, y)$ describes its backscattering properties, arising due to variations in density and elastic modulus.

A common ultrasonic excitation signal for tomographic applications is an impulse. If such an impulse is transmitted from a combined transmitter/receiver transducer, $U(T/R)$, then a simple view of the various interactions may be obtained from a time/amplitude graph, called an *A-scan*. Such a graph could be used to depict both reflection data, received at the original transmitter, or at a secondary receiver, $V(R)$, at the opposite side of the process vessel.

Attenuation from transmitter to receiver will be critically dependent on the material through which the ultrasound travels. It may be modelled by Lambert's law of absorption, which relates intensities at transmitter and receiver (where L represents the total path length from U to V)

$$I_R = I_T \exp\left[-\int_L f(x, y)\mathrm{d}l\right] \tag{4.1}$$

The *time of flight* of the transmitted impulse sensed from the signal at V allows the estimation of the mean speed of sound in the interposing medium along path L. If two-dimensional projection information were available this could be used to estimate $f(x, y)$, through an appropriate tomographic reconstruction algorithm.

Reflection data may also provide an insight into the process. Reflections may arise at material interfaces in the process where there are changes in acoustic impedance. A simple assumption may be made that the reflections return to $U(T/R)$ only through the higher density (liquid) component. Even if some reflections are attenuated by a return transit through a lower density (gas) region along their return path, these will be attenuated relative to those which do not. Under this simplification the speed of sound is that for the liquid, c_l. Thus the time of the observations may then be assumed to be proportional to their distance of travel. Hence for an echo received at time t_r, the distance l_r from the interface to the transducer U is given simply by

$$2l_r = c_l t_r \tag{4.2}$$

The above illustrations are clearly considerably simplified. A more thorough review would include simple transducers, which radiate a wavefront of parallel rays, and more practically include transducers which radiate on a circular front. These have been introduced above as fan-shaped beam transducers, which allow the simultaneous insonification of a large area reducing the serious real-time limitations outlined above.

Circular objects, which could be cross-sections of cylindrical objects, will increase the scattering of the ultrasonic energy, generally reducing both the transmitted and reflected energy in any given direction. Three-dimensional spherical objects may be expected to further increase the scattering, resulting in lower energy incident at the transducers.

In simple terms the intensity of reflection interactions in liquid/gas mixtures can thus be expected to be related to the bubble size. For an incident parallel ray the reflected intensity is greatest when the reflecting surface is flat. The intensity may then be expected to decrease as the radius of curvature of the bubble is increased. Hence smaller bubbles may be expected to be poor reflectors and more difficult to detect.

The interaction of gas voids with ultrasonic energy can also be shown to depend on their size and their resulting natural frequency of vibration, which may occur in several modes (Minnaert, 1933; Gaunaurd and Überall, 1981), and these effects may be expected to introduce errors. The process is further complicated at the micro-scale by the swarm behaviour of bubbles (Nakagawa, 1989). Such resonant effects may offer the opportunity to detect specific sizes and agglomerations of bubbles with ultrasonic transducers of selected frequency. A more thorough examination of the interaction of ultrasonic energy with multicomponent objects is beyond the scope of this chapter.

The realization of ultrasonic process tomography is dependent critically on the design of transducers. There is a wealth of design experience due to the widespread use of transducers for non-destructive testing, flow measurement and for medical applications (Crecraft, 1983). Current products suited to these mature application areas may not be suited to use in process tomography.

An ultrasonic tomography system must insonify the interfaces within the process of interest from several viewpoints, to obtain a number of independent projections and in essence to provide information relating to two dimensions. Simple determinants have been proposed for the assessment of real-time limits for ultrasonic tomography (Hoyle, 1994) where the dimension of time is likely to be critical in dynamic process applications due to the low velocity of sound. Typically a set of fixed identical transducers can be arranged at intervals around the field of interest and excited in sequence. Simple calculations determine the envelope for a sequentially excited, multiple transducer system. For reflection interactions the maximum transit distance is $2L$. If there are N transducer positions, or transducers, then the time taken per complete scan, T, is

$$T = \frac{2LN}{c} \tag{4.3}$$

If the process is dynamic and the features under observation by the sensor system are evolving, then the time, T, will impose an upper limit on the speed of the evolution. If for example the cross-sectional features are in motion at right angles to the imaging plane and, the depth of the observation view is d, the maximum linear speed of evolution, v, is simply:

$$v = \frac{d}{T} \tag{4.4}$$

For a system using multiple transducers, the maximum number is likely to be governed by practical, as well as theoretical considerations: there will be a practical limit in terms of space and cost. The use of a narrow beam offers relatively precise directional information but implies a large number of transducers, or transducer positions. In contrast a wide angle beam offers poor directional information but a wide area of simultaneous insonification. The use of multiple omni-directional transducers has been investigated by Norton and Linzer (1979) who illustrate how multiple projection data are reconstructed.

A transducer employing a fan-shaped beam has been investigated by Moshfeghi (1985, 1986). Wiegand and Hoyle (1989a) have investigated a large variety of possible transducer arrangements through simulations, including systems employing fan-shaped transducers. It is apparent that such transducers are probably near optimal for dynamic process tomography applications where the scan time is critical, such as flow imaging (Plaskowski *et al.*, 1992). A fan-shaped beam can offer the dynamic advantage of wide-angle insonification. However, the corresponding reconstruction algorithm must re-assemble the multiple projection data and estimate intersections which correspond to objects in the view field.

A further refinement of the fan-shaped transducer is one employing *multiple segments* which can be simultaneously triggered in transmit mode, but can be separately interrogated in receive mode, thus providing enhanced directional information. This enhancement, suggested by Flemons (1988), has been examined both through simulation studies (Wiegand and Hoyle, 1989a), and through simple laboratory tests (Wiegand, 1990).

Such investigations indicate that the highest fidelity images are obtained, for a given data collection interval, when using several multiple-segment transducers with a combination of reflection and transmission interaction data. Li and Hoyle (1993) have simulated a range of possibilities. Further improvements in real-time performance is available when multiple transducers can be triggered simultaneously; feasibility studies have been carried out with some success to this end (Li and Hoyle, 1995).

4.4 Tomography using ultrasonic transduction

The use of ultrasonic transducers for tomography rests on the careful design of the transducers to meet the requirements of their intended operating environment. The incorporation of ultrasonic transducers into a process is also an important aspect of its overall system design. The process may have restrictions on the intrusion, or invasion of the transducer (Gai *et al.*, 1989a; Gai, 1990). The number and form of the transducers, their mode of use, their excitation and the capture of the resulting data are critical *architectural* aspects of the design of an ultrasonic process tomography system.

A narrow ultrasonic beam can only observe part of an object field; the use of multiple narrow beams, perhaps applied in sequence would allow the whole object to be insonified from one viewpoint. This is akin to the *B-scan* produced by common medical ultrasound systems using a mechanically scanned, or phased-array, transducer. Alternatively a fan-shaped beam could be used to insonify a wide sector of the object at one time.

Further projections from multiple facets are then necessary to complete the data for reconstruction. In essence the single-facet sensor view considered above must be rotated around the object. In a process tomography application, for practical reasons, this is generally achieved through the provision of multiple sensors which surround the object field.

Either reflection or transmission data may be used in isolation as a basis for a projection. Radon (1917) provided a basis for the description of scalar (two-dimensional) tomographic projections (Deans, 1983; Ramm and Katsevich, 1996). For the attenuation function, $f(x, y)$, introduced above the projection, $p(l, \theta)$, is the sum of line integrals of the normal straight line paths, s (see Figure 4.1)

Thus, the projection for the object function, $f(x, y)$, for one projection angle, ϕ, is:

$$p(l, \phi) = \int_s f(x, y)\,\mathrm{d}s \tag{4.5}$$

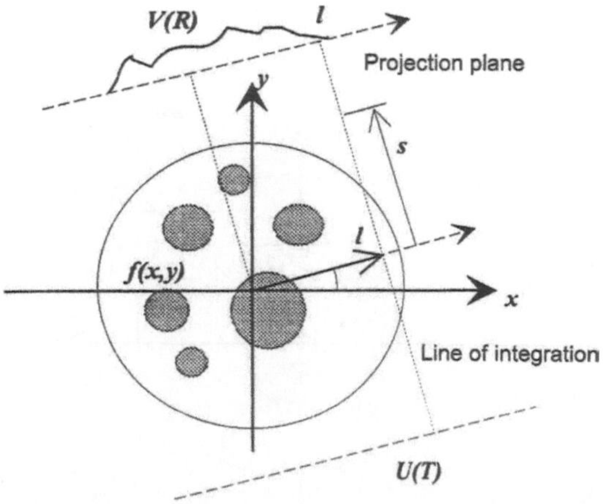

Figure 4.1 Attenuation function projection.

The mapping defined by the projection along all possible lines of integration, S, is the (two-dimensional) *Radon transform*, R. A limited collection of projections is known as a *sample* of the Radon transform. Radon showed that an inverse transform exists such that

$$f = R^{-1}\{R\{f\}\} \tag{4.6}$$

A similar general analysis to the above can be carried out for the reflection function, $g(x, y)$, in which a reflectivity projection is derived for a co-located ultrasonic transmitter and receiver (see Figure 4.2).

The similarity between Figures 4.1 and 4.2 can be clearly seen; the line integration path is now a circular arc rather than a straight line

$$p(q, \phi) = \int_c g(x, y)\, ds \tag{4.7}$$

where the radius of the *projection arc* is:

$$q = \sqrt{r^2 + l^2 - 2rl \cos(\theta - \phi)} \tag{4.8}$$

The value q may be estimated from

$$q = \frac{ct}{2} \tag{4.9}$$

To provide a range of projections (over angle ϕ) a number of transducers will typically be arrayed around the process of interest. In order to increase the volume of data available, in a limited real-time window, it is possible to utilize these other transducers as receivers, in effect as a multiple set: $V(R)$. In this case it has been shown (Norton and Linzer, 1979) that the projection arc for these cases is elliptical rather than circular.

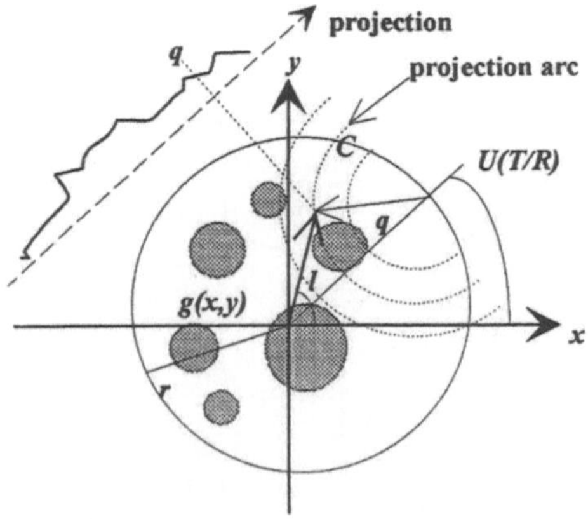

Figure 4.2 Reflection function projection.

A transducer will not deliver an idealized impulse. Thus the measured projection at any point may be considered in relation to the actual time domain function of the pulse. In order to account for this the measured projection at any point is considered in relation to the actual time-domain function of the pulse. In essence the time domain measurement, at a given point in time, will be a composite function of the response function of the transducer and that of the object field. The most convenient representation of this interaction is through the convolution integral. This is the limiting case when an approximation to an ideal impulse is synthesized by a sequence of ideal impulses. Thus the composite result, in this case the measured value, is the integral of effects on the object field of the varying sequence of model impulses. The resulting integral equation may be usefully expressed in the operator form

$$p_{\text{meas}}(q, \phi) = p(q, \phi) \otimes p_{\text{trans}}(q, \phi) \tag{4.10}$$

Although outside the scope of this discussion, the convolution integration operator in the time domain may be simply replaced by a multiplication when the components of the equation are replaced by their respective Fourier transforms. This allows a direct frequency domain evaluation to be carried out (Lynn, 1982).

In practice a further distortion will arise. The actual impulse will be distorted throughout its propagation. The distortion may be considered in terms of the differing attenuations corresponding to the frequency components of the Fourier transform of the impulse. Hence equation (4.10) is only an approximate description of the process.

In both the attenuation case, summarized by equation (4.5) and the reflection case, summarized by equation (4.7), an *inverse algorithm* must be applied, which operates on the multiple projection data, to yield an estimate of the tomographic cross-section of the object. A variety of techniques such as *Fourier inversion* or *filtered backprojection* (Hermon, 1980; Natterer, 1986) exist based on the measurement of the projection data and the above relationships.

The interactions introduced above may be used to gain information relating to the internal nature of a process. The choice will clearly depend on the features of the process of interest but possibilities yield a variety of sensor configurations which may provide raw data for processing to yield both one- and two-dimensional scalar and three-dimensional vector tomographic results. Theoretically all forms may be treated as individual solutions (based on different assumptions) of the same inverse problem of a scattered field produced by an object, or group of objects (Norton and Linzer, 1981; Xu, 1990). The theoretical possibilities which exist may be classified as summarized in Table 4.1 (Hoyle and Xu, 1995).

Table 4.1 Major tomographic modes

	Reflected wave measurement and processing		Attenuated transmitted wave measurement and processing	
Measurement	Amplitude flight time of back-scattered echoes	Amplitude and phase of back-scattered signals	Amplitude flight time of trans-mitted signals	Amplitude and phase of trans-mitted signals
Reconstruction processing	Linear, e.g. back-projection	Non-linear, e.g. approx. inverse soln of wave eqns	Linear, e.g. back-projection	Non-linear, e.g. approx. inverse soln of wave eqns
Tomographic mode	Reflection mode	Reflection mode diffraction	Transmission mode	Transmission mode diffraction

For practical application it is appropriate to consider these in terms of the data that must be sensed, measured and processed in three forms: transmission mode; reflection mode; and diffraction mode.

Transmission-mode tomography was the first mode to be employed in X-ray and optical transmission applications. Early acoustic tomography developed from algorithms developed in these applications, whose underlying assumption was that a transmitted wave travels in a straight line. Hence the sensed wave is related to some characteristic of the medium only along this straight line; as summarized earlier by equation (4.5). In the acoustic case this assumption is clearly an approximation at best, even though reasonably good results have been obtained. Where applications exhibit strongly inhomogeneous components large errors may result.

Corrections to the straight-path approximation can be made using geometrical optical and ray-tracing methods. However, these corrections are difficult to apply to fast real-time applications, such as 100% inspection of food processes, since they demand intensive computation. Furthermore, these corrections neglect diffraction effects (Greenleaf, 1983) which can be very significant in the heterogeneous materials that comprise so many foods. Hence the relatively high-quality performance of transmission-based medical X-ray systems is practically impossible to match in ultrasonic systems due to these intrinsic distortions and lower resolving power.

As summarized above in equation (4.7), the sensed wave at the receiver can be related to the characteristics of the interface which produces a reflection along the straight line outward and return paths; in essence reflection mode tomography. If these paths transit only one of the media, such as a liquid with gas bubbles, then to a reasonable approximation this is homogeneous and the problems arising in transmission systems may be avoided.

A detailed theoretical analysis of reflection mode tomography is offered by Norton and Linzer (1979), based on a single pulse-echo transducer used as a transmitting and receiving element. In their analysis the transducer is assumed to be moved around the object on a circular path in the cross-sectional plane viewing the object field radially. In a practical application a number of transducers could, of course, be employed. Their results demonstrate that, as noted above, the backscattered echoes from a transmitting transducer to a coincident receiver arise from a circular arc at the radius of each object interaction. They also demonstrate the further result noted above that for a transmitter and receiver separated by a constant angle the same perceived echo lies on an elliptical arc.

Medical imaging based on ultrasound is commonly based on reflections. For example a *B-scan* provides multiple view reflection data from a single viewpoint. Although this cannot yield sufficient data to reconstruct the reflectivity function it may be adequate in certain circumstances for processes with some degree of symmetry.

For fine process structures, object inhomogeneity may be comparable in size to the wavelength of the ultrasound. In such cases it is not appropriate to treat propagation along lines or rays as diffraction effects will predominate. In order to indicate the scales involved, for the example of water, the wavelength of sound at a frequency of 1 MHz is approximately 1.5 mm. Energy transmission must then be discussed in terms of wavefronts and the scattering arising from the inhomogeneity; and the interactions can be modelled by the wave equations.

Reconstruction in the diffraction case can be achieved by solving the inverse scattering problem (Mueller *et al.*, 1980). As noted above, most process tomography applications will present a strongly inhomogeneous object field. Lu (1986) and other researchers have presented several new reconstruction algorithms which allow the diffraction mode to be employed in the case of a moderately or strongly scattering medium.

4.5 Applications and limitations

Investigations of ultrasonic process tomography have been limited to a small number of trials across a wide spectrum of techniques and application areas (as reviewed in Hoyle and Xu, 1995). The choice of parameters and mode of operation of ultrasound tomography system is likely to be critical to achieving success. This section deal with a range of previous applications and with their apparent major limitations.

A few authors (Bull, 1993; Chan and Batchelor, 1993) have reviewed the use of ultrasound tomographic techniques, for use in food materials. Bull (1993) considered the use of ultrasound, amongst a range of tomographic modalities and concluded that ultrasound may be appropriate in the processing of liquid foods. A number of authors have applied medical ultrasound tomographic systems to the study of fruit, vegetables and other foods. A good example of this is the work by Chivers *et al.* (1995), who used a medical ultrasound scanner to look for stones in canned peaches. It is fair to say that, while good data have been obtained from this work, the difficulties involved in obtaining these results have been a barrier to the direct transferral of medical ultrasound scanners into food process monitoring.

As noted above, the speed of sound is temperature dependent. Peyrin (1989) has reported investigations of the use of ultrasound for temperature measurement. This principle may have applications where internal temperature is of interest, for example during a heating process. Although far removed from the food sector, work on acoustic transmission techniques have been reported in the estimation of temperature distribution in furnaces (Hoyle and Luke, 1994).

Clearly if temperature measurement is not the underlying aim, variations in temperature in the process may become a potential source of error in reconstruction. If a priori information is available for the temperature variation in the process it may be possible to introduce compensation into the reconstruction algorithm.

The application area of two-phase flow has attracted a range of research, not least because of potential oil industry interest. This work may find parallels in the monitoring and characterization of food process mixtures. Density distributions in two-component flows have been investigated by Fincke (1980); later work by Wolf (1988a, b) has examined the determination of the spatial distribution of gas bubbles in the cross-section of a gas/liquid flow, using transmission mode sensing. Investigations on ultrasound transducers and architectures suitable for two-phase flow imaging were carried out by Gai with others (1989a, b, 1990). He also proposed a novel variant: *time-resolved transmission mode tomography*, which utilizes a timed sample of transmission pulse data arrival rather than the attenuation. Further investigations in the use of transmission tomography for the characterization of two-component bubbly gas/liquid flow have been carried out by Xu *et al.* (1993).

A ultrasonic process tomography system, aimed at real-time flow monitoring, using fan-shaped beam transducers with both single and multiple (three) segments, has been investigated at a laboratory scale by Wiegand and Hoyle (1989a, b). This system exploited both reflection and transmission mode data. It demonstrated real-time image reconstruction performance viable for imaging cross-sections in horizontal flow rates of up to a few metres/second. A physical simulation arrangement was employed in which two transducers were rotated to obtain data as from multiple transducers. An array of parallel processing devices was used for the reconstruction processing (Wiegand and Hoyle, 1991). Further work has shown the complete real-time viability of this general approach using multiple actual transducers (Schlaberg *et al.*, 1996). Typical images from this system with circular reflective objects in water are illustrated in Figure 4.3.

Blackledge (1992) demonstrated that a B-scan system, based on a conventional medical ultrasonic scanner, can produce useful data from a single viewpoint. His preliminary work captures the B-scan as a raster and utilizes standard frame-capture techniques together with conventional medical image-processing algorithms to enhance and analyse the result.

Limited trials have been carried out of process tomography for three-dimensional or vector field imaging. A number of theoretical studies have examined the general problem: Johnson *et al.* (1975, 1977) on the reconstruction of three-dimensional velocity vector field from transmission mode measurements; Kramer (1979) on reconstructing vector fields. Norton (1988) introduced a corollary of the important projection-slice theorem for vector fields. Hauk (1989) and Hauk and Braun (1991) investigated the application of ultrasonic transmission mode tomography to the imaging of vector fields using an array of transducers. Jhang and Sato (1989, 1991) explored a related concept realized through the synthesis of a parallel plane of ultrasound using acrylic lenses. Although this collection of work does elegantly demonstrate the potential power of ultrasonic sensing for three-dimensional imaging the practical problems would be onerous. The trials discussed typically involved pseudo-static data and a small number of transducers, movable in several planes in order to provide the necessary multiple projections.

A number of potential problems must be considered in the application of ultrasonic tomography systems: the low speed of sound; the soft-field effect; and at a more practical level the problem of mounting transducers.

(a) (b)

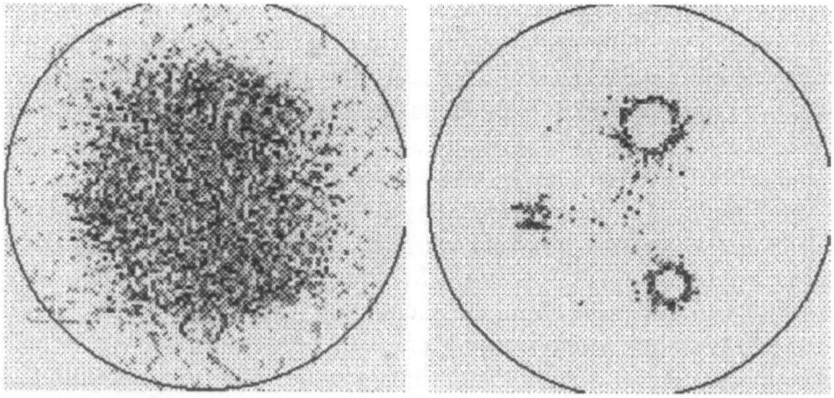

Figure 4.3 Reconstruction image from real-time system. (a) Raw reconstructed image. (b) Threshold image.

In terms of viability for processes having fast dynamics the slow speed of sound is likely to present a critical upper limit. This real-time constraint has two aspects: the relatively trivial engineering problem of ensuring that the real time data can be captured in the available sampling window; and the fundamental systematic limitation on the *maximum speed of evolution* of the process under examination relative to the spatial sampling window. The use of modern digital electronics and high-speed computer systems will provide a solution to the data collection problem. This will probably be relatively straightforward for systems which employ the simplest sequential excitation and sensing regime. More complex systems employing multiple simultaneous sensing will demand more sophisticated engineering solutions, perhaps employing parallel computer systems for real-time data collection and processing (Wiegand and Hoyle, 1989b, 1991).

There is of course no engineering solution to the fundamental limitation imposed by the low speed of sound. However, it is possible to estimate simple upper limits for particular configurations. For example for water-based mixtures at 20°C, the speed of sound in the containing solution is approximately 1500 m/s. Then, for the reflection case, equation (4.3) above gives the data acquisition time needed for one complete *frame* of multiple projections. For a system devised to estimate gas voids in an water-based material, having 30 transducers arrayed around the circumference of pipeline of diameter of 0.2 m, equation (4.3) indicates that the minimum time per scan is 8 ms. A simple system may transmit a pulse from each transducer in sequence. A delay to allow reverberations to decay must be included, otherwise interference with later pulses is likely to result. Experience for a particular system will indicate how long this delay should be. If set equal to the time taken for an echo from the far wall of the vessel, the total frame time will be doubled; in the above case to 16 ms. Equation (4.4) then indicates the real time constraints in terms of the maximum axial flow rate which can be tolerated by the system; in this case for a beam depth of 15 mm this would be approximately 1 m/s. In order to allow objects in the field to be captured in the image frame a value of half the maximum may be taken as a practical limit. This limitation may be expressed in terms of a standard ratio of the speed of sound and maximum sensing range (typically diameter) of the vessel (Hoyle, 1994, 1995).

Many practical process tomography systems will exhibit strong inhomogeneity and where the ultrasound energy has to transit such differing regions the *soft field* nature of scattered ultrasound waves may introduce large errors. These will arise from diffraction and multipath effects. It may theoretically be necessary to use a non-linear reconstruction algorithm in order to reconstruct a cross-section image with reasonable resolution. An alternative is to reply on reflection and blocking for simplicity, where all transits are assumed to be in a single medium.

One of the major potential advantages of an ultrasonic system is that it can be non-invasive and possibly non-intrusive, but these desirable features are not simple to achieve in practice. A key decision is the choice of transducer geometry: narrow or fan-shaped beam acoustic field. If a narrow beam is chosen (in common with medical systems) a large number of views of the object from different viewpoints must be obtained, requiring many transducers which may be impractical. In the second case a small number of fan-beam transducers will cover the area. With few transducers the resulting restricted number of projections will limit the resolution of the image which can be obtained and will result in a *blurred* image.

In general terms a number of factors must be taken into account when designing a transducer system for ultrasonic tomography. Ideally a non-invasive system will be required where transducers are mounted on the outside surface of the process vessel. Factors to be considered then include: the coupling efficiency between the pipe wall and transducer; introduction of the fan-shaped beam or narrow beam into the process via the vessel wall; maximizing the number of views of the imaged object; the number of transducers; the scanning regime; ratio of signal-to-noise; and signal preprocessing. Obviously careful engineering compromises must be made to produce a system able to deliver the process user information required.

A number of key points emerge for critical consideration in ultrasonic tomography systems applied to process applications:

- Potential applications in the food industry must be considered for feasibility in the light of the properties and limitations of ultrasound outlined above.
- A range of theoretical modes have been classified above, the most practicable modes are concerned with reflection and transmission sensing. Similarly in terms of computational simplicity the filtered back projection algorithm is efficient for real-time applications.
- Design of the transducer will be a critical aspect in the composite tomography system. A viable system demands a number of projections which will typically for process applications require a number of transducers distributed around its periphery. The choice of ultrasound frequency, geometry and architecture will depend on the process, for example increased frequency may increase resolution but will increase attenuation. In order to enhance real-time performance multiple simultaneous receiving transducers should be employed.

4.6 References

Asher R.C. (1983) Ultrasonic sensors in the chemical and process industries. *J. Phys. E: Instrum.*, **16**, 959–63.

Blackledge J. (1992) Analysis of two-phase flows using ultrasonic scanning. *Process Industry J.*, May, 27–9.

Braun H. and Hauk A. (1991) Tomographic reconstruction of vector fields. *IEEE Trans. Acoustics, Speech Signal Processing*, **ASSP-39**, 464–71.

Bull C.R. (1993) A review of sensing techniques which could be used to generate images of agricultural and food materials. *Computers Electron. Agric.* **8**.

Chan J.P. and Batchelor B.G. (1993) Machine vision for the food industry, in *Food Process Monitoring Systems*, Editors A.C. Pinder and G. Godfrey, Blackie, London, pp. 58–101.

Crecraft D.I. (1983) Ultrasonic instrumentation – principles, methods and applications. *J. Phys. E Sci. Instrum.*, **16**, 183–9.

Deans S.R. (1983) *The Radon Transform and Some of Its Applications*, Wiley-Interscience.

Fincke J.R. (1980) The application of reconstructive tomography to the measurement of density distribution in two phase flow. *Proc. of 26th International Instrum. Symposium*, ISA, Research Triangle Park, NC, pp. 235–43.

Flemons R. (1988) Ultrasonic transducer and ultrasonic flow imaging method. UK Patent application 8815000.

Gai H. (1990) Ultrasonic techniques for flow imaging: front-end transducers, image reconstruction algorithms and system design. PhD thesis, University of Manchester.

Gai H., Beck M.S. and Flemons R. (1989a) An integral transducer/pipe structure for flow imaging. *Proc. IEEE 3rd International Ultrasonic Symposium*, Montreal.

Gai H., Li Y.C., Plaskowski A. and Beck M.S. (1989b) Flow imaging using ultrasonic time-resolved transmission mode tomography. *IEE 3rd International Conf. on Image Processing and its Applications*, Warwick.

Gaunaurd G.C. and Überall H. (1981) Resonance theory of bubbly liquids. *J. Acoust. Soc. Am.*, **69**, 362–70.

Greenleaf J.F. (1983) Computerized tomography with ultrasound. *Proc. IEEE*, **71**(3), 330–7.

Hauk A. (1989) Ultrasonic time of flight tomography for the non-instrusive measurement of flow velocity fields. *Proc. 18th International Symposium on Acoustical Imaging*, Santa Barbara, CA.

Havelice J.F. and Taenzer J.C. (1979) Medical ultrasonic imaging: an overview of principles and instrumentation. *Proc. IEEE*, **67**(4), 620–41.

Hermon G.T. (1980) *Image Reconstruction From Projections – The Fundamentals of Computerised Tomography*, Academic Press, London.

Hoyle B.S. (1994) Real-time implementation of ultrasonic process tomography. *Proc. 3rd ECAPT Workshop*, Oporto, pp. 359–66.

Hoyle B.S. (1995) Real-time ultrasonic process tomography of flowing mixtures. *J. Part. Part. Systems Characterization*, **12**, 81–6.

Hoyle B.S. and Luke S.P. (1994) Ultrasound in the process industries. *IEE ESEJ*, **3**(3), 119–22.

Hoyle B.S. and Xu L.A. (1995) Ultrasonic sensors, in *Process Tomography – Principles, Techniques and Applications*, Editors R.A. Williams and M.S. Beck, Butterworth-Heinemann, pp. 119–50.

Javanaud C. (1988) Applications of ultrasound to food systems. *Ultrasonics*, **26**, 117–23.

Jhang K.Y. and Sato T. (1989) Flow velocity field tomography using multiple ultrasonic beam detectors and high-order correlation analysis. *J. Acoust. Soc. Am.*, **86**, 1047–52.

Jhang K.Y. and Sato T. (1991) 3-D velocity field measurement using multiple plane detections and high-order correlation analysis. *IEEE Trans. Ultrason. Ferroelec. Freq. Control.* **UFFC-38**, 93–9.

Johnson S.A. (1977) Reconstructing three dimensional fluid velocity fields from acoustic transmission measurements. *Trans Inst. Soc. America*, **16**, 3–15.

Johnson S.A., Greenleaf J.F., Samoya W.A., Duck F.A. and Sjostrand J. (1975) Reconstruction of three-dimensional velocity fields and other parameters by acoustic ray tracing. *Proc. IEEE Symposium on Ultrasonics*, 46–51.

Kramer D. (1979) On the problem of reconstructing images of non-scalar parameters from projections: application to vector fields. *IEEE Trans. on Nuclear Sci.* **NS-26**, 2674–7.

Li W. and Hoyle B.S. (1993) Sensor optimisation for ultrasonic process tomography. *Proc. 2nd ECAPT Workshop*, Karlsruhe, pp. 121–3.

Li W. and Hoyle B.S. (1995) Spectral discrimination in multiple transmitter active real-time ultrasonic tomography. *Proc. 4th ECAPT Workshop*, Bergen, pp. 187–95.

Lu Z.Q. (1986) JKM Perturbation theory, relaxation perturbation theory, and their applications to inverse scattering: theory and reconstruction algorithms. *IEEE Trans. Ultrason. Ferroelec. Freq. Control.*, **UFFC-33**, 722–30.

Lynn P.A. (1982) *An Introduction to the Analysis and Processing of Signals*. Macmillan, London.

Lynnworth L.C. (1989) *Ultrasonic Measurements for Process Control*. Academic Press, London.

McCann, J.D. (1986) A review of ultrasonic techniques and applications, with particular reference to low frequency systems and their use in the food processing industry. AERE Report 11271, Atomic Energy Research Establishment, UK.

McMaster R.C., Editor (1963) *Nondestructive Testing Handbook*, Ronald Press, New York.

Minnaert M. (1933) On musical air bubbles and the sounds of running water. *Phil. Mag.* **16**, 235–48.

Moshfeghi M. (1985) Ultrasonic reflection mode tomography using cylindrically diverging beams. PhD thesis, University of Bristol.

Moshfeghi M. (1986) Ultrasonic reflection mode tomography using fan-shaped beam insonification. *IEEE Trans. Ultrason. Ferroelec. Freq. Control*, **UFFC-33**, 299–314.

Mueller R.K., Kaveh M. and Iverson R.D. (1980) A new approach to acoustic tomography using diffraction techniques. *Acoust. Imaging*, **8**, 615–28.

Nakagawa J. (1989) Measurement of the diameter of bubbles in a bubble swarm in water by the response to a sound pulse method. *Int. Chemical Engineering*, **29**, 729–33.

Natterer F. (1986) *The Mathematics of Computerized Tomography*, John Wiley.

Norton S.J. (1988) Tomographic reconstruction of two-dimensional vector fields: application to flow imaging. *Geophys J. Int.*, **97**, 161–68.

Norton S.J. and Linzer M. (1979) Ultrasonic reflectivity tomography: reconstruction with circular transducer arrays. *Ultrasonic Imaging*, **1**, 154–84.

Peyrin F. (1989) Mapping of internal material temperature with ultrasonic computed tomography, *Proc. Int. Conf. Ultrason.* **83**, 31–6.

Pinder A.C. and Godfrey G. (1993) *Food Process Monitoring Systems*, Blackie, London.

Plaskowski A., Beck M.S. and Thorn R. (1992) *Flow Imaging*, Adam Hilger, Bristol.

Povey, M.J.W. (1989) Ultrasonics in food engineering II. *J. Food Engng*, **9**, 1–20.

Povey, M.J.W. (1995) A bat's eye on food. *Acoustics Bull.* **20**, 27–30.

Povey M.J.W. (1997) *Ultrasonic Techniques for Fluids Characterization*, Academic Press, San Diego, CA.

Radon J. (1917) Über die Bestimmung die Funktionen Durch Irhe Integralwerte Längs Gewisser Mannigfaltigkeiten, *Ber. Sächische Akademie der Wissenschaften* **29**, 262–79.

Ramm A.G. and Katsevich A.I. (1996) *The Radon Transform and Local Tomography*, CRC Press, Boca Raton, FL.

Schlaberg H.I., Yang M. and Hoyle B.S. (1996) Real-time ultrasonic tomography for two-component flows. *Electron Lett.*, **32**, 1571–2.

Silk M.G. (1984) *Ultrasonic Transducers for Non-destructive Testing*, Adam Hilger, Bristol.

Wells P.N.T. (1969) *Physical Principles of Ultrasonic Diagnosis*, Academic Press, London.

Wiegand F. (1990) VLSI parallel processing in flow image based measurement. PhD thesis, University of Leeds.

Wiegand F. and Hoyle B.S. (1989a) Simulations for parallel processing of ultrasound reflection-mode tomography with applications to two-phase flow measurement. *IEEE Trans. Ultrason. Ferroelec. Freq. Control.*, **UFFC-36**, 652–60.

Wiegand F. and Hoyle B.S. (1989b) Real-time parallel processing in industrial flow imaging using transputer arrays, in *CONPAR-88, Parallel Processing and Applications*, Editor C. Jesshope, Cambridge University Press, pp. 482–91.

Wiegand F. and Hoyle B.S. (1991) Development and implementation of real-time ultrasound process tomography using a transputer network. *Parallel Computing* **17**, 791–807.

Wolf J. (1988a) Investigation of bubbly flow by ultrasonic tomography. *Part. and Part. Charac.* **5**, 170–3.

Wolf J. (1988b) Ultrasonic tomography in the field of flow measurement. *Proc. Acoust. Soc. Am. Congress*, Seattle, WA, pp. 1–10.

Xu L. (1990) Ultrasonic computerized tomography and its applications to process monitoring. Internal report, UMIST.

Xu L., Xu L. and Chen Z. (1993) Investigation of transmission-mode UCT system for bubbly gas/liquid distribution monitoring. *Proc. 2nd ECAPT Workshop*, Karlsruhe, pp. 209–12.

5 Particle sizing of food emulsions using ultrasonic spectrometry: principles, techniques and applications

D.J. MCCLEMENTS

5.1 Introduction

Many natural and processed foods consist either partly or wholly of emulsions, or have been in an emulsified state at some time during their existence, e.g. milk, cream, ice-cream, butter, cheese, margarine, salad dressing, cream liqueurs, mayonnaise, soups and sauces (Dickinson and Stainsby, 1982; Dickinson 1992; Walstra, 1996). The rheology, appearance, stability and chemical reactivity of these products is determined by the size of the emulsion droplets they contain. Consequently, it is important to have analytical techniques to provide information about the size of the droplets in food emulsions.

Techniques based on optical microscopy, electron microscopy, light scattering and electrical conductivity have been used to characterize emulsions for many years (Dickinson and Stainsby, 1982). Each of these techniques has its own advantages and limitations, and is therefore most suitable for a particular range of applications. Optical and electron microscopy provides the most direct information about the microstructure of food emulsions, i.e. the dimensions, interactions and spatial distribution of the droplets. Nevertheless, sample preparation is time consuming and laborious, and may cause alterations in the structures being examined. Particle sizing instruments based on light scattering and electrical conductivity are simple to operate, give information about the full droplet size distribution and can carry out a complete analysis in a few minutes or less. The major disadvantage of these techniques is that concentrated samples have to be diluted prior to analysis. This may cause alterations in the structural organization of the droplets, and means that many food samples cannot be analysed *in situ*. Consequently, there is a growing interest in the development of alternative technologies for non-destructively determining the droplet size distributions of concentrated and/or optically opaque emulsions. Techniques are needed for fundamental studies in the laboratory, as well as for on-line sensors in food processing factories.

Nuclear magnetic resonance (NMR) techniques have recently been developed for measuring the droplet size distribution of concentrated and optically opaque emulsions (van den Enden *et al.*, 1990; Soderman *et al.*, 1992; Li and Dougherty, 1993; Lonnqvist *et al.*, 1991; Dickinson and McClements, 1995). These techniques are based on measurements of the restricted diffusion of molecules within emulsion droplets. At present, the widespread use of NMR is limited because the equipment is relatively expensive to purchase, requires highly skilled operators and is not easily adapted to on-line measurements.

It has been recognized for over a century that there is a relationship between the size of the droplets in an emulsion and the scattering of ultrasonic waves. Mathematical theories that relate the ultrasonic properties of an emulsion to its droplet size distribution have been available for many years (Epstein and Carhart, 1953; Waterman and Truell, 1961; Allegra and Hawley, 1972). Even so, it is only recently that the possibility of using ultrasound to determine the droplet size distribution of emulsions has become practically feasible (Javanaud, 1988; Harker and Temple, 1988; Barret-Gültepe *et al.*, 1988; Riebel and Löffler, 1989; McClements and Povey, 1989; McClements, 1991, 1992, 1996; Dickinson and McClements, 1995; Dukhin and Goetz, 1996; McClements and Coupland, 1996). This is because recent advances in microelectronics have led to the widespread availability of relatively low-cost instrumentation for carrying out ultrasonic measurements, as well as computers for rapidly analysing the resulting data. The use of ultrasound to determine the droplet size distribution of an emulsion involves two steps. Firstly, the ultrasonic velocity and/or attenuation coefficient of the emulsion is measured over as wide a range of frequencies as possible. Secondly, these measurements are converted into a droplet size distribution using a suitable mathematical model and a knowledge of the thermophysical properties of the component phases.

A number of manufacturers have recently developed ultrasonic particle sizing instruments that can be used to analyse concentrated food emulsions. It is therefore an appropriate time to review the physical principles of ultrasonic particle sizing, to outline its advantages and limitations over alternative technologies, and to give some examples of its application to food systems. Thus food scientists will be able to decide whether ultrasonic spectrometry is the most appropriate technique for their particular application.

5.2 Measurement techniques

To determine the droplet size distribution of an emulsion using ultrasonic spectrometry it is necessary to measure the ultrasonic velocity and/or attenuation coefficient over a sufficiently wide range of frequencies (typically between 0.1 and 100 MHz). The ultrasonic velocity is equal to the distance that an ultrasonic wave travels through a material in unit time (c = distance/time), whilst the attenuation coefficient is a measure of the decrease in amplitude of an ultrasonic wave per unit distance travelled. The experimental techniques used to measure the frequency dependence of the ultrasonic properties of liquids can be conveniently divided into three categories: *pulse echo*, *through transmission* and *interferometric* methods. The major difference between these techniques is the nature of the ultrasonic signal applied to the sample (Figure 5.1), and the experimental configuration used to carry out the measurements (Figures 5.2 and 5.3).

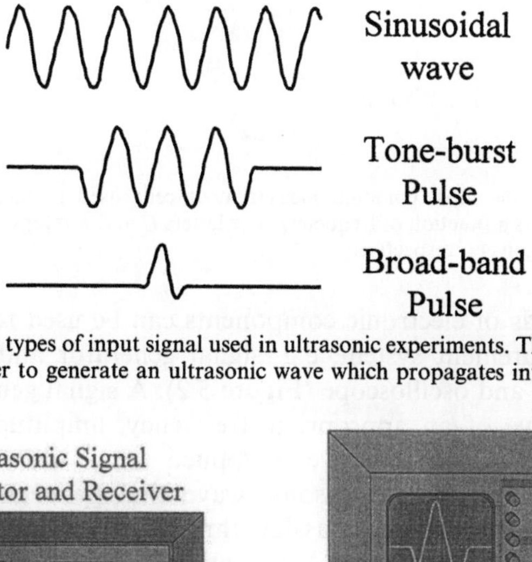

Figure 5.1 Different types of input signal used in ultrasonic experiments. These signals excite an ultrasonic transducer to generate an ultrasonic wave which propagates into a sample.

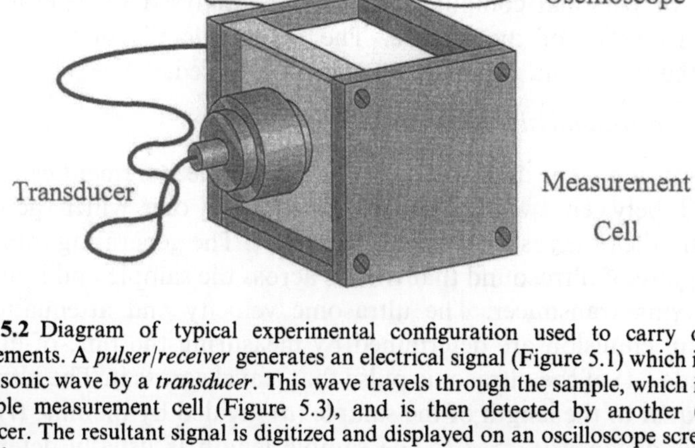

Figure 5.2 Diagram of typical experimental configuration used to carry out ultrasonic measurements. A *pulser/receiver* generates an electrical signal (Figure 5.1) which is converted to an ultrasonic wave by a *transducer*. This wave travels through the sample, which is contained in a suitable measurement cell (Figure 5.3), and is then detected by another (or the same) transducer. The resultant signal is digitized and displayed on an oscilloscope screen, where it is analysed.

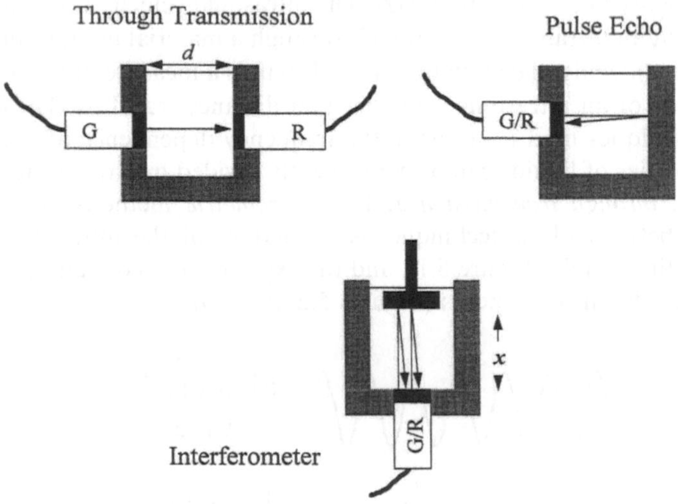

Figure 5.3 Three of the most common measurement cells used to measure the ultrasonic properties of liquids as a function of frequency. The letters *G* and *R* refer to transducers that act as generators and receivers, respectively.

The same kinds of electronic components can be used for most types of ultrasonic measurement system, e.g. signal generator and receiver, ultrasonic transducer and oscilloscope (Figure 5.2). A signal generator produces an electrical wave of an appropriate frequency, amplitude and duration (Figure 5.1). This electrical wave is applied to an ultrasonic transducer which converts it into an ultrasonic wave that propagates through the sample being analysed. After passing through the sample the ultrasonic signal is detected by another (or the same) ultrasonic transducer and converted back into an electrical signal. This electrical signal is digitized by an analogue-to-digital converter, amplified and displayed on the screen of an oscilloscope or personal computer, where it is analysed to determine the ultrasonic properties of the sample. The precise details of this analysis depend on the ultrasonic measurement technique used.

5.2.1 Through transmission techniques

The food emulsion to be analysed is poured into a measurement cell so that it is located between two ultrasonic transducers, one which acts as a generator and the other as a receiver (Figure 5.3). The generating transducer produces a pulse of ultrasound that travels across the sample and is detected by the receiving transducer. The ultrasonic velocity and attenuation coefficient of an emulsion are determined by measuring the time-of-flight (t) and amplitude (A) of the ultrasonic pulse that travel across it. The ultrasonic velocity is equal to the length of the sample (d) divided by the time (t) taken to travel this distance: $c = d/t$. The sample length is usually determined accurately by measuring the time-of-flight through a liquid of known ultrasonic properties (such as distilled water): $d = c_{water}t_{water}$. The attenuation coefficient α is calculated by comparing the reduction in amplitude of a pulse that has travelled through the material being analysed, with that of a pulse that has travelled through a material whose attenuation coefficient is known (usually distilled water): $\alpha_2 = \alpha_1 + \ln(A_R/A_S)/d$, where the subscripts R and S refer to the properties of the reference material and the sample, respectively. In principle, ultrasonic measurements are simple to carry out, but in practice there are a number of factors that must be considered if accurate and reliable measurements are to be obtained. The measurement cell must be carefully designed to minimize temperature variations, reverberation of ultrasonic pulses in cell walls and transducers, diffraction effects and phase cancellation due to non-parallel walls (McClements, 1991).

Two different approaches are available for measuring the frequency dependent ultrasonic properties of emulsions using this technique. In the first approach, a *broad-band* ultrasonic pulse is used, i.e. a single pulse which contains a wide range of different frequencies (Figure 5.1). After the pulse has travelled through the sample it is analysed using a Fourier transform algorithm to determine the values of t and A (and therefore c and α) as a function of frequency (McClements and Fairley, 1992). To cover the whole frequency range (0.1–100 MHz) a number of transducers (typically three or four) with different resonant frequencies are used. In the second approach, a *tone-burst* ultrasonic pulse is used, i.e. a single pulse which contains a number of cycles of ultrasound at a particular frequency (Figure 5.1). In this case the transducer is tuned to a particular frequency and a measurement of the ultrasonic velocity and attenuation is carried out. The transducer is then tuned to another frequency and the process is repeated. Because measurements are carried out separately at a number of different frequencies, this approach is more time consuming and laborious than the one that uses broad-band pulses.

5.2.2 Pulse echo techniques

The frequency dependent ultrasonic properties of a material are measured in almost exactly the same fashion using this technique as the *through transmission technique*, except that a single transducer is used to both generate and receive the ultrasonic pulses (McClements and Fairley, 1991, 1992). The ultrasonic transducer generates a pulse of ultrasound which travels across the sample, is reflected from the back wall of the measurement cell, travels back through the sample and is then detected by the same transducer

(Figure 5.3). The velocity and attenuation coefficient are calculated in exactly the same manner as described for the through transmission technique, except that the pulse has now travelled a distance $2d$ rather than d.

5.2.3 *Interferometric technique*

In an interferometer the emulsion to be analysed is poured into a measurement cell so that it is located between an ultrasonic transducer and a moveable reflector plate (Figure 5.3). A sinusoidal electrical signal of a given frequency (Figure 5.1) is applied to the transducer, where it is converted it into a sinusoidal ultrasonic wave that propagates into the sample (McSkimin, 1964). This wave is reflected back and forth between the reflector plate and transducer, which results in the formation of a standing wave in the emulsion. As the reflector plate is moved vertically through the sample the amplitude of the signal received by the transducer goes through a series of maxima and minima as destructive and constructive interference occurs. The distance x_{max} between successive maxima is equal to half the ultrasonic wavelength ($\lambda/2$) of the sample and so the velocity can be calculated: $c = f\lambda = 2fx_{max}$. The amplitude of the maxima decrease as the distance between the reflector plate and transducer is increased because of attenuation by the sample, imperfect reflection at the boundaries and diffraction. The attenuation coefficient is determined by measuring the amplitude of the maxima as a function of separation between the transducer and the reflector plate for both the sample and a calibration material. The accuracy of the measurements can be improved by measuring the amplitude and distance between a large number of successive maxima. The frequency of the measurement is determined by the resonant frequency of the crystal in the transducer. Crystals can be operated at odd-harmonics of their resonant frequency (f_R, $3f_R$, $5f_R$, etc.), and so measurements can be made over a wide range of frequencies using a single transducer. Nevertheless, measurements must be carried out separately at each individual frequency, which is more time consuming and laborious than for the broad-band pulsed technique described above. Some interferometers use tone-burst signals, rather than continuous sinusoidal waves, so that the ultrasonic energy is not applied continuously to the sample (which may lead to a slight increase in temperature due to absorption of the ultrasound).

5.3 Theory of ultrasonic propagation in emulsions

Once the ultrasonic properties of an emulsion have been measured it is necessary to relate them to the droplet size distribution using an appropriate mathematical method. A number of theories have been derived to describe the propagation of ultrasonic waves through emulsions. These theories are based on a mathematical consideration of the physicochemical processes that occur when an ultrasonic wave propagates through a liquid containing an ensemble of spherical droplets. These processes are of a number of different types, including: (i) *scattering* of the wave into directions different from that of the incident wave; (ii) *absorption* of ultrasonic energy due to thermal conduction and viscous flow; and (iii) *interference* between the waves that travel directly through the droplet, the waves that travel directly through the surrounding medium, and the waves that are scattered. The relative importance of these mechanisms depends on the relative thermophysical properties of the component phases, the frequency of the ultrasonic wave, and the concentration and size of the droplets. The dependence of the ultrasonic properties of an emulsion on the size of the droplets that it contains is the basis of ultrasonic particle sizing.

5.3.1 General theory of ultrasonic scattering

The problem of relating the ultrasonic properties of an emulsion to its droplet size distribution involves two stages. Firstly, the magnitude and phase of the waves scattered from each of the individual emulsion droplets are calculated. Secondly, the magnitude and phase of the wave which emerges from the emulsion (that which is observed in an ultrasonic experiment) is calculated by summing the contributions of the incident wave and the waves that have interacted with the droplets, taking into account their spatial distribution and any droplet–droplet interactions. In dilute systems, each wave is only scattered by a single particle (single scattering), but in concentrated systems a wave might be scattered by a number of different particles (multiple scattering). Mathematical analysis of single scattering is well understood and there is excellent agreement between theory and experimental measurement (Povey, 1997). Multiple scattering effects are more difficult to quantify. Even so, much progress has been made in this area, and it has been shown experimentally that existing ultrasonic theories give reasonable predictions of the ultrasonic properties of many emulsions up to fairly high droplet concentrations (McClements, 1992).

The ultrasonic properties of an ensemble of scatterers are characterized in terms of a complex propagation constant: $K = \omega/c + i\alpha$, where c is the ultrasonic velocity and α is the attenuation coefficient of the scattering material (Epstein and Carhart, 1953; Waterman and Truell, 1961; Allegra and Hawley, 1972). One of the most comprehensive and widely used multiple scattering theories is that derived by Waterman and Truell (1961).

$$\left(\frac{K}{k_1}\right)^2 = 1 + \frac{4\pi N f(0)}{k_1^2} + \frac{4\pi^2 N^2}{k_1^4}\,[f^2(0) - f^2(\pi)] \tag{5.1}$$

where $f(0)$ and $f(\pi)$ are the far-field *scattering amplitudes* of the waves scattered from the individual droplets

$$f(0) = \frac{1}{ik_1}\sum_{n=0}^{\infty}(2n+1)A_n \tag{5.2}$$

$$f(\pi) = \frac{1}{ik_1}\sum_{n=0}^{\infty}(-1)^n(2n+1)A_n \tag{5.3}$$

Here k_1 is the complex propagation constant of the continuous phase ($= \omega/c_1 + i\alpha_1$), N is the number of droplets per unit volume ($\phi = 4\pi r^3 N/3$), ϕ is the disperse phase volume fraction, ω ($= 2\pi f$) is the angular frequency, f is the frequency, $i = \sqrt{-1}$ and r is the droplet radius. The A_n terms are the *scattering coefficients* of the various types of waves scattered from the individual droplets. Mathematical theories for calculating the scattering coefficients of emulsion droplets are available in the literature (Epstein and Carhart, 1953; Allegra and Hawley, 1972). The most general theory calculates the A_n terms by solving a series of linear simultaneous equations at each value of n. Under certain physical situations explicit analytical equations have been derived for calculating the A_n terms (Epstein and Carhart, 1953; Allegra and Hawley, 1972). The actual value of the scattering coefficients depends on the relative thermophysical properties of the component phases, the ultrasonic frequency used and the size of the emulsion droplets. The terms containing N in equation (5.1) account for single scattering effects, whilst those containing N^2 account for (some of the) multiple scattering effects. Multiple scattering becomes increasingly important, and more difficult to describe mathematically, as the concentration of droplets in an emulsion increases. Equation (5.1) can be used to relate the ultrasonic properties of an emulsion (c and α) to its droplet size, once the thermophysical properties of the component phases and the droplet concentration are known: $c = \omega/\mathrm{Re}(K)$, $\alpha = \mathrm{Im}(K)$.

The above equations assume that the contribution of higher-order multiple scattering terms (N^3, etc.) is negligible, and that the emulsion droplets are spherical, do not physically interact with each other and are randomly distributed in space (Waterman and Truell, 1962). These equations may therefore give poor predictions of the ultrasonic properties of concentrated emulsions or emulsions where the droplets are flocculated (McClements, 1992). Recently, attempts have been made to include the effects of droplet flocculation in ultrasonic scattering theory (Al-Nimr and Arpaci, 1993). Further work is needed in this area before ultrasound can be used to provide quantitative information about the nature of droplet–droplet interactions in emulsions, or before these interactions can be adequately accounted for in the existing multiple scattering theory.

5.3.2 *Simplified mathematical models*

Generally, the interactions between an ultrasonic wave and an emulsion droplet are complex, involving various forms of scattering, absorption, reflection, diffraction and resonant behaviour (Riebel and Löffler, 1989). This accounts for the complexity of the general mathematical theory described above. Nevertheless, under certain physical situations the number of interactions involved are reduced and the mathematical description of ultrasonic propagation can be simplified significantly. In practice, it is convenient to divide ultrasonic propagation in emulsions into three categories according to the relationship between the droplet size and the ultrasonic wavelength (Figure 5.4):

- the long wavelength regime (LWR): $r \ll \lambda$
- the intermediate wavelength regime (IWR): $r \sim \lambda$
- the short wavelength regime (SWR): $r \gg \lambda$

Calculations of the extent of each of these regimes for typical emulsion droplet sizes (0.01–100 μm) and ultrasonic frequencies (0.1–100 MHz), indicate that almost all emulsions of practical importance to the food industry fall into either the LW or IW regimes (Figure 5.5).

Ultrasound can be used to determine the size of large droplets (> 1 mm) in the short wavelength regime, but ultrasonic imaging, rather than ultrasonic spectrometry, tends to be used (Gregus, 1980). For this reason, only ultrasonic analysis of emulsions in the LW and IW regimes will be considered.

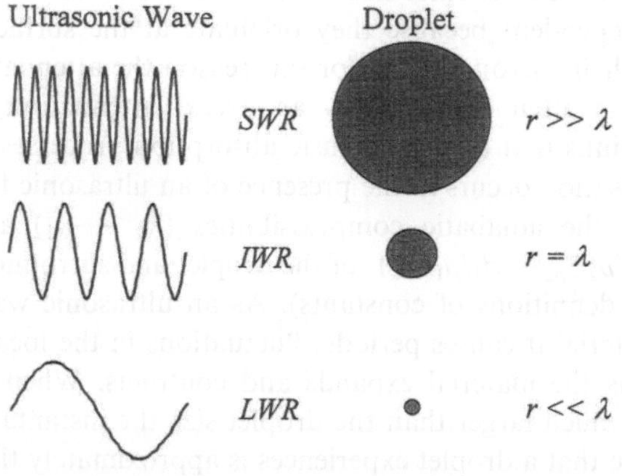

Figure 5.4 The relationship between droplet size and ultrasonic wavelength can be conveniently divided into three regions: long wavelength regime ($\ll \lambda$), intermediate wavelength regime ($\sim \lambda$) and short wavelength regime ($\gg \lambda$).

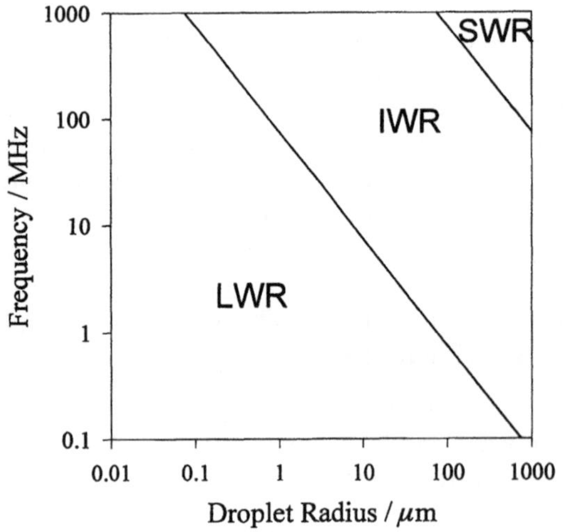

Figure 5.5 Dependence of the long, intermediate and short wavelength regimes on droplet size and ultrasonic frequency for water continuous emulsions. It was assumed that for the LWR, $r < 0.05\lambda$; for the IWR, $0.05\lambda < r < 50\lambda$; and for the SWR, $r > 50\lambda$.

5.3.2.1 Long wavelength regime. Most ultrasonic measurements made on emulsions with droplet diameters less that about 10 μm fall into the LWR. In the LWR only three types of interaction between ultrasonic waves and emulsions are important: intrinsic absorption, droplet pulsation and droplet oscillation. The intrinsic absorption of an emulsion is simply the volume average of the absorptions of each of the individual component phases, and is thus independent of droplet size. The other two forms of interaction are droplet size dependent because they originate at the surface between the droplets and their surroundings. For this reason the attenuation coefficient of emulsions is often expressed as an *excess* attenuation, i.e. the total attenuation minus that due to intrinsic absorption processes.

Droplet pulsation occurs in the presence of an ultrasonic field because of differences in the adiabatic compressiblities ($\kappa_2 - \kappa_1$) and/or thermal properties $(\beta_2/\rho_2 C_{p2} - \beta_1/\rho_1 C_{p1})^2$ of the droplet and surrounding liquid (see Table 5.1 for definitions of constants). As an ultrasonic wave propagates through a material it causes periodic fluctuations in the local pressure and temperature as the material expands and contracts. When the ultrasonic wavelength is much larger than the droplet size the instantaneous pressure or temperature that a droplet experiences is approximately the same as that of the surrounding liquid (Figure 5.4). Nevertheless, the droplet still contracts or expands relative to the surrounding liquid because it has a different adiabatic compressibility (change in volume per unit change in pressure) or

Table 5.1 Selection of the thermophysical properties of some common liquids found in food emulsions at 20°C (taken from Coupland and McClements, 1997). Here c is the ultrasonic velocity, ρ is the density, η is the viscosity, C_p is the specific heat capacity, β is the thermal expansivity, τ is the thermal conductivity and α is the attenuation coefficient. The attenuation data is given at 2 MHz

Liquid	c (m s^{-1})	ρ (kg m^{-3})	η (mPa s)	C_p (J kg^{-1} K^{-1})	β (K^{-1})	τ (J m^{-1} s^{-1} K^{-1})	α (Np m^{-1})
Water	1482.3	998.2	1	4182	2.1	0.59	1.04
Sunflower oil	1471	920.6	54	1980	7.1	0.17	–
Soybean oil	1469.8	922.8	40–60	1917	7.4	0.16	5.5
Rape seed oil	1468.4	911.4	73	–	–	–	5.1
Olive oil	1465.3	915.8	67–82	–	–	0.17	6.5
Cotton seed oil	–	918.7	51	–	7.4	–	–

because it has different thermal properties (change in volume per unit change in temperature). Particle pulsation leads to the generation of a monopole wave which propagates away from the droplet equally in all directions (Figure 5.6). Thus some of the ultrasonic energy is scattered into directions that are different from that of the original wave: this process is referred to as monopole *scattering*. As the droplet pulsates its temperature also fluctuates periodically because of pressure–temperature coupling. This leads to heat flow across the interface between the droplet and the surrounding liquid. If the heat flow out of the droplet is not compensated for by the heat flow into the droplet during a compression–expansion cycle some of the ultrasonic energy is converted to heat: this process is referred to as *thermal absorption*.

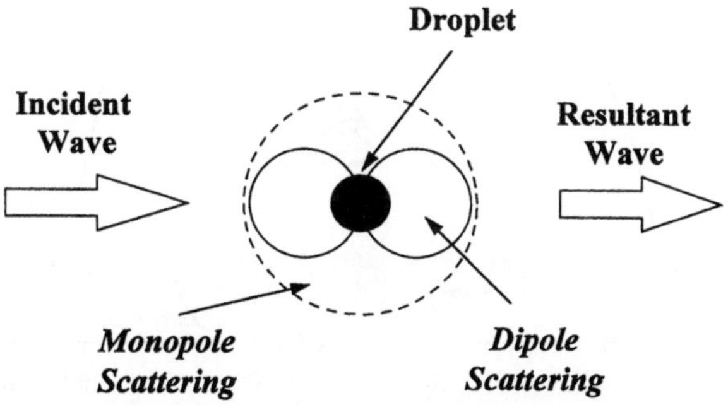

Figure 5.6 The two most important types of interaction between an ultrasonic wave and an emulsion droplet in the long wavelength regime are droplet *oscillation and pulsation*. Pulsation leads to the generation of a monopole wave, whilst oscillation leads to the generation of a dipole wave. Ultrasonic energy is also lost due to thermal and visco-inertial absorption processes associated with the pulsation and oscillation of the droplets.

Droplet oscillation is caused by a density difference between the droplet and the surrounding liquid ($\rho_2 - \rho_1$). In the presence of an ultrasonic wave the droplet moves backwards and forwards relative to the surrounding liquid because it has a different inertia (Figure 5.6). Oscillation of the droplet leads to the generation of a scattered wave which moves away from the droplet with a cosine dependence. Thus some of the ultrasonic energy is scattered into directions which are different from that of the original wave: this process is referred to as *dipole scattering*. As the droplet oscillates in the ultrasonic field its movement is damped because of the viscous drag of the surrounding liquid. These frictional losses mean that some of the ultrasonic energy is converted to heat: this process is referred to as *visco-inertial absorption*.

The effects of the pulsation and oscillation of droplets in an ultrasonic field on the ultrasonic properties of an emulsion are contained in the *scattering coefficients* A_0 and A_1, respectively. All higher order scattering coefficients are negligible in the LWR and so equations (5.1) to (5.3) can be simplified

$$\left(\frac{K}{k_1}\right)^2 = 1 - \frac{i4\pi N(A_0 + 3A_1)}{k_1^3} - \frac{48\pi^2 N^2 A_0 A_1}{k_1^6} \tag{5.4}$$

In addition, explicit analytical expressions for the A_0 and A_1 terms can be used in the LWR, rather than having to solve the full set of complex linear simultaneous equations needed in the general solution (McClements, 1996).

The dependence of the ultrasonic velocity and attenuation coefficient on droplet size and ultrasonic frequency in the LWR is illustrated in Figure 5.7 for a typical emulsion (10 wt% *n*-hexadecane oil-in-water emulsion at 20°C).

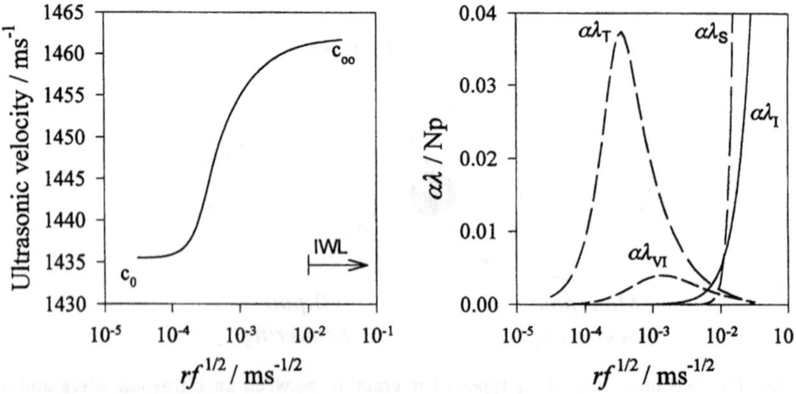

Figure 5.7 Variation of the ultrasonic velocity and attenuation coefficient of an emulsion with droplet size and frequency in the long wavelength regime. Values were calculated using equation (5.4) and the A_0 and A_1 terms for a 10 wt% *n*-hexadecane oil-in-water emulsion with a droplet radius of 1 μm.

The precise shape of the curves actually depends on the relative physicochemical properties of the component phases. In the LWR the ultrasonic properties of an emulsion are usually plotted against $rf^{1/2}$ because results from emulsions with different combinations of particle size and frequency fall on the same curve. The ultrasonic velocity increases from a constant value (c_0) at low $rf^{1/2}$ values to a another constant value (c_∞) at high $rf^{1/2}$ values, and can be described by an equation of the form (McClements and Coupland, 1996)

$$c = c_0 + (c_\infty - c_0)f(f, r) \tag{5.5}$$

where $f(f, r)$ is a function which depends on the ultrasonic frequency and droplet size, and varies from 0 (at low $rf^{1/2}$ values) to 1 (at high $rf^{1/2}$ values). The overall attenuation in an emulsion can be conveniently divided into four contributions: intrinsic absorption (α_I), (monopole + dipole) scattering losses (α_S) visco-inertial absorption (α_VI) and thermal absorption (α_T)

$$\alpha_\mathrm{TOTAL} = \alpha_\mathrm{I} + \alpha_\mathrm{S} + \alpha_\mathrm{VI} + \alpha_\mathrm{T} \tag{5.6}$$

Expressions for the different terms can be found in the literature (McClements, 1996). There is a maximum in the attenuation per cycle at intermediate values of droplet size and frequency (Figure 5.7). For food emulsions the major cause of attenuation in the LWR is thermal absorption, because the density difference between the droplets and surrounding liquid is so small that the visco-inertial losses are negligible. Scattering losses and intrinsic absorption only make a significant contribution to the overall attenuation at higher frequencies, and become the dominant form of attenuation when the IWR is entered.

In the LWR, the ultrasonic velocity of an emulsion is only sensitive to droplet size in the region $10^{-4} < rf^{1/2} < 10^{-2}$ (Figure 5.7). If an instrument is used that can measure the ultrasonic velocity over the frequency range 0.1–100 MHz it is possible to analyse droplets with sizes between 0.01 and 30 μm. This range can be extended by using attenuation measurements, either alone or in conjunction with velocity measurements.

5.3.2.2 Intermediate wavelength regime, Emulsions that contain fairly large emulsion droplets (typically $> 10\,\mu$m) tend to fall in the IWR, especially when relatively high ultrasonic frequencies are used (> 10 MHz). Interactions between ultrasonic waves and emulsion droplets are most complicated in this regime, and many different types of scattered wave are generated. Consequently, it is necessary to include many more scattering coefficients (A_n) in the calculation of the ultrasonic properties of an emulsion, than the two (A_0 and A_1) required in the LWR. Scattering of the ultrasonic waves dominates the other types of interaction in the IWR, and so simplified expressions for the A_n terms can be used to calculate the ultrasonic properties of an emulsion from equation (5.1) (McClements 1996). The number of A_n terms needed to calculate the ultrasonic properties of an emulsion varies with frequency, and is determined at each frequency by incrementally increasing the value of n and calculating the complex propagation constant K until it does not change any more. As many as 20 or 30 A_n terms may be needed to calculate the ultrasonic properties in the IWL.

The dependence of the ultrasonic velocity and attenuation coefficient of an emulsion on droplet size and frequency in the IWR is shown in Figure 5.8. As the droplet size increases relative to the ultrasonic wavelength the attenuation coefficient increases until it reaches a maximum when $r \sim \lambda$. Above this value the attenuation coefficient decreases, and at high ratios of r to λ it becomes negligible. The ultrasonic velocity decreases initially as the particle size increases and then increases dramatically when $r \sim \lambda$, until it reaches a constant value (equal to the velocity in the pure continuous phase) at high ratios of r to λ.

The ultrasonic properties of emulsions are particularly sensitive to droplet size in the region $0.1 < r/\lambda < 50$ in the IWR (Figure 5.8). If an instrument is used which can measure the ultrasonic velocity and attenuation coefficient over the frequency range 0.1–100 MHz it is therefore possible to analyse droplets with sizes between about 1 μm and a few meters. Thus by utilizing measurements in both the LWR and IWR it is possible to cover the whole range of droplet sizes that are important in food emulsions.

5.3.3 Polydispersity

Food emulsions do not contain emulsion droplets that all have the same size, but rather contain a range of different droplet sizes, which may be more or less broad. The above equations must therefore be modified

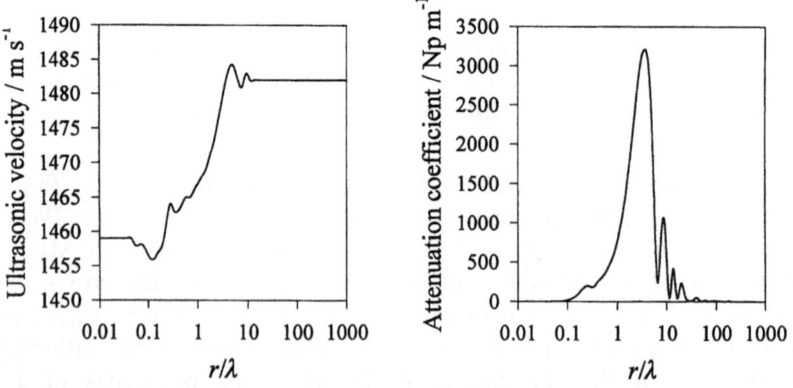

Figure 5.8 Variation of the ultrasonic velocity and attenuation coefficient of an emulsion with droplet size and frequency in the intermediate wavelength regime. Values were calculated using equation (5.1) and the A_n terms for a 10 wt% n-hexadecane oil-in-water emulsion with a droplet radius of 50 μm.

$$\left(\frac{K}{k_1}\right)^2 = 1 + \frac{4\pi \sum_j N_f f_f(0)}{k_1^2} + \frac{4\pi^2}{k_1^4} \sum_j N_f^2 (f_j{}^2(0) - f_j{}^2(\pi)) \tag{5.7}$$

Here $f_j(0)$ and $f_j(\pi)$ are the scattering amplitudes of droplets with radius r_j, and N_j is the number of droplets with radius r_j.

5.4 Conversion of ultrasonic measurements to droplet size distributions

Emulsions are analysed using ultrasonic spectrometry by measuring their ultrasonic velocity and/or attenuation coefficient as a function of frequency, and then finding the droplet size distribution that gives the best fit between the experimental measurements and predictions made using multiple scattering theory. There are two main approaches to solving this inverse scattering problem: model-independent inversion and model-dependent inversion. In the model-dependent inversion method it is assumed that the particle size distribution of an emulsion can be modelled using a relatively simple mathematical model, e.g. log-normal

$$P(r) = \frac{1}{x_g \ln \sigma_g \sqrt{2\pi}} \exp - \left[\frac{\ln^2 \sigma_g}{2}\right] \exp - \left[\frac{(\ln r - \ln x_g)^2}{2\ln^2 \sigma_g}\right] \tag{5.8}$$

Here x_g and σ_g are the mean and standard deviation of the droplet radius. The droplet size distribution can then be calculated from experimental ultrasonic measurements using a microcomputer. Initially, a guess at the mean droplet size (x_g) and standard deviation (σ_g) is used to calculate the ultrasonic properties of the emulsion using the multiple scattering theory. The predicted values are then compared with the experimental values, and the sum of the squares of the differences (SSD) is calculated. For ultrasonic velocity measurements as a function of frequency

$$SSD = \sum_i [c_{\mathrm{THEORY}}(f_i) - c_{\mathrm{EXPT}}(f_i)]^2 \tag{5.9}$$

whilst for attenuation measurements

$$SSD = \sum_i [\alpha_{\mathrm{THEORY}}(f_i) - \alpha_{\mathrm{EXPT}}(f_i)]^2 \tag{5.10}$$

where $c_{\mathrm{THEORY}}(f_i)$ and $\alpha_{\mathrm{THEORY}}(f_i)$ are the predicted ultrasonic velocity and attenuation coefficient, and $c_{\mathrm{EXPT}}(f_i)$ and $\alpha_{\mathrm{EXPT}}(f_i)$ are the experimentally measured values at frequency f_i. The mean and standard deviation are then varied until the computer finds the minimum in SSD. Once the values of x_g and σ_g have been determined then the particle size distribution $P(r)$ can be calculated using equation (5.8). The model-dependent method has the advantage that only a small number of variables need to be determined during the inversion procedure: e.g. x_g and σ_g. Problems may arise if the assumed model is not a reasonable representation of the system being studied. In this case it may be possible to use a different type of model droplet size distribution, or to increase the complexity of the model, e.g. by assuming a bimodal distribution. Ideally, one would like to determine the droplet size distribution without making any a priori assumptions about its shape. Although this is possible, the solution is extremely sensitive to small changes in the experimental measurements and so there may be more than one droplet size distribution that gives a good fit between the theory and experimental data (Riebel and Löffler, 1989). If this situation does occur it should be possible to ascertain which droplet size distribution is the correct one by measuring the ultrasonic velocity and the attenuation coefficient (rather than just one or the other). Another potential problem with the model-independent method is that the computation time required to determine the droplet size distribution is much longer than the model dependent method, because the values of many more variables must be determined.

5.5 Thermophysical properties of component phases

One of the major factors holding back the widespread use of ultrasonic spectrometry for characterizing food emulsions is the fact that a large amount of information is needed about the thermophysical properties of the component phases before one can convert ultrasonic measurements into droplet size distributions, e.g. ultrasonic velocities, density, viscosity, specific heat capacity, thermal conductivity, thermal expansivity and attenuation coefficient. In a recent paper, we have tabulated the thermophysical properties of a large number of edible oils that are relevant to ultrasonic characterization of emulsions (Coupland and McClements, 1997). A selection of some of these properties is included in Table 5.1. It can be seen that most food oils have quite similar thermophysical properties, and therefore we expect that the ultrasonic properties of emulsions manufactured from different oils will have quite similar ultrasonic properties, which will facilitate the ultrasonic analysis of their properties.

5.6 Application of ultrasonic spectrometry to food emulsions

The ultrasonic velocity and attenuation coefficient of a series of 20 wt% corn oil-in-water emulsions stabilized by Tween 20 were measured at 30°C using a Frequency Scanning Pulse Echo Reflectometer (McClements and Fairley, 1991, 1992) in the frequency range 1–5 MHz. The droplet size distribution of the emulsions was then determined using a computer program written in our laboratory. This program finds the droplet size distribution (assuming the emulsions have a log-normal distribution) that gives the best-fit between the experimental measurements and the velocity and attenuation coefficient predicted by multiple scattering theory. Good agreement was found between the droplet size distributions determined using the ultrasonic technique and those determined by laser-light scattering (Figure 5.9).

5.7 Advantages and limitations of ultrasonic spectrometry

The major advantages of ultrasonic spectrometry over other methods of particle sizing are that it is non-destructive, non-invasive, can be used in concentrated and optically opaque emulsions, is relatively inexpensive and measurements are rapid. In addition, it can easily be adapted for on-line measurements which is particularly useful for monitoring food processing operations. One of the major limitations of the ultrasonic technique is that it cannot be used to study emulsions which contain small gas bubbles. This is because the gas bubbles scatter the ultrasound so effectively (even at very low concentrations) that the ultrasonic signal is completely attenuated

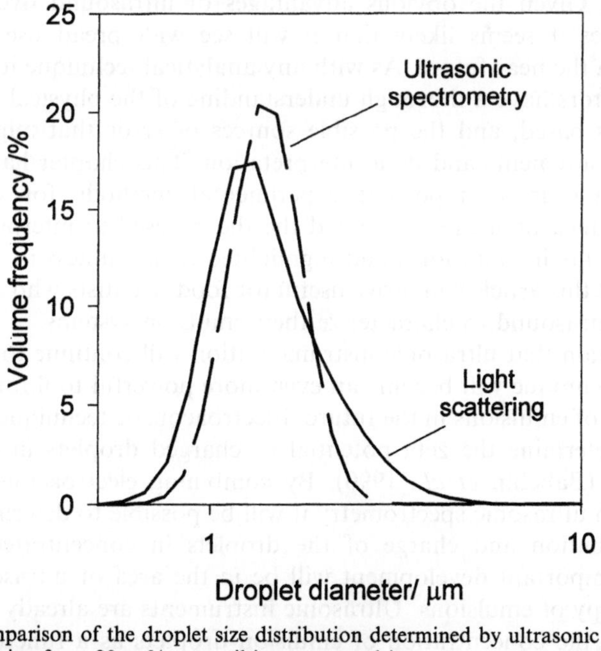

Figure 5.9 Comparison of the droplet size distribution determined by ultrasonic spectrometry and light scattering for a 20 wt% corn oil-in-water emulsion.

(Gaunuard and Überall, 1981). Ultrasound also has limited application for analysis of very dilute emulsions (<0.5 wt%), because the change in ultrasonic properties with droplet size becomes of the same order as the experimental error. At present ultrasound also has limited application to very concentrated (>30 wt%) or aggregated systems. It is possible to propagate ultrasound through these emulsions, but the theories that relate the measurable ultrasonic properties to the droplet size distribution are still poorly developed because of the mathematical complexity of accounting for particle–particle interactions in flocculated or concentrated systems. Nevertheless, significant progress is being made in this area, and it is likely that new theoretical developments will extend the use of ultrasound to higher droplet concentrations in the near future. In summary, ultrasonic spectrometry is most useful for studying concentrated or optically opaque materials, or when on-line measurements are required, whereas light scattering or electrical conductivity measurements are preferable for dilute emulsions (ϕ < 0.1 wt%).

5.8 Conclusions

Ultrasonic spectrometry is a particle sizing technology that has finally come of age. Analytical instruments are now commercially available that are capable of measuring the droplet size distribution of concentrated food emulsions. Given the obvious advantages of ultrasound over alternative technologies it seems likely that it will see widespread use in the food industry in the near future. As with any analytical technique it is important that operators have a thorough understanding of the physical principles on which it is based, and the possible sources of error that might adversely affect measurements and data interpretation. This chapter has highlighted some of the most important experimental methods for carrying out ultrasonic measurements, presented the theory used to interpret ultrasonic measurements in emulsions, and highlighted its usefulness in practice. It is hoped that this article will prove useful for food scientists who are interested in using ultrasound to characterize their emulsion systems.

It is certain that ultrasonic instrumentation will continue to develop and that the technique will become an even more powerful tool for probing the properties of emulsions in the future. Electroacoustic techniques are already used to determine the zeta potential of charged droplets in concentrated emulsions (Babchin *et al.*, 1990). By combining electroacoustic measurements with ultrasonic spectrometry it will be possible to determine both the size distribution and charge of the droplets in concentrated emulsions. Another important development will be in the area of ultrasonic imaging spectroscopy of emulsions. Ultrasonic instruments are already available for measuring the concentration of emulsion droplets as a function of height

(Howe *et al.*, 1986; Povey, 1995). By measuring the frequency dependence of the ultrasonic properties of an emulsion at each height it should also be possible to determine the size of the droplets, thus providing food scientists with a powerful tool for studying kinetic phenomenon, such as creaming and coalescence.

Acknowledgements

This material is based on work supported by the Cooperative State Research, Education, and Extension Service, US Department of Agriculture, under Agreement No. 95-37500-2051. The author also thanks Dr John Coupland for valuable advice and discussions, and for providing the information about the thermophysical properties of the oils, and Jochim Meier for providing Figure 5.2.

5.9 References

Allegra J.R. and Hawley S.A. (1972) Attenuation of sound in suspensions and emulsions: theory and experiments. *J. Acoust. Soc. Am.*, **51**, 1545–64.

Al-Nimr M.A. and Arpaci V.S. (1993) Acoustical properties of interacting particles. *J. Acoust. Soc. Am.*, **93**, 813–18.

Babchin A.J., Chow R.S. and Sanatzky R.P. (1990) Electrokinetic measurements by electroacoustical methods. *Adv. Coll. Int. Sci.*, **30**, 111–51.

Barret-Gültepe M.A., Gültepe M.E., McCarthy J.L. and Yeager E.B. (1988) The determination of particle-size distribution of perfluorochemical emulsions by ultrasonic measurement. *Biomat. Art. Cells Art. Org.*, **16**, 691–92.

Coupland J.N. and McClements D.J. (1997) A compilation of some physical properties of liquid edible oils. *J. Am. Oil Chem. Soc.* Submitted.

Dickinson E. (1992) *An Introduction* to *Food Colloids*, Oxford University Press, Oxford.

Dickinson E. and Stairisby G. (1982) *Colloids in Foods*, Applied Science, London.

Dickinson E. and McClements D.J. (1995) *Advances in Food Colloids*, Blackie, London.

Dukthin A.S. and Goetz P.J. (1996) Acoustic spectroscopy for concentrated polydisperse colloids with high density contrast. *Langmuir*, **12**, 4987–97.

Epstein P.S. and Carhart R.R. (1953) The absorption of sound in suspensions and emulsions. I. Water fog in air. *J. Acoust. Soc. Am.*, **25**, 533–65.

Gaunaurd G.C. and Überall H. (1981) Resonance theory of bubbly liquids. *J. Acoust. Soc. Am.*, **62**, 362–70.

Gregus P. (1980) *Ultrasonic Imaging: Seeing by Sound*, Focal Press, New York.

Harker A.H. and Temple J.A.G. (1988) Propagation and attenuation of ultrasound in suspensions. *J. Phys. D: Appl. Phys.*, **21**, 1576–88.

Howe A.M., Mackie A.R. and Robins M. M. (1986) Technique to measure emulsion creaming by velocity of ultrasound. *J. Disp. Sci. Tech.*, **7**, 231–43.

Javanaud C. (1988) Applications of ultrasound to food systems. *Ultrasonics*, **26**, 117–23.

Li B. and Dougherty E.R. (1993) Size distribution estimation in process fluids by ultrasound for particle sizes in the wavelength range. *Optical Engng*, **32**, 1967–80.

Lonnqvist I., Khan A. and Soderman O. (1991) Characterisation of emulsions by NMR methods. *J. Colloid lnt. Sci.*, **144**, 401–11.

McClements D.J. and Povey M.J.W. (1989) Scattering of ultrasound by emulsions. *J. Phys. D: Appl. Phys.*, **22**, 38–47.

McClements D.J. (1991) Ultrasonic characterisation of emulsions and suspensions. *Adv. Coll. Int. Sci.*, **37**, 33–72.

McClements D.J. and Fairley P. (1991) Ultrasonic pulse echo reflectometer. *Ultrasonics*, **29**, 58–62.
McClements D.J. and Fairley P. (1992) Frequency scanning ultrasonic pulse echo reflectometer. *Ultrasonics*, **30**, 403–5.
McClements D.J. (1992) Comparison of multiple scattering theories with experimental measurements in emulsions. *J. Acoust. Soc. Am.*, **91**, 849–53.
McClements D.J. (1994) Ultrasonic determination of depletion flocculation in oil-in-water emulsions containing a non-ionic surfactant. *Coll. Surf. A*, **90**, 25–35.
McClements D.J. (1996) Principles of ultrasonic droplet size determination in emulsions. *Langmuir*, **12**, 3454–61.
McClements D.J. and Coupland J.N. (1996) Theory of droplet size distribution measurements in emulsions using ultrasonic spectroscopy. *Coll. Surf. A*, **117**, 161–70.
McSkimin H.J. (1964) In *Physical Acoustics*, Vol. 1A, Editor W.P. Mason, Academic Press, New York, p. 271.
Povey M.J.W. (1995) In *New Physico-chemical Techniques for the Characterization of Complex Food Systems*, Editor E. Dickinson, Blackie, London, pp. 196–213.
Povey M.J.W. (1997) *Ultrasonic Techniques for Fluids Characterization*, Academic Press, San Diego, CA.
Riebel U. and Loffler F. (1989) The fundamentals of particle-size analysis by means of ultrasonic spectrometry. *Part. Part. Syst. Charact.*, **6**, 135–43.
Soderman O., Lonnqvist I. and Balinov B. (1992) In *Emulsions – A Fundamental and Practical Approach*, Editor J. Sjoblom, Kluwer, Dordrecht, p. 239.
Van den Enden J.C., Waddington D., Van Aalst H., Van Kralingen C.G. and Packer K.J. (1990) Rapid-determination of water droplet size distributions by Pfg-Nrnr. *J. Colloid Int. Sci.*, **140**, 105–13.
Walstra P. (1996) In *Food Chemisty*, 3rd edn, Editor O.R. Fennema, Marcel Dekker.
Waterman P.C. and Truell R. (1961) Multiple scattering of waves. *J. Math. Phys.*, **2**, 512–37.

6 Power ultrasound in food processing – the way forward

T.J. MASON

6.1 Introduction

The use of ultrasound within the food industry has been a subject of research and development for many years and, as is the case in other areas, the sound ranges used can be divided basically into diagnostic and power ultrasound (Figure 6.1). Diagnostic applications involving the high-frequency end of ultrasound are covered in other chapters of this book, in this and succeeding contributions the emerging study of power ultrasound in food processing will be explored. In this case the mechanical and chemical effects of cavitation are important and applications are very wide ranging (Table 6.1). The majority of studies are restricted to the range of 20–40 kHz, i.e. the usual ranges employed for power ultrasonic equipment used in cleaning, cell disruption and plastic welding.

This chapter will review the basics of power ultrasonic equipment and survey some of the areas in which it has shown potential for future development.

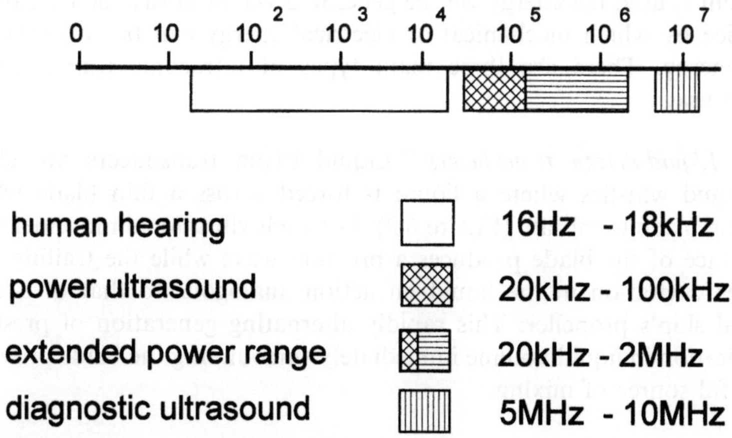

Figure 6.1 The frequency ranges of sound.

Table 6.1 Some uses of power ultrasound in food processing

Mechanical effects
Crystallization of fats, sugars, etc.,
Degassing
Destruction of foams
Extraction of flavourings
Filtration and drying
Freezing
Mixing and homogenization
Precipitation of airborne powders
Tenderization of meat
Chemical and biochemical effects
Bactericidal action
Effluent treatment
Modification of growth of living cells
Alteration of enzyme activity
Oxidation
Sterilization of equipment

6.2 Power ultrasound: generation and equipment

In order to appreciate the potential for the uses of power ultrasound in processing technology it is important to understand firstly the methods by which it can be generated and secondly how the acoustic energy, once generated, can be introduced to a process (Mason, 1991).

6.2.1 Methods of generating power ultrasound

The first requirement for investigations on the use of ultrasound in food processing is a source of ultrasound and whatever type of commercial instrument is used the energy will be generated via an ultrasonic transducer – a device by which mechanical or electrical energy can be converted to sound energy. There are three main types of ultrasonic transducer in common usage.

6.2.1.1 Liquid-driven transducers. Liquid-driven transducers are effectively liquid whistles where a liquid is forced across a thin blade which causes the blade to vibrate (Figure 6.2). For each vibrational movement the leading face of the blade produces a pressure wave while the trailing face generates cavitation in the liquid an action analogous to that of a conventional ship's propeller. This rapidly alternating generation of pressure and cavitation in liquid volume immediately surrounding the blade provides a powerful source of mixing.

6.2.1.2 Magnetostrictive transducers. Magnetostrictive transducers are electromechanical devices which use magnetostriction, an effect found in some ferromagnetic materials, e.g. nickel or iron which change in dimension

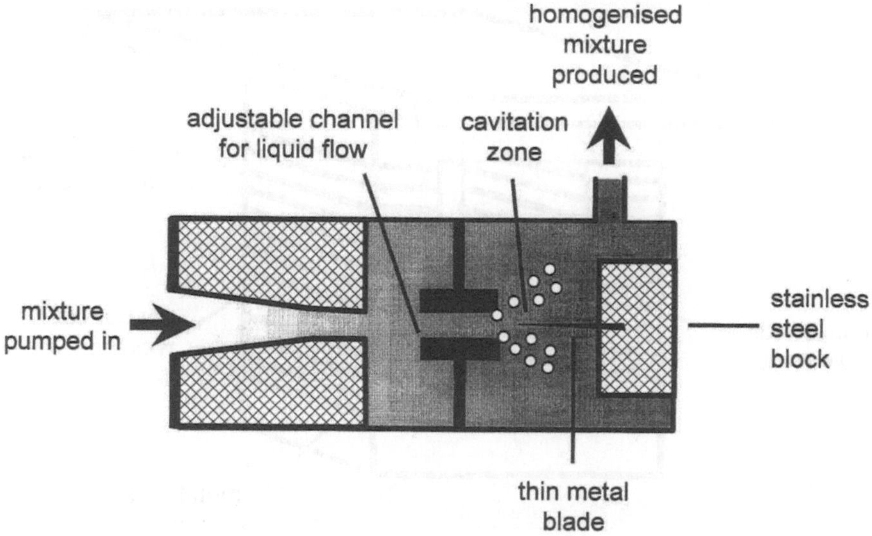

Figure 6.2 Liquid-driven transducer.

on the application of a magnetic field. This type of transducer may be considered to be a form of solenoid with a magnetostrictive material as the core. The core is assembled as a laminate of many layers of thin nickel (or nickel alloy) plate and in its simplest form this comprises of a closed square loop of core material with coils wound around two opposite sides (Figure 6.3). Applying a current to the coil produces a reduction in dimension of the core (magnetostriction) and consequently a reduction in the dimensions of the transducer. Switching off the current results in a return to the original dimensions. Repeated rapid switching on and off of the current generates the mechanical vibrations required. The dimensions of the transducer must be accurately designed so that the whole unit resonates at a precise sound frequency.

There are two disadvantages with this type of transducer. Firstly the frequency range is restricted to below 100 kHz and secondly the system is only about 60% electrically efficient with losses occurring through heating. For this latter reason external cooling is often required. The major advantages are that the system is of an extremely robust construction and that very large driving forces can be obtained.

6.2.1.3 Piezoelectric transducers. Piezoelectric transducers are the most common devices employed for the generation of ultrasound and utilize ceramics containing piezoelectric materials such as barium titanate or lead metaniobate. The piezoceramic element commonly used in ultrasonic cleaners and for probe systems is produced in the form of a disk with a

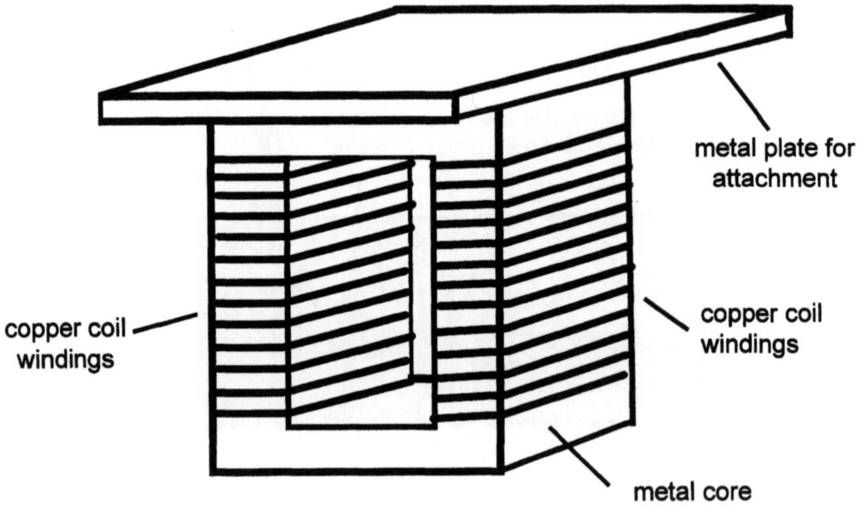

Figure 6.3 Construction of magnetostrictive transducer.

central hole. Ceramic transducers are potentially brittle and so it is normal practice to clamp them between metal blocks (the front and back masses). This serves both to protect the delicate crystalline material and to prevent it from overheating by acting as a heat sink. Usually two elements are combined (known as a 'sandwich' construction) so that their overall mechanical motion is additive (Figure 6.4). Piezoelectric transducers are better than 95% electrically efficient and can operate over the whole ultrasonic range.

6.2.2 *The design of power ultrasonic systems*

The liquid whistle is a very simple and robust device but its operation depends on pumping a fast flowing liquid through an orifice and across a blade.

For this reason applications of the whistle to processing are restricted to operations involving mixing and homogenization although it should be emphasized that for this purpose it is excellent.

The majority of devices for the generation of power ultrasound are based on electroacoustic systems, i.e. piezoelectric or, less commonly, magnetostrictive transducers. Whichever of these two transducer types is used the equipment needed to deliver the ultrasonic energy to the liquid system under investigation will comprise of three essential parts (Figure 6.5):

- A generator to convert mains electricity into the required high frequency alternating current to drive the transducer assembly.

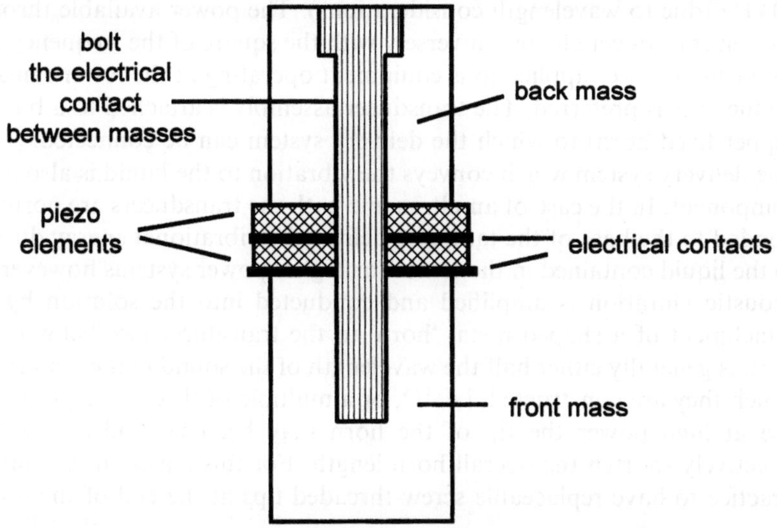

Figure 6.4 Piezoelectric sandwich transducer.

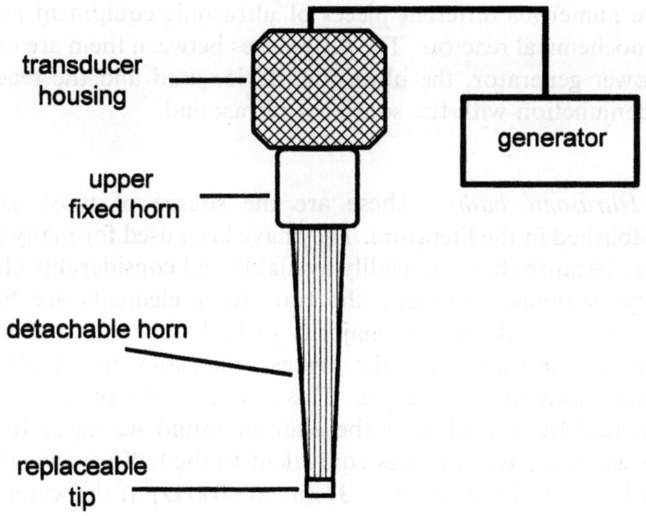

Figure 6.5 Ultrasonic probe system.

- The transducer element itself which converts the high-frequency alternating current into mechanical vibrations. The majority of present day power ultrasonic equipment is based on piezoelectric technology and the shape and dimensions of the sandwich transducer assembly is dependent on its working frequency thus a 20 kHz assembly is twice the length of a

40 kHz (due to wavelength considerations). The power available through a given transducer changes inversely with the square of the frequency and so for high power applications equipment operating at the lower range of frequencies is preferred. The transducer assembly is attached to a booster (upper fixed horn) to which the delivery system can be connected.

- The delivery system which conveys the vibration to the liquid is also a key component. In the case of an ultrasonic bath the transducers are normally bonded to the base of the tank and deliver the vibrational energy directly to the liquid contained in the tank. For higher power systems however the acoustic vibration is amplified and conducted into the solution by the attachment of a shaped metal 'horn' to the transducer (see below). The horn is generally either half the wavelength of the sound in the material of which they are constructed, i.e. l/2, or a multiple of this. After prolonged use at high power the tip of the horn may become eroded and thus effectively shorten the overall horn length. For this reason it is common practice to have replaceable screw-threaded tips at the end of the horn.

6.2.3 *Types of ultrasonic reactor*

There are numerous different pieces of ultrasonic equipment available for use as sonochemical reactors. The differences between them are in the design of the power generator, the ultrasonic device used and the reactor or cell used in conjunction with the source of ultrasound.

6.2.3.1 Ultrasonic baths. These are the source of most experimental results published in the literature. They have been used for many years in the laboratory because they are readily available and considerably cheaper than probe type systems. Generally the transducer elements are fixed to the underside of the tank and the majority of baths operate at around 40 kHz (Figure 6.6). For such baths the highest intensities are obtained at fixed levels throughout the bath depth. This is due to the presence of standing waves created by reflection of the emitted sound waves at the liquid/air interface separated by distances equivalent to the half wavelength of sound in the bath liquid (for water, $\lambda = 37$ mm at 40 kHz). If the water level in the tank falls below λ it is no longer possible to generate high intensities.

There are many different options available for ultrasonic baths which permit a range of different parameters to be studied, these include:

- Thermostatically controlled heating.
- Frequency sweeps; these produce a more uniform cavitation field, by avoiding standing wave zones which would otherwise restrict cavitation.
- Variable power.
- Either on–off actuation which provides bursts of intermittent power or power pulsing. This is often useful for degassing since if the power is

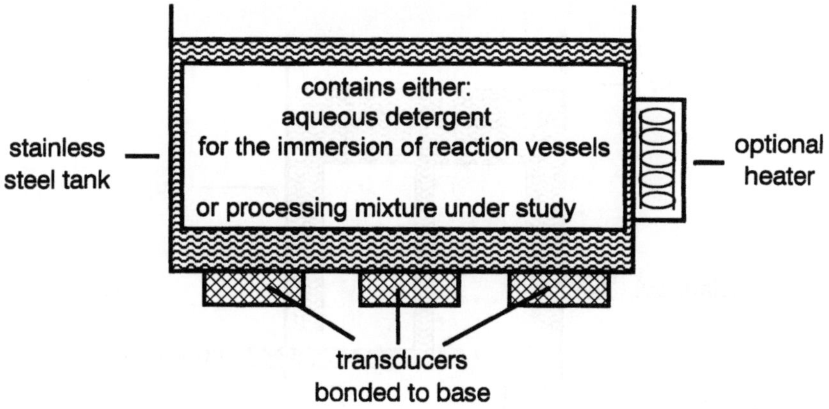

Figure 6.6 Ultrasonic bath.

stopped during part of the time any degassed bubbles can rise to the surface. Without this, the bubbles may be trapped in the standing wave field within the liquid and are responsible for ultrasound energy absorption.
- Electronic timers which can provide variable lengths of treatment and off periods.

Standard ultrasonic baths are of rather low power in order to avoid cavitation damage to the tank walls and the power density is low because the volumes of liquid in the tanks are generally large.

An alternative design to the bath is the cup-horn (Figure 6.7). Essentially this can be considered to be a very high-power ultrasonic bath because in this system the emitter surface, linked to a powerful transducer, is directly in contact with the solution. In the cup-horn configuration power transmission patterns are also influenced by experimental conditions and the level of the liquid is of paramount importance. Standing waves are only created if the emitted waves meet a liquid/air interface or reflecting surface at a distance of $\lambda/2$ from the source. In this situation very high activity zones are generated in the nodes of the standing wave.

6.2.3.2 Ultrasonic 'probe' systems. In order to amplify the acoustic energy generated from a transducer it is normal practice to attach to the transducer a 'horn' as described in Section 6.1.2.3. Some common designs for acoustic horns are shown in Figure 6.8 indicating the amplification factors possible (Perkins, 1990). The horns are generally a half wavelength (or multiple thereof) in length.

The amplitude gain generated by the system depends on the shape of the final horn, for a uniform cylinder there is no gain in amplitude but the horn acts as an extender for the transference of the acoustic energy. The gain, for

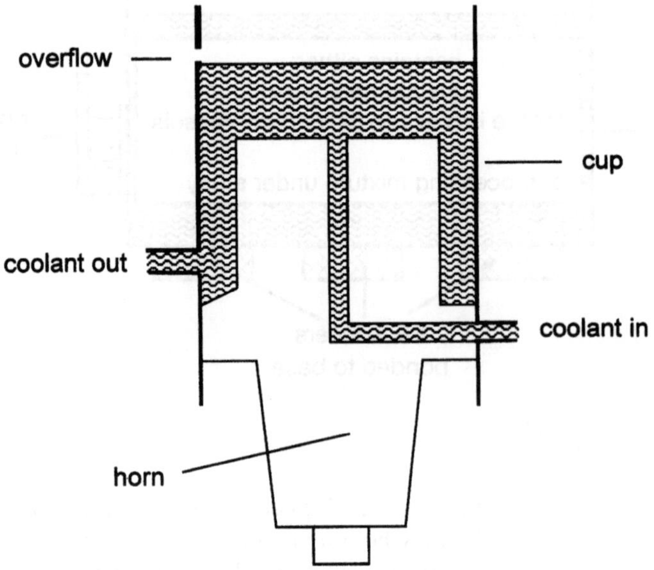

Figure 6.7 The cup horn.

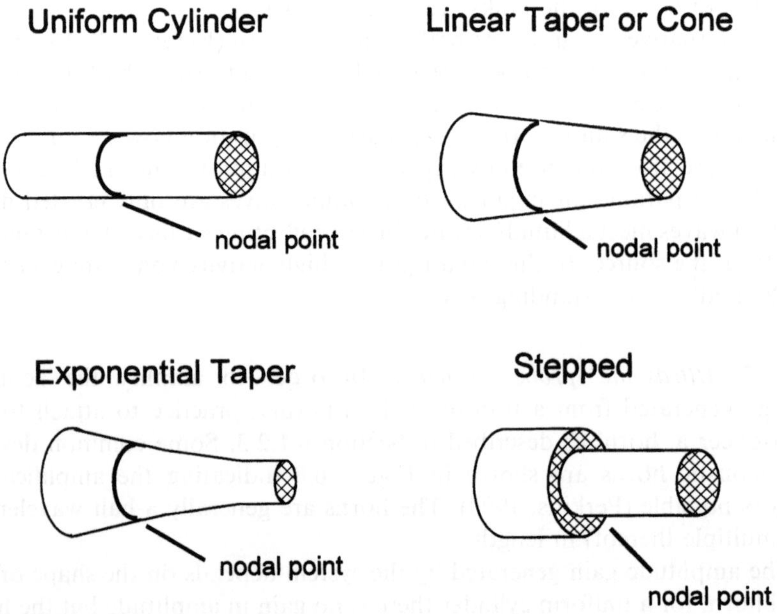

Figure 6.8 Shapes of acoustic horns.

a given form of the amplifier, is defined by a specific ratio of the diameter of the two faces of the horn (*D* is the driven face; *d* is the emitting face). For an exponential or linear tapered horn, this ratio is D/d but for a stepped horn it is $(D/d)^2$. Thus horn magnification is always higher for the stepped horn. However, to avoid damage to the horn material caused by internal stress, the D/d ratio must not be made too large.

The source of vibration and the load on which the source is acting are interdependent (Shoh, 1971) and so to obtain efficient energy transfer there should be an acoustic 'match' between these two elements. In practical terms, the highest energy transferable from the source depends on two factors:

- The characteristics of the materials used to construct the transducer assembly and the horn. For the best transmission of the vibration, highly resilient materials are required. Titanium or aluminium alloys are chosen because of their high mechanical fatigue resistance. Aluminium alloys are not ideal in situations in which they are in contact with cavitating liquids because of their susceptibility to erosion (and chemical reactivity) and so titanium alloys are chosen.
- The area of the emitting face. The most efficient acoustic coupling is achieved using small faced emitters, but at high amplitudes cavitation bubbles generated at the surface will limit power transmission to the load.

6.2.3.3 Equipment involving parallel vibrating plates. This design is probably the best option for on-line use, the product being treated homogeneously during its transfer through the reactor whose walls are vibrating. When the plates are close together the attenuation of the sound by the liquid is minimal and standing waves patterns are not present. The advantage of vibrating plates on each side of the liquid compared to a single vibrating surface is that the vibrational energy, before being absorbed by the liquid, is reverberated by the opposite plate. In this way the mechanical effect is maximized.

Different, but overlapping, patents claim the same general cell shape involving parallel vibrating walls for different purposes. The RUM (Reverberating Ultrasonic Mixer) invented by the Holl Research Group (1978), with its two irradiation frequencies of 20 and 16 kHz produces a 'beat' frequency of 4000 Hz. The two different frequencies create a larger number of different cavitating bubbles than a single frequency. Operating on this principle the Nearfield Acoustic Processor (NAP) was developed some years ago by the Lewis Corporation and is currently marketed by Advanced Sonic Processing Systems (263 Bacon Pond Road, Woodbury, CT 06798, USA). This processor is driven by magnetostrictive transducers, is very robust and can cope with large throughput.

6.2.3.4 Radial vibrating systems. An ideal way of introducing ultrasound into a liquid flowing in a pipe would be to use the pipe itself as a vibrating source. Such systems are capable of handling high flow rates and viscous materials. The cross-sectional geometry is an important factor in such devices. A cylindrical resonating pipe will provide a focus of ultrasonic energy in the centre of the tube as will a hexagonal pipe. Thus relatively low power at the perimeter inner surface will give high energy in the centre, this reduces erosion problems at the emitting surface. Cylindrical tube reactors with transducers directly bonded to the outer surface have been developed by a number of groups. The Battelle resonating pipe (Battelle, Columbus Laboratories, 505 King Avenue, Columbus, OH 43201, USA) is 6 inches in diameter and operates at 25 kHz and has been used for the degassing of oils.

With a transducer attached a metal cylinder can be considered to be a tubular resonator which will vibrate in the radial mode with the vibrational nodes and antinodes distributed at $\lambda/2$ intervals along the cylinder. There are two basic approaches for the generation of radial power ultrasound:

- One is to make the tube essentially a 'horn' attached to a transducer. Telsonic have a tube which has a single transducer at one end and the other end of the tube is made a null point as opposed to a maximum vibration position for a horn (Telsonic AG, Industriestrasse, CH-9552, Bronshofen, Switzerland) In this system the acoustic energy will be emitted as radial vibrations. When the tube is in the form of a solid bar and has a transducer at each end the radial emissions are stronger. This latter device is known as the Push–Pull (Martin Walter Ultraschalltechnik GmbH, Hardstrasse 13, Postfach 6, Ortsteil Conweiler, D-7541 Straubenhardt 5, Germany). With this system it is possible to use a combination of frequencies which allows the standing waves to be distributed over a larger volume (Walter, 1992).
- When the transducer is attached directly to the wall of the tube, i.e. at right angles to its axis, again the acoustic energy is emitted radially and an example is the Sodeva Ultrasonic Tube (Sodeva, Bat. MODUL , BP 299, Savoie Technolac, 73375 Le Bourget du Lac, France). This is constructed in the form of a hollow tube and liquids can be treated by flowing either inside or outside the vibrating structure.
- If the transducer, or transducers, are simply driving a ring of metal rather than a tube then the ring itself becomes the source of very high energy acoustic radial vibrations. This 'ring' can be used as the die to assist with the cold drawing of metal pipes or simply as the source for liquid sonication (FFR Ultrasonics, Unit 15, Queniborough Industrial Estate, Melton Road, Queniborough, Leicestershire LE7 8FP, UK).

6.3 Applications of power ultrasound in food processing

6.3.1 *Oxidation processes*

A good example of potentially useful ultrasonically enhanced oxidation is in the ageing of fermented products such as wines and spirits where limited oxidation is potentially useful for flavour development and early maturation of the product. The effect of irradiation with 1 MHz ultrasound has been shown to alter the alcohol/ester balance in such products to give the appearance of an aged product (Ishimori *et al.*, 1981). This seems to be particularly successful for whisky where a patented method has been described which reduces the normally prolonged ageing time to less than one year by applying ultrasound to the liquor in a standard barrel (Rosenfeld and Schmidt, 1984).

6.3.2 *Enzyme reactions*

It has been known for many years that ultrasound can be employed as a method of inhibiting enzyme activity. Some 60 years ago Chambers (1937) reported that pure pepsin was inactivated by sonication probably as a result of cavitation. Enzyme inactivation through sonication is also considered to be responsible for the inhibition of sucrose inversion (Crawford, 1955).

Peroxidase, which is found in most raw and unblanched fruit and vegetables, is particularly associated with the development of off-flavours and brown pigments. The effect of ultrasound (20 kHz, 371 W cm^2) on peroxidase:sigma-P8000 dissolved in 0.1 M potassium phosphate at 20°C buffer pH7 has been reported (Figure 6.9) (Wiltshire, 1992). The activity of the peroxidase was estimated by the Bergmeyer method involving measurements of the decomposition of hydrogen peroxide with guaiacol as hydrogen donor through the rate of colour development at 436 nm. The original activity of peroxidase was progressively reduced by 90% as ultrasound was applied over a 3-h period.

A convenient generalization concerning the effect of ultrasound on enzyme activity has been suggested in which oxidases are usually inactivated by sonication whilst catalases are only affected at low concentrations (Naimark and Mosher, 1953). Reductases and amylases however appear to be highly resistant to sonication.

6.3.3 *Stimulation of living cells*

The production of yoghurt is an increasingly important process in the food industry. Recent investigations have shown that the use of ultrasound as a processing aid can lead to a total reduction of up to 40% in production time (Cordemans, 1995). A very important finding in this work was that sonication reduced the normal dependence of the process on the origin of milk.

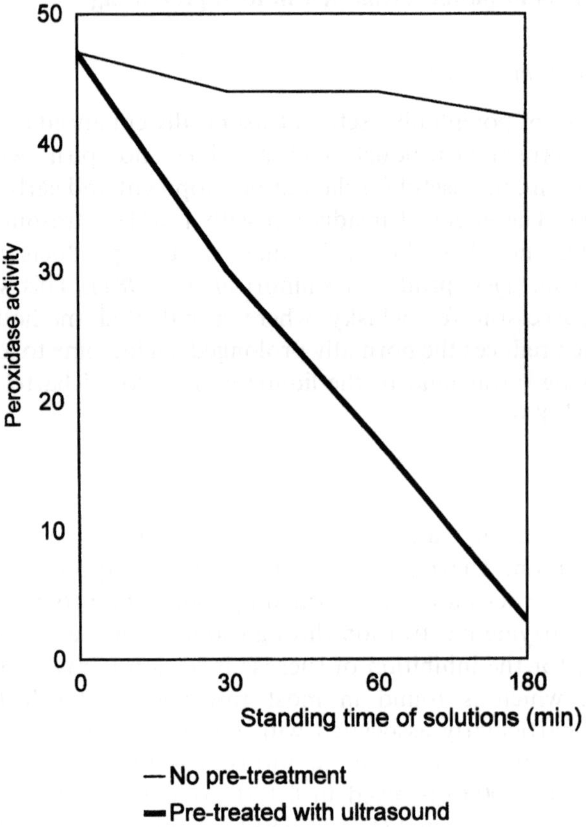

Figure 6.9 Ultrasonic control of peroxidase activity.

Further it appeared to improve both the consistency and the texture of the product. Ultrasonic irradiation has also been used in the production of lactose-reduced yoghurt (Toba, 1990). Sonication improves the action of the lactobacillus by nearly 50% and also results in a sweetening effect without any increase in calorific value.

Ultrasonically stimulated seed germination offers the possibility of increased productivity for large scale farm crops and in more general horticulture. Agricultural crop yields are dependent on the quality of the plant variety and on the percentage seed germination and growth. The effect of the suspension media on the ultrasonically stimulated germination of seeds from a temperate *Cymbidium* species has been studied. Pretreatment with 0.1 M KOH for 0.5 h or ultrasonic waves for 4 h resulted in the best germination but the ultrasonic treatment induced quicker germination and faster rhizome growth (Choi and Chung, 1991).

More recent Russian studies have applied sonication under dry conditions which may be carried out up to several months before actual sowing (Abramov, 1994). The subsequent handling and packaging of the seeds is perfectly conventional with no special precautions required. The treatment unit itself uses ultrasonic equipment operating at 20 kHz with a vibrational amplitude of between 1–40 μm and is able to handle up to 600 kg h^{-1} with the possibility of installation in a continuous packing line. Examples of the use of this process include ultrasonic treatment leading to a three-fold enhancement in sunflower seed germination in soil and a ten-day reduction in the ripening time of tomatoes.

6.3.4 Sterilization

One of the major long-established industrial applications of power ultrasound is for cleaning and it has proved to be an extremely efficient technology. Ultrasound is particularly useful in surface decontamination where the inrush of fluid which accompanies cavitational collapse near a surface is non-symmetric (Figure 6.10). The surface itself restricts the inrush of fluid from that side and so a jet of liquid is formed from the resultant major flow from the other side of the bubble. This powerful jet will dislodge dirt and bacteria from surfaces. The particular advantage of ultrasonic cleaning in this context is that it can reach crevices that are not easily reached by conventional cleaning methods. Indeed a general patent has been

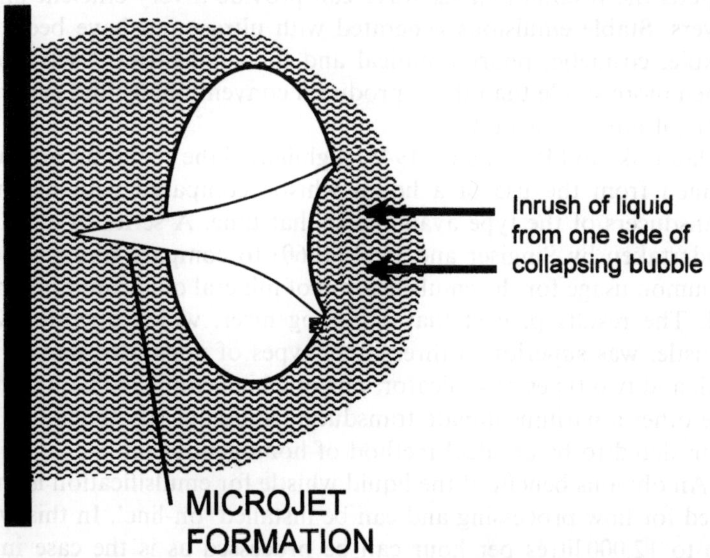

Figure 6.10 Cavitation collapse near a solid surface.

applied relating to the use of ultrasound as a method of pasteurization, sterilization and decontamination of instruments and surfaces used within the medical, surgical, dental and food processing industries (Boucher, 1980).

One recent example of surface cleaning in food technology is the use of a combination of a bactericide and ultrasound to decontaminate hatchery eggs (Slapp, 1995). In hatcheries the eggs are normally laid 'clean' but a number (some 10%) have surface contamination having found their way outside of the nest. Dirty eggs are removed from the hatchery because all egg shells are porous and so any contamination could penetrate to the inside and kill the developing embryo. The result is that these 'dead' eggs can explode after about 10 days. Ultrasonically assisted cleaning produces decontamination of 'clean' eggs which is more efficient than currently used spraying or gassing technique and, more significantly, can provide decontamination of the 'dirty' eggs which allows them to be used in the hatchery.

Power ultrasound is also capable of enhancing the effects of chemical biocides, this is in part due to the break-up and dispersion of bacterial clumps which then renders the bacteria more susceptible to chemical attack. Such effects can be of great use in the sterilization and disinfection of food materials particularly where ultrasound is used in conjunction with a conventional sterilization technology.

6.3.5 Ultrasonic emulsification

One of the earliest uses of power ultrasound in processing was in emulsification. If a bubble collapses near the phase boundary of two immiscible liquids the resultant shock wave can provide a very efficient mixing of the layers. Stable emulsions generated with ultrasound have been used in the textile, cosmetic, pharmaceutical and food industries. Such emulsions are often more stable than those produced conventionally and generally require little, if any, surfactant.

Janovski and Pohlmann (1948) highlighted the economic advantages to be gained from the use of a liquid whistle compared with magnetostrictive transducers of the type available at that time. A series of experiments was undertaken by Singiser and Beal (1960) to compare four methods then in common usage for the emulsification of mineral oil, peanut oil and safflower oil. The results proved that a homogenizer, which operated via a liquid whistle, was superior to three other types of apparatus namely a colloidal mill and two types of sonicator, one of which employed a quartz crystal and the other a barium titanate transducer. Subsequently the liquid whistle was considered to be an ideal method of homogenization.

An obvious benefit of the liquid whistle for emulsification is that it can be used for flow processing and can be installed 'on-line'. In this way volumes up to 12,000 litres per hour can be processed as is the case in the manufacture of such items as fruit juices, tomato ketchup and mayonnaise.

A whistle type device has been used for the production of cheese (Gaffney, 1996). The action of the ultrasound was to coagulate various combinations of concentrated proteins and oils to produce soft cheeses. It is claimed that the process also results in reduced waste and lower energy costs.

6.3.6 Extraction

The mechanical effects of ultrasound provide two major benefits which increase the classical extraction procedures

- a greater forced penetration of solvent into cellular materials
- improved mass transfer to and from interfaces
- the disruption of biological cell walls on the surface and within the plant material to facilitate the release of contents.

The extraction of sugar from sugar beets has been found to be improved with power ultrasound as a result of these effects (Chendke and Fogler, 1975). Ultrasonically assisted extraction can also be applied to the production of medicinal compounds such as helicid, berberine hydrochloride and bergenin from Chinese plants (Zhao *et al.*, 1991). Thus helicid, which is normally extracted by refluxing in ethanol, can be obtained in a 50% higher yield in half the extraction time at room temperature using ultrasound.

Protein extraction from defatted soya-beans has been studied by Wang (1975). A continuous process was developed involving sonication of the slurry by a 550 W probe operating at a 20 kHz frequency which resulted in an efficiency of extraction which was better than that available using any previously available technology. This was scaled up to pilot-plant for the extraction of soya bean protein (Moulton and Wang, 1982). A two-fold improvement in the extraction of protein from algae has also been reported (Cordemans, 1995).

The extraction of tea solids from leaves is commercially important because it provides a starting point for the production of instant tea. Instant tea is a powder derived from pure tea infusion from which water has been removed by spray drying. The use of ultrasound improves extraction at 60°C by nearly 20% (Mason and Zhao, 1994). The effectiveness of ultrasonic extraction is greater than normal thermal extraction and requires less time. The majority of material is extracted in the first 10 min of sonication.

6.3.7 Meat products

Closely linked with extraction is the methodology employed for the production processed meats where power ultrasound aids in the extraction of protein (Vimini *et al.*, 1983). Generally this involves tumbling the meat particles with an aqueous liquor containing salt. Ultrasound disrupts the meat myofibril and this releases a sticky exudate which binds the meat together. The binding strength, water holding capacity, product colour and yields were examined after treatment either with salt tumbling, sonication or both. Samples which received both salt treatment and sonication were superior in all qualities. Similar results were obtained from a study of the effect of sonication on cured rolled ham (Reynolds *et al.*, 1978). Ultrasonic treatment enhanced the extraction of myofibrillar proteins leading to an increase in the strength of the reformed meat.

A traditional method of tenderization of meat is by mechanical pounding which makes poorer quality meat more palatable. Power ultrasound has also been found to be useful for this process. In a pilot study involving sirloin steak the connective tissues were clearly seen to be reduced when the meat had been subjected to treatment at room temperature at 40 kHz using $2\,W\,cm^{-2}$ for 2 h (Roberts, 1991).

6.3.8 Crystallization

Power ultrasound has proved to be extremely useful in crystallization processes. It serves a number of roles in the initiation of seeding and subsequent crystal formation and growth. It also has a property which is particularly beneficial in processing application namely that the cleaning action of the cavitation effectively stops the encrustation of crystals on cooling elements and thereby ensures continuous efficient heat transfer. It is of considerable practical importance to be able to control the point at which crystallization occurs in any large-scale production process. Often crystallization can occur in an uncontrolled manner simply due to a slight drop in temperature or pressure. On the other hand to control the initiation of crystallization is often difficult since problems may occur due to incorrect external factors such as temperature and pressure settings. Sonication is thought to enhance both the nucleation rate and rate of crystal growth in a saturated or supercooled medium by producing fresh and/or more nucleation sites in the medium. This may be due to cavitation bubbles acting as nuclei for crystal growth or by the disruption of seeds/nuclei already present within the medium thus increasing the number of nuclei present in the solution/melt.

The scale-up of ultrasonically assisted crystallization was achieved successfully some years ago in the production of a crystalline drug (Midler, 1970). The ultrasound was introduced through an array of horns at the base of the crystallization tower and served two purposes. Firstly the saturated liquid was ultrasonically seeded as it entered the base of the tower and these small seed crystals rose with the liquid flow and continued to grow until they were large enough to start to sink under the influence of gravity. These larger crystals were fragmented by cavitation as they sank downwards moving closer to the ultrasonic sources. The fragments produced were small enough to rise once again in the flow and provide large numbers of particles which themselves could act as seeds for further nucleation.

One very important area related to crystallization in the food industry is the formation of ice crystals during the freezing of water. The quality of 'fresh' thawed foods preserved through freezing can be somewhat disappointing in terms of texture. This is particularly true of soft fruits such as strawberries. The problem arises because the small ice crystals which are formed initially inside of the cellular material of the food continue to grow. As these crystals increase in size they break some of the cell walls leading to a partial destruction of the structure of the material. There is a considerable 'dwell time' between the initiation of crystallization (usually at about −3°C and complete freezing at which point the temperature of the whole item can fall (Figure 6.11). Under the influence of ultrasound a much more rapid and even seeding occurs and this leads to a much shorter dwell time (Acton and Morris, 1992). In addition, since there are a greater number of seeds the final size of the ice crystals is smaller and cell damage is reduced. In a somewhat related field the effect of ultrasound on the production of ice lollipops has been investigated (Wiltshire, 1992). Sonication generates ice crystals of a significantly reduced size with a more even distribution through the solid. This gives rise to a lollipop which is much harder than the conventional product which makes it less customer friendly in that it is far less easy to bite. A true advantage however seems to be that it adheres much more strongly to its supporting wooden stick.

6.3.9 Degassing

The cavitational effects which are the basis of sonochemistry are also the reason for the extremely effective uses of ultrasound for the degassing of liquids. Any dissolved gases or gas bubbles in the medium act as nuclei for the formation of cavitation bubbles. Such bubbles are not easily collapsed in

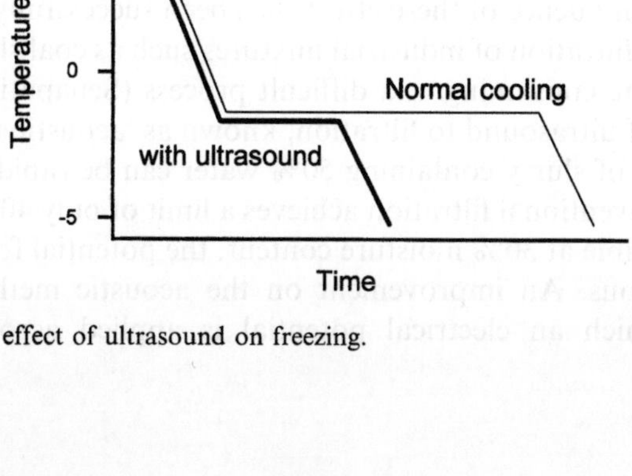

Figure 6.11 The effect of ultrasound on freezing.

the compression cycle of the wave due to the fact that they contain gas and they will continue to grow on further rarefaction cycles, filling with more gas and eventually floating to the surface. Since the rarefaction cycles are taking place extremely rapidly (around 40 000 times per second using an ultrasonic bath) the bubbles grow so quickly that degassing appears to occur almost instantaneously.

The removal of unwanted air or gas is a important in many food processes and it can be extremely difficult especially in very viscous liquids such as chocolate. Ultrasonically assisted degassing is particularly rapid in aqueous systems and can be used to remove any dissolved gas down to a very low level and is of great benefit in situations requiring the rapid and controlled removal of a gas or gases from a system. However ultrasound has also been employed to degas liquids as viscous as molten glass (Spinosa and Ensminger, 1987) and as dense as molten metals during solidification (Eskin, 1996).

6.3.10 *Acoustically aided filtration*

The requirement to remove suspensions of solids from liquids is common to many industries including chemical, engineering as well as food. This separation can be either for the production of solids-free liquid or to isolate the solid from its mother liquors. Ultrasonic filtration of particulate matter from a liquid is now arousing some interest since the rate of flow through a filter can be increased substantially on application of ultrasound.

Conventionally, membranes of various sorts have been employed ranging from the simple filter pad through semipermeable osmotic type membranes to those which are used on a size-exclusion principle for the purification of polymeric materials. Unfortunately the conventional methodologies often lead to 'clogged' filters and, as a consequence, there will always be the need to replace filters on a regular basis. There are two specific effects of ultrasonic irradiation which can be harnessed to improve the filtration technique: (i) sonication will cause agglomeration of fine particles (i.e. more rapid filtration), and at the same time, (ii) will supply sufficient vibrational energy to the system to keep the particles partly suspended and therefore leave more free 'channels' for solvent elution.

The combined influence of these effects has been successfully employed to enhance vacuum filtration of industrial mixtures such as coal slurry, which is a particularly time consuming and difficult process (Senapati, 1991). With the application of ultrasound to filtration, known as 'acoustic filtration', the moisture content of slurry containing 50% water can be rapidly reduced to 25% whereas conventional filtration achieves a limit of only 40%. Since coal slurry is combustible at 30% moisture content, the potential for this process is clearly enormous. An improvement on the acoustic method has been developed in which an electrical potential is applied across the slurry mixture while acoustic filtration is performed (Muralidhara *et al.*, 1985). The filter itself is made the cathode while the anode, on the top of the slurry, functions as a source of attraction for the predominantly negatively charged particulate material. The additional mobility introduced by the electric charge 'electro-acoustic filtration' increases the efficiency of drying of 50% coal slurry by a further 10%. When applied to fruit extracts and drinks this technique has been used to increase the apple juice extracted from pulp. Where conventional belt vacuum filtration achieves a reduction in moisture content from an initial value of 85% to 50%, electro-acoustic technology achieved the significantly lower figure of 38%.

6.3.11 Acoustic drying

Acoustic drying is of potentially great commercial importance. Sonically enhanced drying can be carried out at lower temperatures than conventional methodology which reduces the probability of oxidation or degradation in the material. Additionally unlike high-velocity gas drying the material is not blown about or damaged. In 1964 Soloff (1964) reported a study of acoustic drying of a number of materials using a rotary drier with a sonic source of 169 dB at 10.9 kHz. The comparative results in terms of increased throughput due to the application of sonics is shown are Table 6.2.

Sugar crystals can be dried to a moisture content which is 50–75% less than can be obtained in the same time as conventional processing using ultrasound (Boucher, 1971). Similar technology reduced the drying time of fermentation sediment while keeping the temperature below 40°C (Boucher, 1959). The effect of sonication during the drying process of shelled corn and whole and crushed wheat increased the rate for all samples with the effect being more pronounced at lower temperatures (Huxsoll and Hall, 1970). A 130% drying rate increase was observed at 21°C when sonic irradiation was applied. At 63°C the rate was increased by 66% whereas at 79.5°C only a 6% rate increase was noted.

The drying of coal particles in air can be substantially accelerated using ultrasound (Fairbanks, 1974). Warm air at 65°C was passed over the drying surface of the coal and ultrasound was applied above the surface of the coal

Table 6.2 Improved rotary drying induced by sound (10.9 kHz, 169 dB)

Material for drying	Initial moisture (%)	Target moisture (%)	Increased throughput (%)
Orange crystals	3.5	1.8	375
Grated cheese	16.8	5.9	40
Gelatin beads	12.9	3.7	83.3
Rice grains	27.6	14.5	122.2

particles. Not only was the rate of drying increased but also the moisture content of the final dried sample was reduced. This is thought to be initially due to a reduction of pressure above the sample to encourage water loss.

Ensminger (1988) examined the effects of sonic irradiation during the drying of food materials using electro-osmosis. This involves the removal of water by encouraging water molecules to pass through porous membranes by the application of a direct current electrical potential. However, during such a process electrolysis of the water may occur which involves the formation of hydrogen at the cathode and oxygen at the anode. The presence of these gases reduces the electrode potential and restricts movement of water molecules. Application of ultrasound aids the removal of these gases from the electrodes and encourages higher potentials and faster drying rates.

Ultrasound can be employed to increase the rate of heat transfer between a solid heated surface and a liquid (Fairbanks, 1974). The ultrasound can either be introduced to the liquid itself or by vibrating the solid heated surface. It is thought that cavitation aids in the disruption of the boundary layer and as the temperature of the liquid increases the cavitation threshold is reduced. Heat transfer is increased by approximately 30–60%.

6.3.12 *Effect on rice grains*

When particles of material in a liquid suspension are subjected to sonication a number of physical and mechanical effects can result. Large particles are subject to surface erosion (via cavitational collapse in the surrounding liquid) or particle size reduction (due to fission through interparticle collision or the collapse of cavitation bubbles formed on the surface).

If the particles subjected to sonication are rice grains in water then some destruction of the surface 'shell' or even grain fragmentation might be anticipated. Either of these effects would result in a faster release of starch during subsequent cooking leading to a shorter period to form a gel. The effect of low-power ultrasound on white and brown rice in terms of both surface effects (by optical microscopy) and reduced cooking time (through gel formation at 80°C have been investigated (Paniwnyk, 1993). In the case of white rice the surface abrasion caused through sonication of rice grains (Figure 6.12a) can seen after only 5 min (Figure 6.12b) where the inner starch is revealed as lighter coloured parts of the grain. This is followed by fragmentation after 10-min treatment (Figure 6.12c). These results are confirmed by the reduced gel times observed for both white and brown rice (Table 6.3). In the case of brown rice the effects of sonication are ameliorated to some extent by the tougher outer shell on the grains.

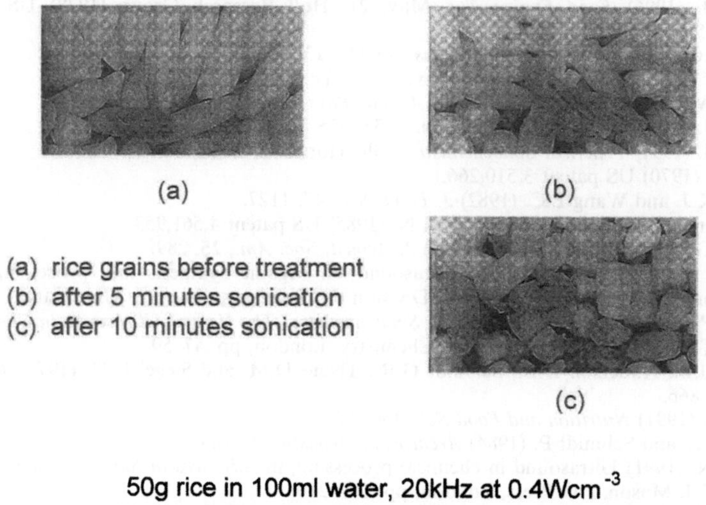

Figure 6.12 Effect of ultrasound on rice grains in water.

Table 6.3 Effect of ultrasonic treatment on the gel time of rice (50 g rice in 100 cm^3 water sonicated for 10 min at 25°C)

Type of rice	Normal gel time at 80°C (min)	Gel time after sonication	Acoustic power
White	15	Very fast	0.35 $W cm^{-3}$
Brown	25	5	0.7 $W cm^{-3}$

6.4 References

Abramov O.V. (1994) Institute of General and Inorganic Chemistry, 31 Leninski Prospect, 117905 Moscow, Russia, personal communication.

Acton E. and Morris G.J. (1992) W.O. 99/20420.

Boucher R.M.G. (1959) *Ultrasonic News*, **3**, 14.

Boucher R.M.G. (1980) US patent 4,211,744.

Boucher R.M.G. (1971) US patent 3,175,255.

Chambers L.A. (1937) *J. Biol.Chem.*, **117**, 639.

Chendke P.K. and Fogler H.S. (1975) *Ultrasonics,* 31.

Choi S.O. and Chung J.D. (1991) *J. Korean Soc. Horticultural Sci.*, **32**, 525.

Cordemans E. (1995) Information provided by Undatim Ultrasonics, 7 rue du Bosquet, Louvaine le Neuve, Belgium.

Crawford A.E. (1955) *Ultrasonic Engineering with Particular Reference to High Power Applications*, Butterworths, Oxford.

Ensminger D. (1988) *Drying Tech.*, **6**, 473.

Eskin G.I. (1996) Degassing, filtration and grain refinement processes of light alloys in a field of acoustic cavitation. In *Advances in Sonochemistry*, Vol. 4, Editor T.J. Mason, JAI Press, London, pp. 101–60.

Fairbanks H.V. (1974) *Ultrasonics*, 260.

Gaffney B. (1996) *Food Engineering*, May, 21. Holl Research Group (1978) US patent 4,071,225.
Huxsoll C.C. and Hall C.W. (1970) *Trans. ASAE.*, **13**, 21.
Ishimori Y., Karube I. and Suzuki S (1981) *J. Molec. Catal.*, **12**, 253.
Janovski W. and Pohlmann R.Z. (1948) *Angew. Phys.*, **1**, 22.
Mason T.J. and Zhao Y. (1994) *Ultrasonics*, **32**, 375.
Mason T.J. (1991) *Practical Sonochemistry*, Ellis Horwood, Chichester.
Midler M. (1970) US patent 3,510,266.
Moulton K.J. and Wang L.C. (1982) *J. Food. Sci.*, **47**, 1127.
Muralidhara H., Parekh B. and Senapati N. (1985) US patent 4,561,953.
Naimark G.M. and Mosher W.A. (1953) *J. Acoust. Soc. Am.*, **25**, 289.
Paniwnyk L. (1993) The effect of ultrasound on organic synthesis and processing from laboratory to large scale. PhD thesis, Division of Chemistry, Coventry University.
Perkins J.P. (1990) Power ultrasound. In *Sonochemistry: The Uses of Ultrasound in Chemistry*, Editor T.J. Mason, Royal Society of Chemistry, London, pp. 47–59.
Reynolds J.B., Anderson D.B., Schmidt G.R., Theno D.M. and Siegel D.G. (1978) *J. Food. Sci.*, **43**, 866.
Roberts T. (1991) *Nutrition and Food Sci.*, **130**, 17.
Rosenfeld E. and Schmidt P. (1984) *Archives of Acoustics*, **9**, 105.
Senapati N. (1991) Ultrasound in chemical processing, in *Advances in Sonochemistry*, Vol. 2, Editor T.J. Mason, JAI Press, London, pp. 187–210.
Shoh A. (1971) *Chemical Engineering; Progress Symposium Series.* **109**, 126–134.
Singiser R.E. and Beal H.M. (1960) *J. Am. Pharm. Assoc., (Scient. Ed.)*, **49**, 482.
Slapp P. (1995) BP 95 00587 2.
Soloff R.S. (1964) *J. Acoustic Soc. Am.*, **36**, 961.
Spinosa E.D. and Ensminger D.E. (1987) *Glass Industry*, 10.
Toba T. (1990) *J. Sci. Food Agri.*, **52**, 403–7.
Vimini R.J. Kemp J.D. and Fox J.D. (1983) *J. Food. Sci.*, **48**, 1572.
Wang L.C. (1975) *J. Food. Sci.*, **40**, 549.
Wang L.C. (1981) *J. Agric. Food. Chem.*, **29**, 177.
Wiltshire M. (1992) Presented at Sonochemistry Symposium, Royal Society of Chemistry Annual Congress, Manchester.
Zhao Y., Bao C. and Mason T.J. (1991) *Ultrasonics International '91 Conference Proceedings*, Butterworths, p. 87.

7 Some applications of air-borne power ultrasound to food processing

J.A. GALLEGO-JUÁREZ

7.1 Introduction

It is well known that power ultrasound is used to produce permanent effects in the medium where it is applied. These effects are mainly due to a series of phenomena which accompany the high ultrasonic intensities, the most important of which are radiation pressure, streaming, cavitation, steady forces action, heat, interface instabilities, friction, diffusion, agitation and mechanical rupture.

These phenomena are used to enhance a wide range of processes which basically depend on the irradiated medium. In fact, a typical characteristic of high-intensity ultrasonic waves is the ability to produce different effects in different media in such a way so that, sometimes, these effects seem to be opposite. This is, for example, the case when using power ultrasound in liquid suspensions to separate aggregated particles which contrasts with the application of ultrasound to gas suspensions to produce particle agglomeration. The explanation for this apparently contradictory behaviour can be found in the different medium where the ultrasonic energy is acting and, consequently, in the different mechanism is active. In liquids, the majority of these effects are associated with cavitation; a phenomenon which is not possible to induce in gases or solid media.

Foodstuffs are generally inhomogeneous complex materials which often contain gas, liquid and solid phases. Therefore, it is frequently difficult to understand the different mechanisms provoked by the action of ultrasonic energy. In fact, many aspects of ultrasonic food processing are not well known and this situation limits the possibilities of application in a field which appears to be very promising. A strong research effort on the basic mechanisms involved in any specific application is needed for an efficient development in this area.

The future of power ultrasonics in food processing is also linked to the development of adequate industrial equipment. In fact, the commercial introduction of a new technology depends, in addition to its capability for solving real problems, on the design, operational simplicity, efficiency and low energy consumption of the equipment used for the development of the process.

Air-borne power ultrasound represents an example of a field of application to food processing which has not been sufficiently exploited. This is probably because of the problems already mentioned, i.e. the lack of a thorough understanding of the basic mechanisms involved and the difficulties in power ultrasonic technology. In fact, large-scale industrial applications require efficient transmission of energy and capacity in the acoustic generators to sonify large volumes.

This chapter deals with a new air-borne power ultrasound technology and its application to food processing in two different areas: food dehydration and defoaming

7.2 The new air-borne power ultrasound technology

The most important problems to solve in air-borne ultrasonic technology are related to the efficient generation of ultrasonic energy in gases. In fact, the gaseous media, due to their low density, present a very low specific acoustic impedance and a high acoustic absorption. Therefore, for an efficient generation and transmission of energy it is necessary to achieve ultrasonic generators covering the following requirements: good impedance matching with the air, large amplitude of vibration, high directional or focused radiation, and high power capacity.

There are few commercial power generators for air-borne ultrasound and almost none of them cover all the above-mentioned requirements. For many years the most common transducers for use in gases have been aerodynamic systems of various kinds, such as whistles and sirens, in which the acoustic energy is provided by a gas jet (Hartmann, 1939; Allen and Rudnick, 1947). Although it is possible to yield large acoustic powers with these generators, the efficiencies obtained are low, the sounds emitted are complex and they present poor directivities and difficulties in reaching ultrasonic frequencies (Greguss, 1964).

Over the past few years, a new type of sonic and ultrasonic power generator for air-borne ultrasound has been developed (Gallego-Juárez *et al*, 1989). It comprises a stepped-plate transducer and an electronic unit for driving the transducer. The structure of the transducer is shown in Figure 7.1. Essentially, it consists of an extensive flexural vibrating plate of stepped profile driven at its centre by a piezoelectrically activated vibrator. The extensive surface of the plate produces high radiation resistance and power capacity and offers the vibrating system good impedance matching with the medium. The special profile of the radiating plate permits the control of the vibration amplitude and the radiation pattern in such a way that very

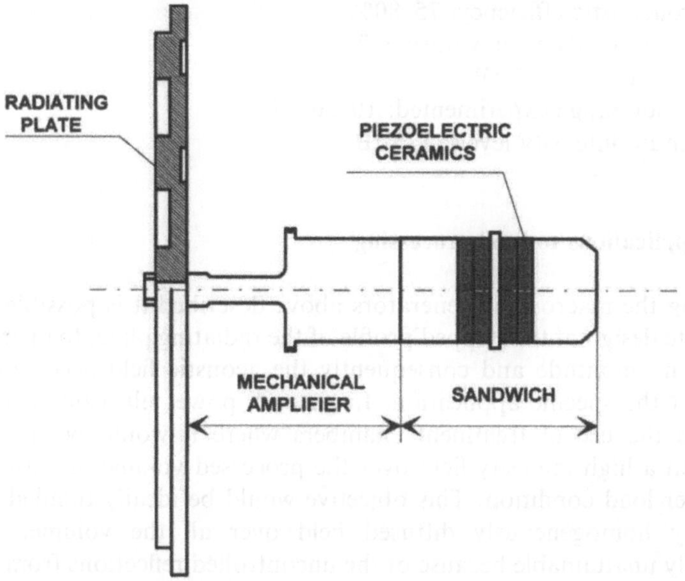

Figure 7.1 Structure of the transducer.

directional or focused radiation can be obtained. In fact, a flat plate radiator vibrating in its flexural modes presents, in general, a poor directivity due to phase cancellation. Nevertheless, if the surface elements, vibrating in counterphase on the two sides of the nodal lines, are alternately shifted along the acoustic axis direction by a quantity equal to half a wavelength of the sound in the propagation medium, the radiation produced will be in phase (piston-like radiation) across the whole beam. Following a similar procedure it is possible, with adequate displacements of the different plate zones, to obtain any acoustic field configuration. Focused radiators have also been designed and constructed.

Figure 7.2 shows the directivity pattern of one of these directional transducers with a radiating plate of 32 cm diameter in comparison with the directivity pattern of a flat-plate transducer. From this it can be seen a beam width at 3 dB of less than 2° is obtained. Figure 7.3 shows the acoustic field along the axis of a focused stepped-plate radiator.

A prototype of this transducer is shown in Figure 7.4. Different versions of these transducers and the corresponding electronic driving units have been designed, constructed and tested for the frequency range of 10–50 kHz.

The electronic units are special power generators provided with a system to drive the transducer at resonance during operation.

The general performance obtained with these new macrosonic generators can be summarized as follows:

- electroacoustic efficiency: 75–80%
- directivity (3 dB beamwidth): < 2
- power capacity: 1–2 kW
- frequency range experimented: 10–50 kHz
- maximum intensity levels: 175 dB

7.3 Applications to food processing

By using the macrosonic generators above described it is possible, with an adequate design of the stepped profile of the radiating plate, to distribute the vibration amplitude and consequently the acoustic field according to the needs of the specific application. In general, power ultrasound processing requires the use of treatment chambers where it would be necessary to maintain a high-intensity field over the processed volume or surface under whatever load condition. This objective would be ideally fulfilled by high-intensity homogeneously diffused field over all the volume, which is generally unattainable because of the uncontrolled reflections from different boundaries. Additionally, this type of operation would require very high power input per unit volume. In practice, standing wave patterns of different nature will be established in the processing volume. For a surface processing a focusing radiation should be useful.

For both volume and surface treatment, the present power generators appear to be very suitable tools. In fact, by using directional or focused stepped-plate radiators sound pressure levels higher than 16 dB have been achieved under various conditions.

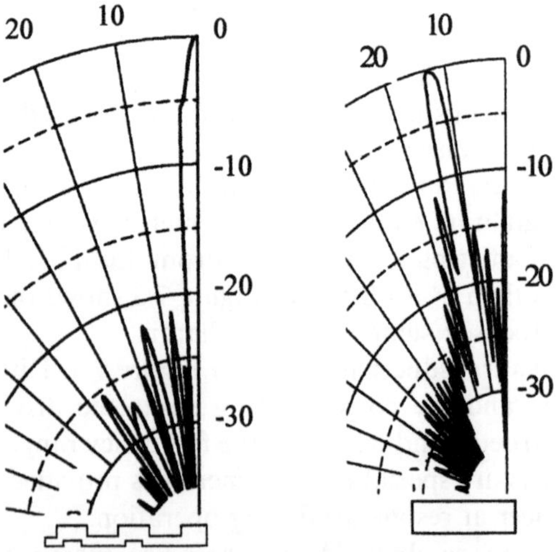

Figure 7.2 Comparison between directivities of a stepped-plate transducer and a flat-plate transducer (five nodal circles plate, $r = 16\,cm$, $f = 19382\,Hz$).

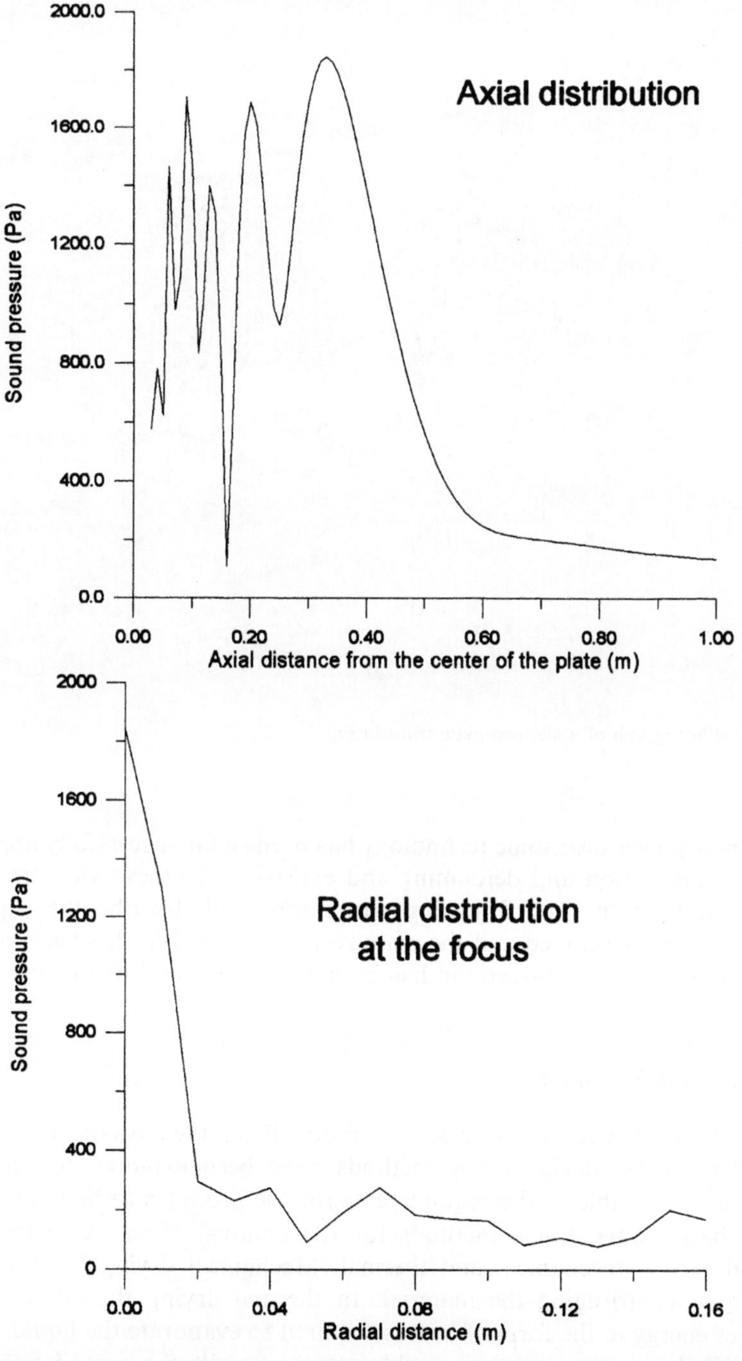

Figure 7.3 Acoustic field of a focused stepped-plate transducer (seven nodal circles stepped-plate $r = 24\,\text{cm}$, $f = 17485\,\text{Hz}$).

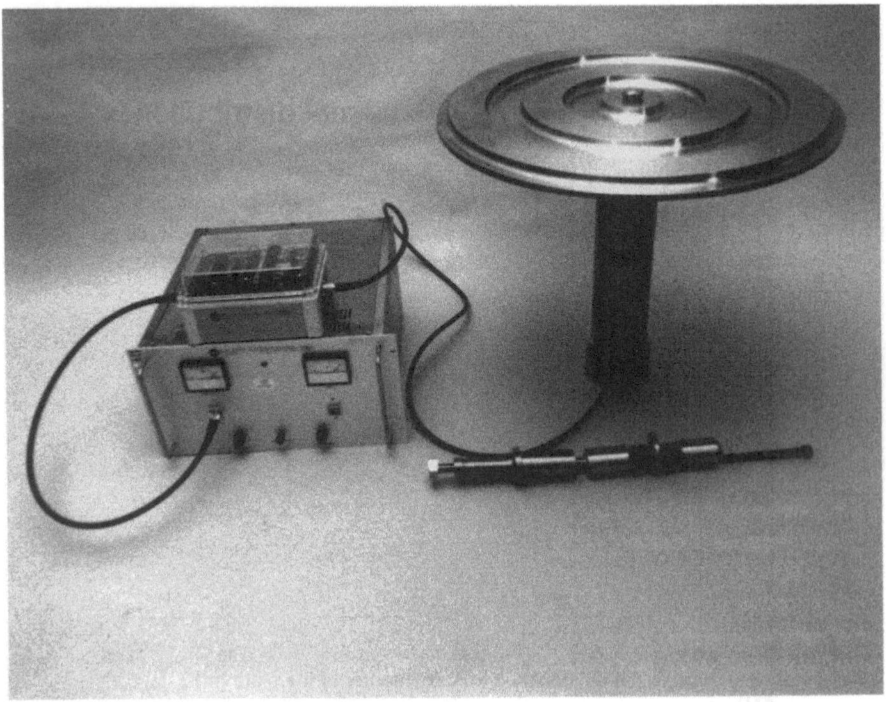

Figure 7.4 Photograph of a stepped-plate transducer.

This new power ultrasonic technology has been so far successfully applied to food dehydration and defoaming and extensions to these are currently under investigation. The following paragraphs will describe the experimental systems developed and the main results obtained in these and other areas where air-borne ultrasound has been applied to food processing.

7.3.1 Food dehydration

Dehydration is a method for preserving foods. From the days of sun-drying to modern freeze drying many methods have been explored to achieve demoisturize suitable to the requirements for the products to be treated.

Two basic conventional methods for the removal of moisture from a material exist – mechanical and thermal. Mechanical drying is based on pressing or centrifuging the material. In thermal drying the addition of whatever energy in the form of heat is required to evaporate the liquid. The first method removes moisture weakly attached while the second one can provide a more complete drying of the product.

The drying method which will be suitable for any specific application is related to the way in which the liquid to be removed is attached to the solid material to be dried. In general, three types of attachment can be considered: chemical, mechanical and physicochemical. The last two types can be treated by conventional drying techniques. The mechanical attachment is that of porous materials containing wetting and capillary moisture. The physicochemical attachment is associated with adsorption and osmotic moisture. In capillary-porous materials both of these kinds of attachment are present. For food dehydration, the present conventional systems employs two main procedures: hot-air drying and freeze-drying. Hot-air drying is a widely used method, but it can produce deteriorative changes in the food. Instead, in freeze-drying, where food pieces are first frozen and then ice sublimates, the product deterioration is negligible but the process is expensive.

High-intensity sonic and ultrasonic energy has been used to increase the drying rate of materials. The acoustically assisted hot-air drying process permits the use of lower temperatures and may be useful for drying heat-sensitive materials (Fairbank, 1975; Seya, 1970).

When the material is completely covered with moisture and the evaporation takes place only at the surface it dries at a constant rate which depends on the diffusion through the surface of water vapour into the air, the moisture content, the relative velocity of the drying air and the surface temperature of the material (Soloff, 1994). This period of constant rate of drying is followed by a period where the rate of drying falls, when the material is no longer completely covered with moisture. High-intensity air-borne ultrasound introduces pressure variations at gas–liquid interfaces, and therefore increases the evaporation rate of moisture. During the negative phase of the pressure cycle the moisture is removed and it does not re-enter during the positive pressure phase. This mechanism is more efficient in the constant-rate period of drying when the surface of the material is completely covered with moisture. In a forced-air drying system, the effect of the air velocity is to influence the heat and mass transfer. The acoustic energy produces an oscillating-velocity effect which can increase the drying rate at stable air velocity. In addition, high-intensity air-borne ultrasound causes microstreaming at the interfaces. This reduces the diffusion boundary layer, increases mass transfer and accelerates diffusion (Arkhangel'skii and Statnikov, 1973). Therefore, the application of air-borne acoustic energy can positively contribute in the drying process. Nevertheless, the use of this technique has been very limited, probably because of the insufficient improvements obtained in the drying rates and/or the practical difficulties to meet large-scale needs of industry.

When high-intensity ultrasonic energy is applied directly to the material to be dried, it travels through the solid medium causing a rapid series of alternative contractions and expansions, in a similar way to a sponge when it is squeezed and released repeatedly (sponge effect). The alternating stress makes dewatering easy by creating microscopic channels for water movement (Muralidhara *et al.*, 1985).

In the following sections the application of stepped-plate macrosonic generators for vegetable dehydration will be described. Two experimental studies are presented: forced-air drying assisted by air-borne ultrasound and a new drying procedure in which the ultrasonic vibration is directly applied to the material together with a static pressure (Gallego-Juárez *et al.*, 1995a, b).

7.3.1.1 Forced-air dehydration assisted by air-borne ultrasound. The experimental set-up designed and constructed for air-borne ultrasonic dehydration is shown in Figure 7.5. It mainly consists of a hot-air generator, a stepped-plate power ultrasonic transducer with the corresponding electronic generator and a flat plate parallel to the ultrasonic radiator, acting as a reflector for the formation of a standing wave and also as sample holder. In addition, complementary sets of equipment for measuring temperature, air flow velocity and weight were used. The acoustic field was previously measured and then controlled through the values of the applied current and voltage.

The experimental tests mainly consisted of measuring the water content of food samples after different times of application of high-intensity ultrasonic fields in combination with forced-air at various temperatures and flow velocities. In all experiments, the frequency was kept constant at about 20 kHz, while different sound pressure levels were applied. The water content of the samples was measured by weighing them. The food samples to be dried were carrot slices. The slices were either of square section (12×12 mm) or of circular section (14 mm in diameter) with 2, 4 and 8 mm in thickness.

A summary of the results is presented in Figures 7.6 and 7.7. Figure 7.6 shows the results obtained with forced-air at 60, 90 and 115°C with and without ultrasound (155 dB). In all cases the size of the samples was 12 × 12 mm square section, 2 mm thick. As can be seen, the effect of the ultrasonic radiation is significant at low air temperature and diminishes when the temperature increases. At the highest temperature (115°C) the ultrasonic effect was negligible and there were no appreciable differences in the curves.

In order to see the influence of the acoustic pressure level some tests were done at higher intensity (163 dB). Figure 7.7 summarizes the results of these tests. Comparison between Figures 7.6 and 7.7 shows that the effect of increasing the acoustic pressure is scarcely appreciable and occurs mostly when the moisture is lost rapidly (i.e. the first 10 min), and diminishes as the drying curves start to level off.

Other tests were done by increasing the air flow velocity up to 3 m/s. The results obtained showed that at this forced-air speed the differences with and without ultrasound are less significant (Figure 7.8).

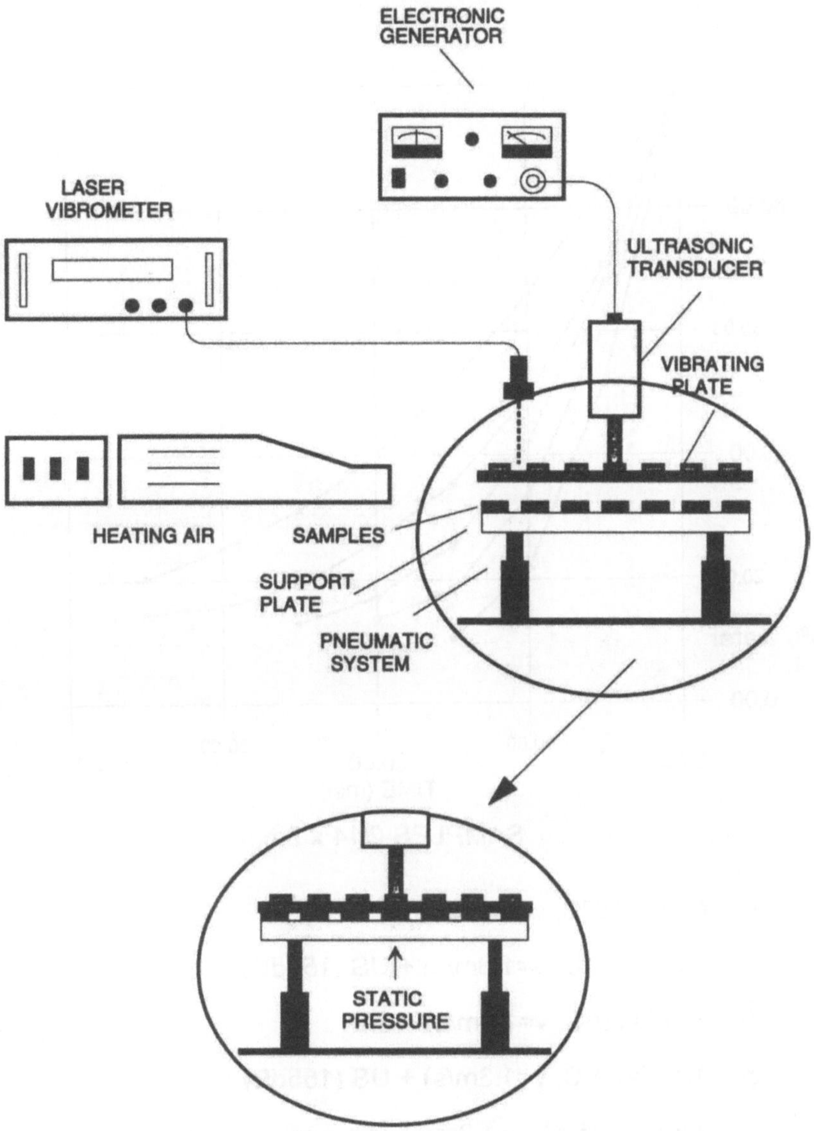

Figure 7.5 Experimental set-ups for forced-air drying assisted by air-borne ultrasound.

In conclusion, it can be stated from the results of the present and previous investigation that the application of air-borne ultrasound can be useful in increasing the efficiency of forced-air drying processes. Nevertheless, the improvement seems to be relatively limited and the use of air-borne ultrasound could be restricted to specific products and/or operations.

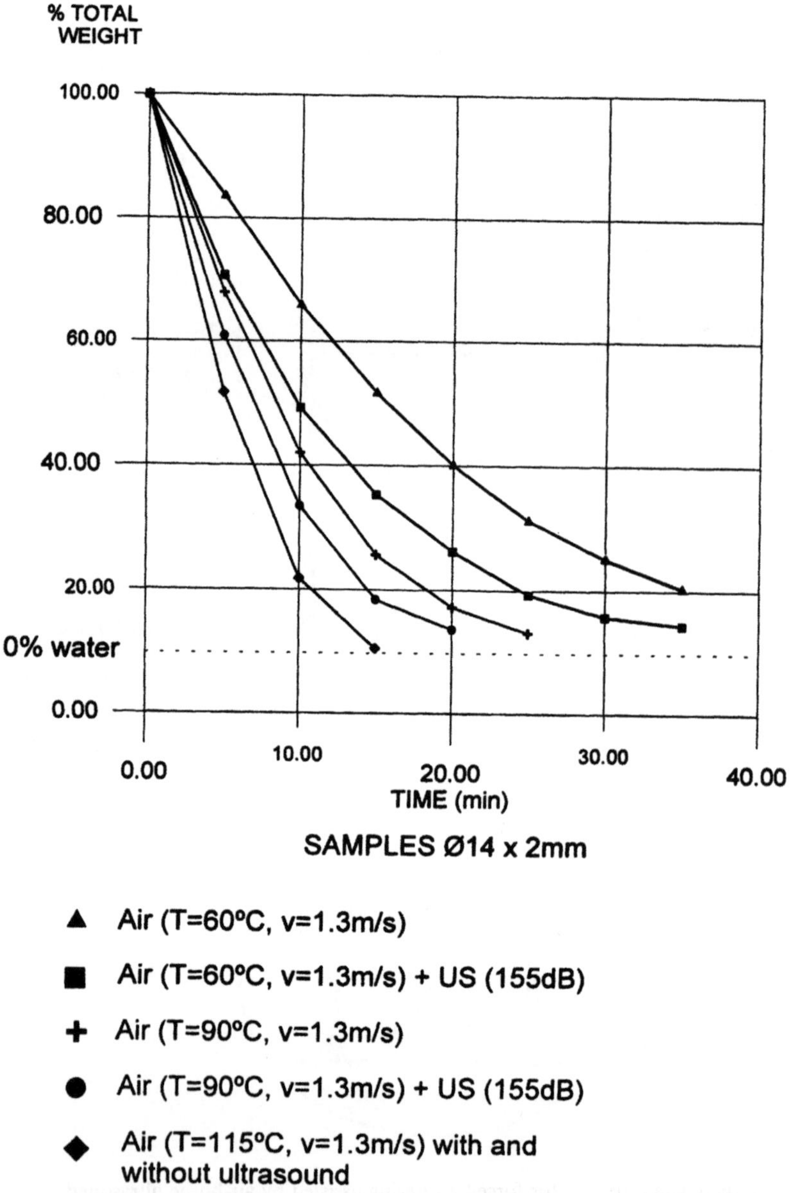

Figure 7.6 Forced-air dehydration curves with and without air-borne ultrasound (155 dB).

7.3.1.2 Dehydration by ultrasonic vibration and static pressure. In order to increase the ultrasonic effect on food dehydration a new procedure was developed and tested. It mainly consisted in applying the ultrasonic vibration in direct contact with the samples and together with a static

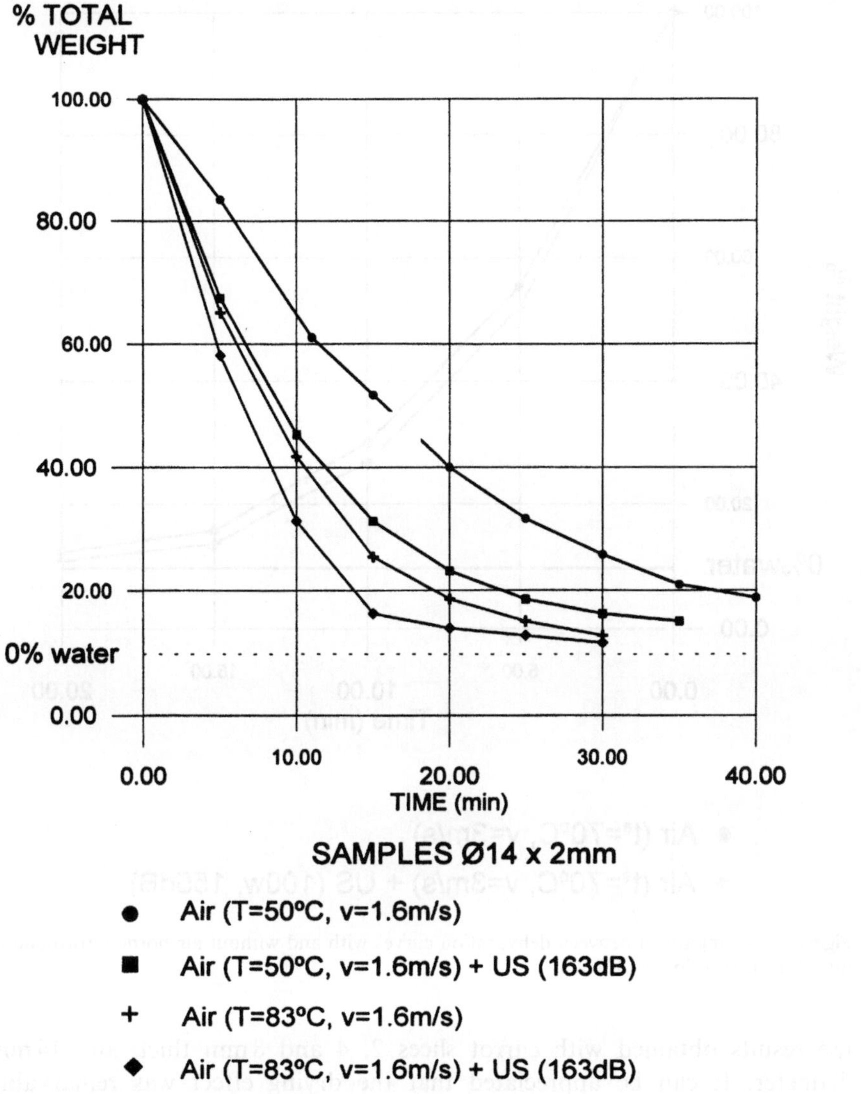

Figure 7.7 Forced-air dehydration curves with and without air-borne ultrasound (163 dB).

pressure. To that purpose, the experimental set-up presented in Figure 7.5 was slightly modified. The samples were placed on the surface of the transducer radiating plate and they were kept there during treatment by applying a static pressure on them (see the detail in Figure 7.5). An air flow at 1 m/s and 22°C was also applied to facilitate the removal of moisture. The drying effect was measured following the previous procedure, i.e. by weighing the samples at different times during treatment. Figure 7.9 shows

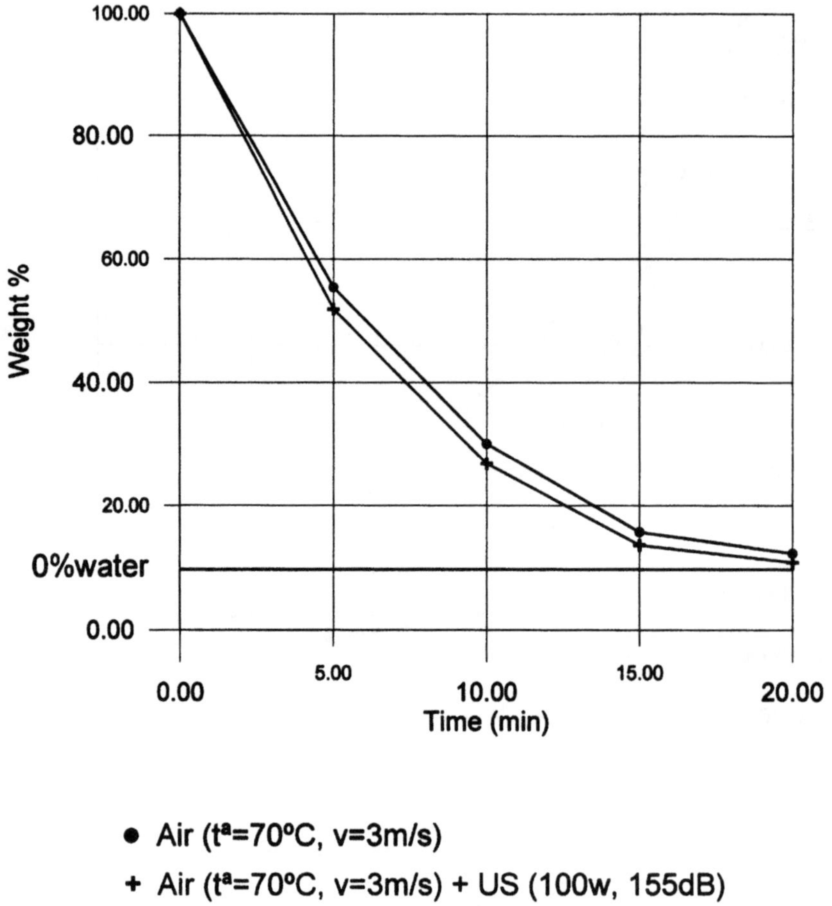

Figure 7.8 Comparison between dehydration curves with and without air-borne ultrasound at forced-air flow velocity of 3 m/s.

the results obtained with carrot slices 2, 4 and 8 mm thick and 14 mm diameter. It can be appreciated that the drying effect was remarkably improved. In fact, the drying process is not only much quicker and less energy consuming than the forced-air drying (with and without airborne ultrasound) but it is more powerful: the final moisture content could be less than 1%. In addition, due to the processing time and low temperature of the air flow, the product qualities are well preserved.

In conclusion, the application of ultrasonic vibration in contact with food samples has demonstrated to be very effective method for food dehydration. The main mechanisms involved in this new procedure seem to be the above-mentioned 'sponge effect' and the creation of dewatering channels by the acoustic stress.

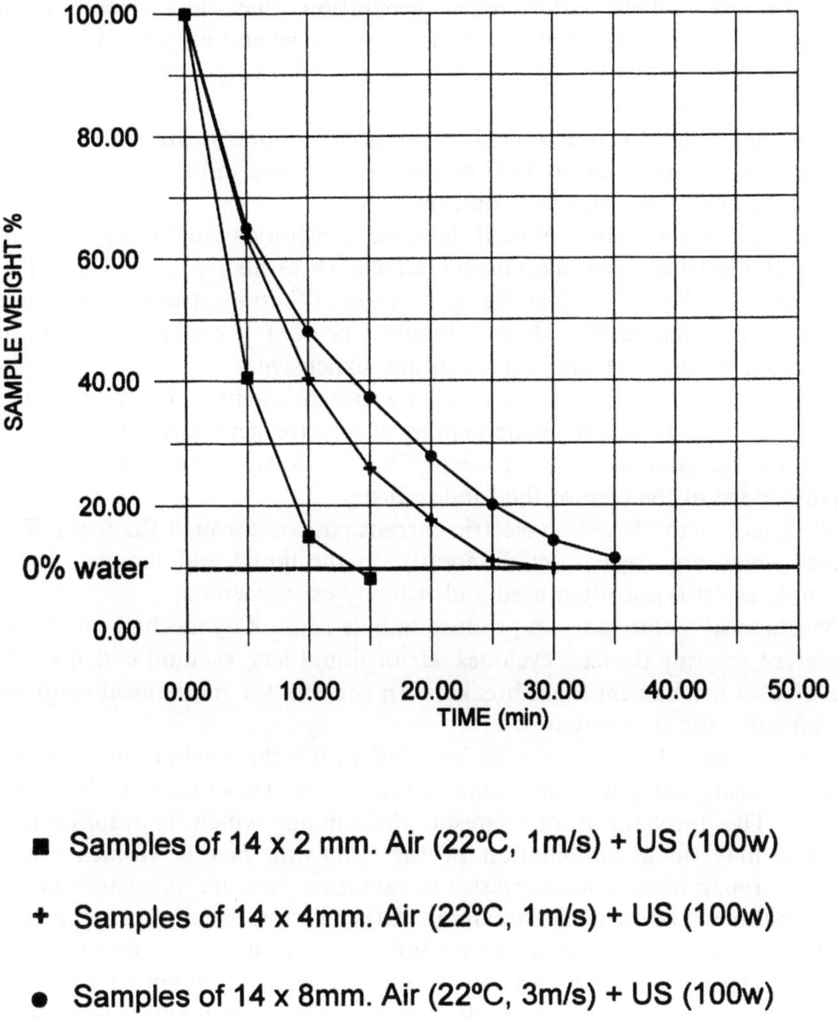

Figure 7.9 Dehydration curves by ultrasonic vibration and static pressure.

7.3.2 Defoaming

Foam is a dispersion of gas in liquid in which the distances between the individual bubbles are very small. In a foam system the volume ratio of gas to liquid is very great and the bulk density approaches that of a gas.

Foams are frequently produced during various manufacturing processes. Dispersions of gases in liquids may be produced by injection of gas into a body of liquid or, if the gas is already dissolved in the liquid, by any procedure (heating, depressing, etc.) which favours the coalescence of the gas molecules. This latter foam generation may also result from vaporization of the liquid phase. Another foam generation procedure is by means of chemical or biological agents which following a certain reaction, will release a gas.

Generally foams cause difficulties in process control and in handling equipment. A typical example is in the fermentation industry where foam represents one of the biggest problems.

There are several conventional defoaming methods employing thermal, chemical electrical and mechanical effects (Viesturs *et al.*, 1982). The thermal methods are based on the application of thermal stresses to produce the rupture of the bubbles. It then involves heating or cooling the foam in some way, which is generally difficult and expensive.

Chemical defoaming methods employ antifoam agents which decrease the surface tension. The chemical defoaming agents are, generally, very effective but they contaminate the process. This obviously represents a great disadvantage in the case of the food industry.

Electrical methods use an electric current passing through the foam. The forces which are created act differently on the liquid and the gas. These methods are still not often used and scarcely experimented.

Mechanical methods try to produce bubble rupture by mechanical shock. Different rotating devices, cyclones, air or liquid jets, vacuum systems, etc. are used as mechanical foam breakers. In general, the mechanical methods are effective for coarse foams.

The ultrasound method can be included within the mechanical methods. High-intensity sonic and ultrasonic waves are a clean means of breaking foams. The mechanism of acoustic defoaming, which is insufficiently known, may be a combination of the following factors (Boucher and Weiner, 1963): high acoustic pressures, radiation pressure, resonance of the foam bubbles and acoustic streaming. Other factors, such as atomization from the film surface, may also have influence on this phenomenon.

The potential use of high-intensity ultrasound for defoaming has been known for many years. Nevertheless, a few acoustic defoamers have been reported in the technical literature but apparently none of them were used in large-scale industrial plants. The majority of these defoaming devices are based on aerodynamic acoustic sources of various types such as the Hartmann whistle (Hartmann, 1939), the rotary siren (Allen and Rudnick, 1947), etc. The main difficulties with these systems come from noise problems (they generally work at sonic rather than ultrasonic frequencies), the need for high air generation capacity, the control and sterilization of the air flow and high energy consumption.

A new ultrasonic defoamer was developed by using the stepped-plate transducer to generate air-borne ultrasound. This is a powerful and compact piece of piezoelectric equipment with no moving parts, there is no air flow required to produce the sound and so there is no interference with the process and it is readily sterilizable. It has been successfully applied to the control of excess foam produced on high-speed canning lines and in the dissipation of foams in reactors.

In canning lines, the air-borne ultrasonic radiation is focused on the working area where the foam excess is quickly destroyed to avoid liquid losses (Figure 7.10). The new system was applied successfully to control foam in the filling operation of cans with a well-known commercial beverage at a speed of more than 20 cans per second. Two focused transducers working at 20 kHz were used in parallel for an efficient and quick operation in order to cover the can surface with high-intensity sound pressure levels (about 165 dB). The power applied to each transducer was only of 150 W. Then the energy consumption was only of about 5 mW h/can.

Figure 7.11 shows the application of the new system for the treatment of foam in reactors. This technique was tested in a beer fermenter on a pilot scale. The rate of foam breaking was of about 200 l/min with an input power on the transducer of 300 W; the energy consumption was of about 30 W h/m^3. In recent tests on an industrial scale (vessels of 6 m in diameter) these figures were improved significantly. The application and extension of this new technique is currently under development.

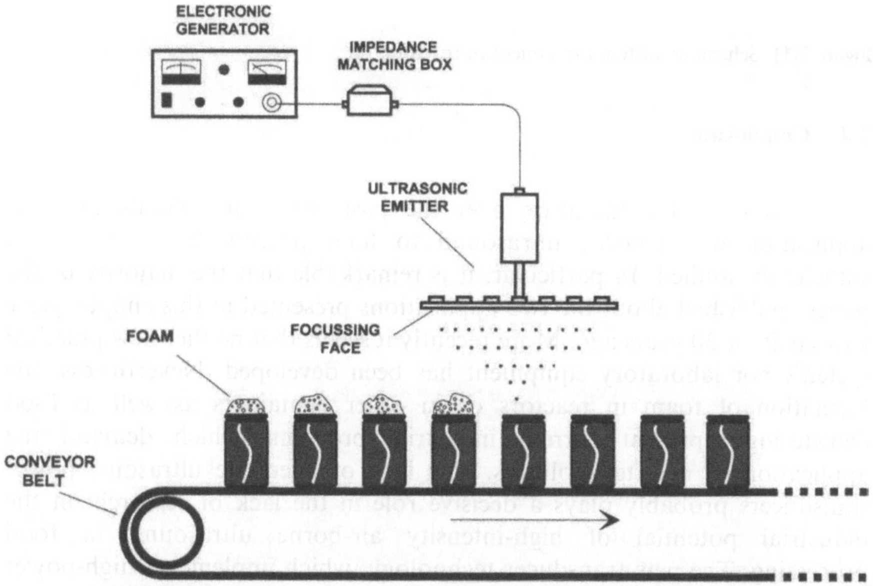

Figure 7.10 Schematic of ultrasonic defoaming in canning lines.

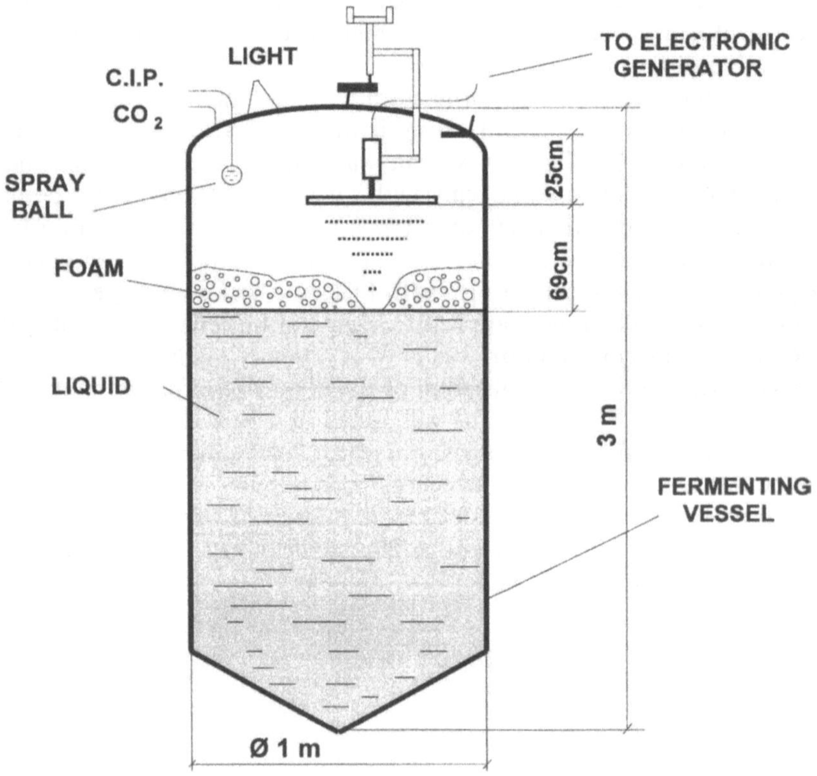

Figure 7.11 Schematic of foam treatment in reactors.

7.4 Conclusion

An analysis of the literature over the past few years reveals that the application of air-borne ultrasound to food processing has not been sufficiently studied. In particular, it is remarkable that the majority of the papers published about the two applications presented in this chapter were written 20 or 30 years ago. More recently it seems that neither new practical systems nor laboratory equipment has been developed. Nevertheless, the formation of foam in reactors or in other containers as well as food dewatering represent current industrial problems which demand the application of new technologies. The lack of adequate ultrasonic power transducers probably plays a decisive role in the lack of research on the industrial potential of high-intensity air-borne ultrasound in food processing. The new transducer technology, which implements high-power capacity, efficiency and directivity, has opened up renewed possibilities for industrial applications. In fact, from the results of the two applications to food processing described previously, it seems possible to draw a positive conclusion about the potential of the stepped-plate power generator for large-scale applications. Its suitability for tailoring the acoustic field by modifying the steps size makes this technology a basic tool for present and future developments of high-intensity ultrasonics in fluids and multiphase media.

In food processing, the new technology could be applied in some other different fields such as agglomeration of micron particles, solid–gas and solid–liquid separation, gas sterilization, mass transfer, etc.

7.5 References

Allen C.H. and Rudnick I. (1947) A powerful high frequency siren. *J. Acoust. Soc. Am.*, **19**, 857–65.

Arkhangel'skii M.F. and Statnikov Y.G. (1973) Diffusion in heterogeneous systems, in *Physical Principles of Ultrasonic Technology*, Vol. 2, Editor L.D. Rosemberg, Plenum Press, New York.

Boucher R.M.G. and Weiner A.L. (1963) Foam control by acoustic and aerodynamic means. *Br. Chem. Engng*, **8**, 808–12.

Fairbank H.V. (1975) Applying ultrasound to continuous drying process, in *Ultrasonic International 1975 Conference Proceedings*, IPC Science and Technology Press, Guildford, pp. 43–5.

Gallego-Juárez J.A., Rodriguez-Corral G., San Emeteno J.L. and Montoya-Vitini F. (1989) Electroacoustic unit for generating high sonic and ultrasonic intensities in gases and interphases, Spanish Patent 8903371, European Patent No. EP 450.030.A1, 1991, USA Patent No. 5,299,175, 1994.

Gallego-Juárez J.A., Rodriguez-Corral G. and Galvez-Moraleda J.C. (1995a) High intensity ultrasound for food dehydration, *Proceedings of the World Congress on Ultrasonics 95*, Gesellschaft fur angewandte Ultraschallforschung (GEFAU), Duisburg, pp. 761–4.

Gallego-Juárez J.A., Yang T., Vázquez Martinez F., Galvez-Moraleda J.C. and Rodriguez-Corral G. (1995b) Procédé et dispositif de désydratation, Belgium Patent No. 9500420.

Greguss P. (1964) The applications of air-borne and liquid-borne sounds to industrial technology. *Ultrasonics*, 5–10.

Hartmann J. (1939) Construction, performance and design of the acoustic air-jet generator. *J. Scient. Instru.*, **16**, 140–9.

Muralidhara H.S., Ensminger D. and Putman A. (1985) Acoustic dewatering and drying (low and high frequency): state of the art review. *Drying Technol.*, **3**, 529–66.

Seya K. (1970) Macrosonic drying, in *Proceedings of the First International Symposium on High-power Ultrasonics*, IPC Science and Technology Press, Guildford, pp. 136–40.

Soloff R.S. (1994) Sonic drying. *J. Acoust. Soc. Am.*, **36**, 961–5.

Viesturs U.E., Kristapson M. Z. and Levitans E.S. (1982) Foams in microbiological processes. In *Advances in Biochemical Engineering*, Editor A. Fiechter, Springer, Berlin.

8 The importance of power ultrasound in cleaning and disinfection in the poultry industry – a case study

T. QUARTLY-WATSON

8.1 Introduction and background

Poultry processing factories are highly automated and may produce from several hundred thousand to a million birds a week. The products have a limited shelf life before they become unfit for consumption. This is determined by the rate that bacteria build up on the carcasses and this is influenced by operations within the complete production and delivery chain. All the elements from slaughter to transport and storage in the shops need to be both clean and cool.

Keeping the factory and transport cool is well within the normal capabilities of suppliers. However, keeping the equipment which comes into contact with the poultry sterile poses more problems. The actual physical shape of the shackles which are used to suspend and transport the poultry and the baskets which are used to store them, render them difficult to clean quickly. In the past continuous cleaning methods have relied on rotating brushes and sprayed water. End-of-shift methods generally involve coating the shackles with a chemical foam and then using high-pressure water sprays to remove the residue. This has proved to be time consuming and bacteriological testing has shown both methods are frequently ineffective.

To extend the shelf life by minimizing cross-contamination each bird must be presented with a clean and disinfected shackle or basket. Effective cleaning using traditional methods can only be accomplished by stopping the line. This is usually done between shifts which means that there has been a potential build up of bacteria on the product with a corresponding reduction in shelf life.

Banham Poultry Ltd asked Production Line Cleaning Ltd to come up with a method which enables effective cleaning to be performed continuously during the production process. The solution involved the use of power ultrasound. Tests with the Stericlean system (developed by Production Line Cleaning) demonstrate that each shackle and basket is kept visibly and bacteriologically clean on a continuous basis, thus permitting each bird to be presented with an uncontaminated shackle or basket. This significantly reduces the risk of cross-contamination and leads to both a longer shelf life and minimization of the post-shift cleaning.

8.2 Factory production requirements

The poultry processing industry needs to operate in a way that maximizes sales and profitability while meeting requirements which include:

- reasonable shelf life: having a shelf life longer than eight days
- meeting health and safety requirements: veterinary inspection, noise pollution, equipment handling.
- minimizing ownership costs: labour, consumables (water, disinfectant – both acquisition and disposal, utilities), inventory.

8.3 Production process

There are three main processes which are used in a typical poultry processing factory:

- The evisceration process in which the birds are suspended from shackles, stunned, killed, scalded, defeathered and then eviscerated
- the weighing process in which the birds are sorted by weight
- the maturation process in which the birds are either stored in baskets or suspended from shackles in a chilled environment for a number of hours prior to packing.

In addition, there may be further processing during which the birds are cut into portions.

8.3.1 Conveyors

During the above processes the birds are hung from shackles which form part of a conveyor. Typical conveyors have the following characteristics:

- shackle spacing: 0.15–0.6 m
- shackle weight: 0.25–0.75 kg
- shackle material: plastic or stainless steel
- conveyor line speed: 2000–9000 shackles per hour
- conveyor length: 150–2000 m
- shackles per conveyor line: 700–10 000
- conveyor repetition frequency: 5 min to several hours.

Where birds are stored in baskets the cleaning rate can be between 200 and 1200 baskets per hour and the basket weight 2–10 kg (plastic–stainless steel).

8.3.2 Problems with conventional cleaning

The difficulty of manually cleaning a line which contains 10 000 shackles and is 2 km long is readily apparent. To do it several times a day so that a clean shackle is presented to every bird is not possible. At the other end of the scale for a 150-m line, a shackle would have to be cleaned every 5 min which is equally impossible. Further, because of the physical shape of shackles and baskets, it is difficult to ensure that the cleaning process is totally effective.

The costs associated with conventional cleaning come from three principal sources: (a) labour, (b) water acquisition and disposal, and (c) disinfectant acquisition. Therefore, if any change in method is to be introduced it must be shown that:

- it is *effective*, i.e. each shackle or basket is sterile
- the *lifetime costs* of the equipment are *acceptable* for the poultry producers in terms of purchase, running and maintenance charges.

8.4 Ultrasonic cleaning

A number of systems have been made by Production Line Cleaning Ltd to meet the requirements of cleaning shackles, stainless steel baskets and plastic baskets. The core of the system is a heated stainless steel tank which contains a number of ultrasonic transducers and a dilute sanitizing solution. The items to be cleaned are passed through the tank and subjected to the ultrasound as they pass in front of the transducers. The ultrasound in conjunction with the sanitizer, removes the fat and blood deposits and sterilizes the item. Tests show that the combined effect of ultrasound and sanitizer are extremely effective. The exact mechanism of bacteria destruction is unclear but it appears that the bacteria are damaged by the ultrasound in such a way that a low concentration of sanitizer is able to destroy them.

Additional modules allow for mechanical handling of the baskets, a pre-wash if there are large pieces of contaminant, and an air dried if needed.

8.4.1 Design variables

The major design variables include the type of contamination, the type of material to be cleaned and the conveyor line speed. Clearly, there is an interdependency between them. In addition to the actual effectiveness of the cleaning process the ancillary costs of heating, water use and disposal, sanitizer use and disposal and noise must be considered.

8.4.2 Type of contamination

These include fats, blood, skin, meat and lime scale.

- *Tank temperature.* The type of contamination will influence the tank temperature to be used. There is a trade-off between the tank temperature and the effectiveness of the ultrasound. Maximum cavitation occurs at about 65–70°C falling to about half when the temperature is either lowered to 40°C or raised to 85°C. Experience has shown that in order to obtain the greatest effect in ultrasonic cleaning the following guidelines apply: to clean *fat* the temperature should be > 60°C and to clean *blood* the temperature should be < 45°C. There is no advantage in heating above about 65°C as the costs involved in heating are not offset in any other way. Where both blood and fat are present, the lower temperature is generally more effective.
- *Sanitizer type.* Lime-scale contamination typically found in hard water areas requires a different sanitizer from food contaminants. It is often practical to periodically change the sanitizer for an acid solution to remove lime-scale.
- *Prewash.* Large lumps of debris are not removed effectively by ultrasound. If they are a significant problem, they need to be removed through a prewash using a water deluge.
- *Ultrasonic power.* There is always a threshold figure for 'watts per litre' which must be exceeded before any effective cleaning can take place. This is partially dependent on the type of contamination present.
- *Immersion time.* The immersion time required is determined to a large extent by the type of contamination present.

8.4.3 Type of material to be cleaned

In most food processing factories the material is either plastic or stainless steel and this determines several factors.

- *Sanitizer type.* The type of sanitizer used is particularly important for plastic. Stainless steel is much more tolerant to a wide range of sanitizers.
- *Ultrasonic power.* The surface of plastic tends to become much more scored than stainless steel. Greater power is needed to clean scored surfaces.
- *Immersion time.* On average plastic needs three times greater immersion time than stainless steel.

8.4.4 Method of operation

The shackles or baskets are guided through the Stericlean tank by introducing a dip section into an existing conveyor system or by purpose building a transport mechanism. The shackles or baskets are submerged in the tank for a time ranging from between about 10 s and 1 min depending on the material and contaminant as discussed above. The water in the tank has sanitizer added, generally at a concentration of between 1% and 2%. The sanitizer is normally disposed of in the drainage system.

At the exit end of the tank, water is continuously injected to create a debris-free surface area through which the clean shackles or baskets can pass without becoming contaminated. The tank water is continuously filtered and circulated through the tank to distribute the heat and remove debris. Water levels are controlled automatically. Some water is lost by being 'dragged' out of the tank with shackles or baskets. This is replaced and the sanitizer is added at the same time through a metering system. After an initial period the tank water and sanitizer are replaced every two to three days. Should the water supply be cut off for some reason and the level drop beneath a predetermined point, a sensor cuts off the heating and ultrasonics to prevent damage. This is because the ultrasonic transducers must only be active when submerged.

Heating may by a steam coil, gas or electricity. Steam or gas are generally preferred as being cheaper options.

Ultrasonic power levels typically vary between 2–3 kW for a small single line shackle tank and 18 kW for a dual line, plastic basket tank. Although a lower frequency may be slightly more effective in cleaning, Production Line Cleaning has standardized on transducers which operate at 38 kHz to minimize noise. The transducers have a design life in excess of 5000 h.

8.5 Results

8.5.1 Cleanliness testing requirements

In all food processing factories veterinary inspection is part of the day-to-day operation. Tests on baskets and shackles are made on an ongoing basis and examples of both total viable count (TVC) testing done by the factories and Biotrace testing done independently by Biotrace Ltd have demonstrated the effectiveness of the ultrasonic cleaning system.

8.5.2 TVC analysis

Some examples from Banham Poultry Ltd, Pic Pic and Proconar are given in Table 8.1.

Table 8.1

Source of sample	Before ultrasonic cleaning	After ultrasonic cleaning
Plastic tray	530	120
Plastic tray	2310	220
Plastic tray	1420	100
Steel basket	7280	80
Steel basket	6240	10
Steel basket	12480	< 10
Steel shackle	300	0
Steel shackle	320	0
Steel shackle	220	1

8.5.3 *Biotrace analysis*

A Biotrace Uni-Lite Xcel test was employed which uses the ATP bioluminescence technique for detecting organic residues. Independent testing was used in the case of both plastic and stainless steel. Results are given in Table 8.2.

Table 8.2 Measurements are in relative light units (RLU)

Source of sample	Before ultrasonic cleaning	After ultrasonic cleaning
Plastic tray	725	160
Plastic tray	681	59
Plastic tray (badly worn)	> 500,000	717
Plastic link belt	2485	52
Plastic link belt	3087	32

8.5.4 *Reliability*

The equipment has proved very reliable with only normal preventative maintenance being necessary. Transducer life for the earlier installed equipment is now in excess of 6000 h without failure.

8.5.5 *Cost benefit*

Ultrasound cleaning exceeds the mandatory cleanliness requirements. This is achieved together with savings in labour time, reduced sanitizer and water consumption, reduced water disposal costs and high reliability and therefore it has proved to be a very cost effective solution.

8.6 Conclusion

The use of power ultrasound has proved to be extremely effective both in sterilization and cost terms. It has been shown to give a significantly better performance than conventional techniques. The risk of cross-contamination has been markedly reduced and this has lead to a longer shelf life for the product.

9 The principles of cavitation

T.G. LEIGHTON

9.1 Introduction: the characterization of cavitation

Acoustic cavitation has in the past been defined in various ways. The area of concern of this book is the use of ultrasound in food processing, so that most relevant applications of cavitation will involve ultrasonic fields of high acoustic pressure amplitude. The bubble is driven into some form of response by an externally applied acoustic pressure, and it is useful to consider cavitation to be the 'activation' of a gas inclusion in a liquid by an acoustic field. This does not therefore include the emission of sound from a gas bubble on entrainment in a liquid which, however, shall be discussed briefly as it is of interest to a range of applications, including industrial injectors, and because it illustrates the existence of the bubble resonance.

In liquids of density ρ, with ambient pressure p_0, a gas bubble of radius R_0 has a well-defined resonance frequency f_0 equal to:

$$f_0 = \frac{\omega_0}{2\pi} = \frac{1}{2\pi R_0}\sqrt{\frac{3\kappa p_0}{\rho}} \tag{9.1}$$

where κ, the polytropic index, varies between unity and γ, the ratio of the specific heats of the gas at constant pressure and constant volume. (The expression is valid for air bubbles in water, if the radius R_0 is greater than around 10 μm. When the ambient pressure is 1 atmosphere, the expression reduces to $f_0 R_0 \approx 3$ Hz m. For more general expressions, see Leighton 1994a.) Such bubbles pulsate as lightly damped oscillators, emitting sound. This can be observed when a gas bubble is injected into water (Minnaert, 1933; Leighton and Walton, 1987). However there is a range of other entrainment modes, as can be illustrated from the scenarios involving liquid impact: waterfalls (Leighton *et al.*, 1995a); rainfall over water (Pumphrey and Crum, 1990); or breaking ocean waves (Medwin and Beaky, 1989). Upon entrainment, the bubbles undergo shape and volume changes (Leighton *et al.*, 1991). Except in certain circumstances (Longuet-Higgins, 1989), the shape changes are not efficient radiators of sound (Strasberg, 1956). It is the volumetric pulsations which generate the characteristic 'signature' typical of an entrained bubble, i.e. an exponentially decaying sinusoid, the frequency of which indicates the bubble size (Minnaert, 1933; Strasberg, 1956). A few milliseconds after injection these passive emissions have decayed to below the level of the noise. However the bubble may still be driven into oscillation by an externally applied sound field, and it is such behaviour which is the main concern of this chapter.

There is a wide range of phenomena which arise as a result of the pulsations or surface perturbations of the gas–liquid interface of the gas inclusion (Leighton, 1994a). Driven by a sound field, the bubble can undergo relatively stable, low energy oscillations (called 'stable' or 'non-inertial' cavitation), which can give rise to acoustical effects (including resonance and non-linear acoustic scattering). Optical and electrical effects may also be observed (Leighton, 1994b). The bubble may also undergo the higher-energy phenomenon of 'inertial' (formerly known as 'transient') cavitation. There exists a threshold, based on parameters relating to the sound field and the bubble, above which inertial cavitation will occur. It is relatively simple with a low-amplitude acoustic field to cause only non-inertial cavitation to occur in a given bubble population. However, if the aim is to generate inertial cavitation through the use of a higher acoustic pressure amplitude, then in an acoustic field the threshold conditions are usually exceeded only for some of the bubbles present. [As the acoustic frequency increases, the tendency is to reduce the proportion which undergo inertial cavitation (see Section 9.5.1).] The result is that both inertial and non-inertial cavitation will occur throughout the cavitation field, and will influence one another (Leighton, 1995a). Non-inertial cavitation can cause a range of acoustic, chemical, biological and erosive effects (Leighton, 1995b). Theoretical descriptions of cavitation have been available since the start of the century. It was the presence of erosive pits in propellers which inspired Lord Rayleigh (1917) to develop his pioneering analysis for the collapse of an empty spherical cavity under a static pressure. Coupling this energetic collapse phase with the explosive growth phase of a bubble (Blake, 1949), Noltingk and Neppiras (1950) characterized a particular type of cavitation, in which appropriately small bubbles in sufficiently strong sound fields undergo growth to many times their original size, and then a subsequent rapid collapse. It has since been verified that such events concentrate the acoustic energy, generating high gas temperatures (Neppiras and Noltingk, 1951), shock waves in both the gas (Vaughan and Leeman, 1986) and the surrounding liquid (Rayleigh, 1917), and free radicals (Suslick *et al.*, 1986). The latter may be involved in sonochemical reactions (Suslick *et al.*, 1986), and may generate sonoluminescence (Leighton, 1994a). Flynn (1964) distinguished the so-called 'transient' cavitation with which such manifestations are associated, from the less energetic 'stable' cavitation, where the bubble pulsates about an equilibrium radius over many acoustic cycles. Perhaps not surprisingly, therefore, this field which grew from the observation of erosive damage in propellers developed into a problem coupling fluid dynamics with acoustics, and indeed acoustic techniques feature amongst the most common ways in which cavitation is initiated and measured (Leighton, 1995b).

Though in common use the distinction between *transient* and *stable* (or *inertial* and *non-inertial*) cavitation is inexact, many formulations have concentrated on attempting to distinguish the threshold values of acoustic frequency, pressure amplitude, and initial bubble radius corresponding to the transition between the types of cavitation (Neppiras, 1980; Apfel, 1981; Flynn and Church, 1988; Holland and Apfel, 1989; Leighton, 1994a). However, once the threshold has been exceeded, the extent to which cavitational effects are generated (the 'activity' of the cavitating field) is a key parameter (see Section 9.6).

With rare exceptions (Gaitan and Crum, 1990), whilst theory has tended to concentrate on the dynamics of single bubbles, in practice when inertial cavitation causes a given effect, it is generally *populations* of bubbles that are involved (Leighton, 1995a). Whilst the exploitation of cavitation in many common applications (such as cell killing, ultrasonic cleaning, cavitation erosion and sonochemistry) relies on inertial cavitation (Leighton, 1995b), the application of these effects depends upon the characteristics of the *bubble population* and its interaction with the sound field. These two key factors may be influenced by other forms of cavitation, affecting nucleation and the relatively low-energy phenomena of bubble migration, dissolution and exsolution (Leighton, 1995a).

An experimental characterization of cavitation which is to be of use to common applications must therefore account for the following problems. First, whilst most applications involve bubble populations, the models and theories overwhelmingly rely on single-bubble descriptors, descriptors which are themselves only approximate. Second, when inertial cavitation occurs in a sound field and produces a range of effects (chemical, erosive, biological or acoustic), non-inertial cavitation may also occur, potentially not only influencing the inertial cavitation and its consequences, but also producing effects of its own.

9.2 Types of cavitation

Conceptualizations of bubble dynamics begin with models of isolated spherical bubbles. These models in turn become modified to allow the incorporation of experimentally observed phenomena. In this way have arisen the well-known classifications of transient and stable cavitation (Neppiras, 1980; Apfel, 1981; Leighton, 1994a). An impression of the range of bubble behaviours is shown in Figure 9.1. Neither of the phenomena shown in the first two rows (slow dissolution or growth of a bubble) will give

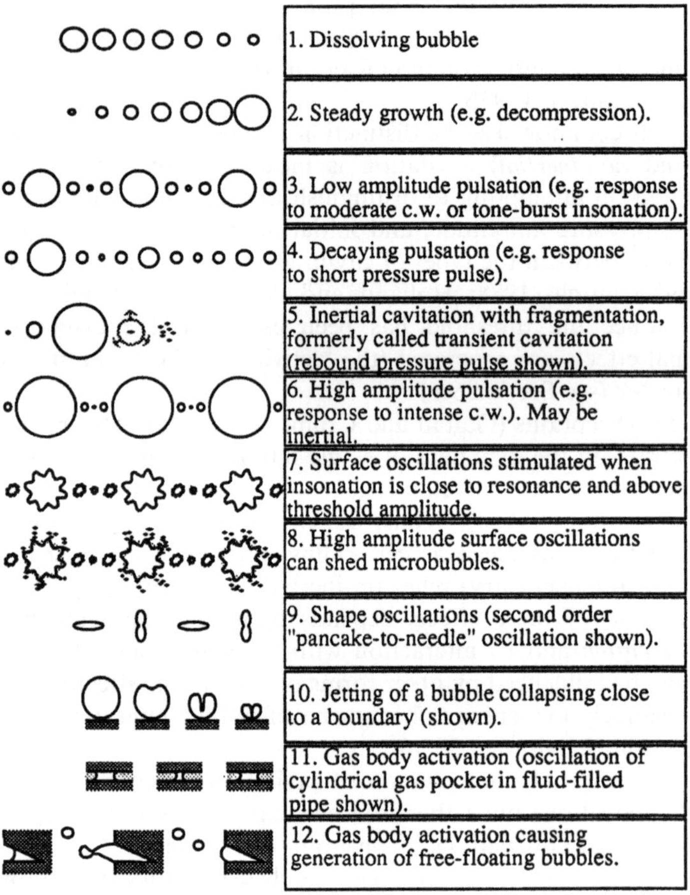

Figure 9.1 A schematic illustration of a range of bubble behaviour. Rows 3–12 are effects through which the bubble can be expected to be acoustically active. The expansion ratios drawn for these bubbles are exaggerated to more clearly illustrate the pulsations.

rise to acoustic emission. However, pulsation will give rise to such emission, and as such can be detected acoustically at a distance. As described in the previous section, the spherical bubble is, to a first approximation, an oscillator with a single degree of freedom, with a resonance frequency f_0 which is determined by the ratio of the stiffness to the inertia. The stiffness is determined by the compressibility of the gas content within the bubble, whilst the inertia is invested principally in the surrounding liquid, which must move if the bubble wall does (Leighton, 1994a).

As an oscillator, the bubble can be driven into pulsation by an incident sound field. If it is driven such that the amplitude of oscillation of the bubble wall is low (for example, the incident sound field has a low acoustic pressure amplitude, or the bubble has a size which differs greatly from that required for resonance with the driving frequency, bearing in mind that we are at present considering only models of bubbles which remain spherical at all times), then the bubble will scatter the sound field. If an ultrasonic beam is incident on the bubble, acoustic energy will be converted to heat through viscous and thermal damping mechanisms associated with the bubble motion (Devin, 1959; Eller, 1970) and scattered out of the beam (to first order as a spherical re-radiation) by the bubble. Row 3 of Figure 9.1 shows just such an oscillation. Despite damping, the illustrated bubble pulsations are shown not to decrease in amplitude, suggesting a continuous wave or tone-burst insonation, rather than a short pulse (which is the case illustrated in row 4). If pulsations of this type are of sufficient amplitude, this form of cavitation can strongly influence the effects of power ultrasound. First, such bubbles can directly generate physical and biological effects. These are usually associated with acoustic radiation forces. For example, cell disruption may be brought about because the bubbles can travel rapidly through the liquid under the influence of acoustic radiation forces (Section 9.4), generating hydrodynamic stresses which can, for example, produce haemolysis (Miller and Williams, 1989). Second, such bubbles can strongly influence both the population of nuclei available to seed inertial cavitation, and the acoustic pressure amplitudes to which those nuclei are subjected (see Section 9.5 and Leighton, 1995a).

It is well known that when ultrasound of sufficient intensity passes through a liquid containing bubbles which initially have radii that lie within a critical size range (Section 9.5), such gas inclusions may nucleate inertial cavitation (Neppiras, 1980; Apfel, 1981; Walton and Reynolds, 1984; Apfel and Holland, 1991). Inertial cavitation is typified by the sudden expansion and then rapid collapse of the bubble (row 5 of Figure 9.1). During the rapid collapse, the gas within the bubble is compressed. It may be heated to high temperatures (Walton and Reynolds, 1984; Suslick *et al.*, 1986; Kamath *et al.*, 1993), and gas shocks may propagate (Wu and Roberts, 1993). Electrical discharge and other effects may be associated with this phenomenon (Margulis, 1985; Lepoint *et al.*, 1993). Mechanisms for the generation of sonoluminescence based on all these, and yet more (Crum and Cordry, 1993; Eberlein, 1996), processes have been proposed.

Certainly the production of free radicals and electronically excited species has been associated with the extreme conditions that occur within the bubble. These can cause chemical and biological effects. However, the cavitation can also affect the medium through a range of associated mechanical processes [and indeed careful studies can distinguish, for example, the relative contributions of chemical and mechanical mechanisms to bioeffect (Miller and Thomas, 1993a, b; Miller *et al.*, 1995)].

One of the most notable such mechanisms occurs immediately after the collapse. The bubble rebounds, emitting a pressure pulse (shown figuratively in Figure 9.1, row 5 by arrows) into the liquid (Rayleigh, 1917; Noltingk and

Neppiras, 1950; Neppiras and Noltingk, 1951; Plesset and Prosperetti, 1977; Leighton, 1994a). This may cause mechanical damage. If clouds of bubbles are present, they may collapse co-operatively, enhancing this effect (Vyas and Preece, 1976). After the emission of such a pressure pulse, the bubble may fragment or repeat the growth/collapse cycle a number of times. For many years it was thought that the type of bubble activity which gave rise to such phenomena as sonoluminescence would involve fragmentation of the bubble after one (as shown in Figure 9.1, row 5) or a few cycles, reinforcing the labels of 'transient' cavitation and, for less energetic bubble oscillations, 'stable' cavitation. Following the discovery that in specialized conditions the bubble can pulsate for thousands of cycles (in a manner similar to row 6 of Figure 9.1), emitting a sonoluminescent flash at each collapse (Gaitan and Crum, 1990; Barber *et al.*, 1992), this terminology clearly needed replacing. Since models of the bubble collapse had shown that, to achieve the energy concentration expected to generate what had been known as 'inertial cavitation' the inertia of the liquid had to play a dominant role during the collapse, the terms 'inertial' and 'non-inertial' replaced 'transient' and 'stable'.

The generic oscillation shown in row 6 of Figure 9.1 is a high-amplitude pulsation of a spherical bubble (Figure 9.2). Depending on the amplitude, such oscillation may be inertial (as described above for specialized conditions) or non-inertial, but of high amplitude. If non-inertial, there are a number of interpretations of how this situation, which differs in wall oscillation amplitude from that shown in row 3, arises. It might, for example, occur if rows 3 and 6 show the same bubble, but in row 6 the acoustic pressure amplitude of the driving field is greater. Through such high-amplitude pulsations a second mechanism for mechanical damage may occur, resulting from the hydrodynamic shear stresses close to the bubble wall. The high amplitude pulsation indicates good coupling of the bubble to the sound field, and the bubble can move rapidly under acoustic radiation forces (see Section 9.4), traversing the medium as described in the previous section. Radiation forces are particularly strong on resonant bubbles, and indeed an increased amplitude of wall oscillation might be expected if the bubble is closer to resonance size. This gives a second source for the difference between rows 3 and 6: the bubbles in these rows now might be in the same sound field, but with the bubbles in row 6 being closer to resonance. However, as the bubble is more closely driven to resonance, other effects begin to occur. Most notable of these are surface waves (Figure 9.3b), which visually cause a 'shimmer' to appear on the surface of the bubble (Kornfeld and Suvorov, 1944; Neppiras, 1980; Phelps and Leighton, 1997), and acoustically may be detected through combination-frequency techniques (Phelps and Leighton, 1996). Such surface waves are illustrated in row 7 of Figure 9.1. They can be associated with an erratic 'dancing' translational motion (Crum and Eller, 1969), and at high amplitude

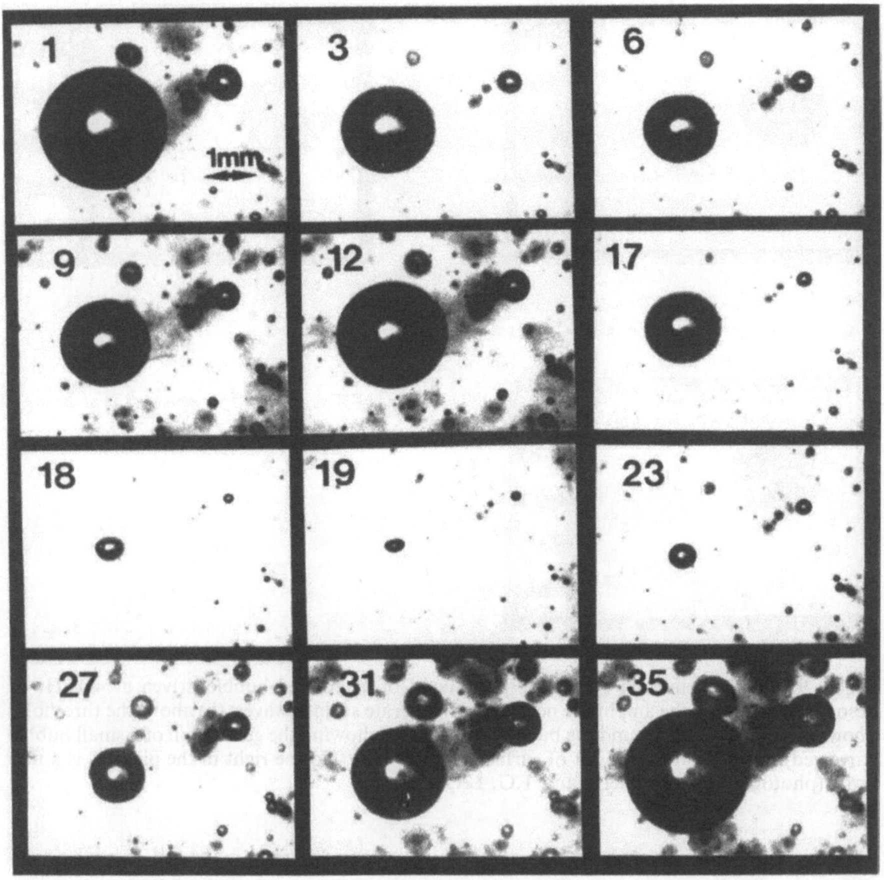

Figure 9.2 The oscillations of air bubbles in glycerol. The bubbles are driven by an inertially generated 100 Hz pressure field, of amplitude 3900 Pa, with 600 Pa static pressure. A selection of frames is shown from a sequence of 35 consecutive frames, filmed at 2000 frames per second. They illustrate the periodic unit in the motion of the largest bubble. The bubble contracts from maximum size in frame 1 to a minimum in frame 6, before expanding again to a second maximum (frame 12), then collapsing to a second minimum (frame 19), and finally expanding to reach, in frame 35, the same size as it had in frame 1. The second collapse is far more rapid than the first. After Leighton *et al.* (1990a). Reprinted by permission from the *European Journal of Physics*, vol. 11, pp. 352–358; Copyright 1990 IOP Publishing Ltd.

microbubbles might break off from the tips of the surface waves (Figure 9.1, row 8; Figure 9.3c; Leighton, 1995a). Surface waves can be associated with local circulation currents in the liquid, called microstreaming, which is a third mechanism by which an erosive effect can arise (Ahmad *et al.*, 1987).

Other departures from spherical symmetry include shape oscillations (Figure 9.1, row 9), found particularly in larger bubbles, where the tendency of surface tension to promote sphericity, is weaker (Strasberg, 1956; Longuet-Higgins, 1992a; Leighton, 1994a). If extreme, such shape oscil-

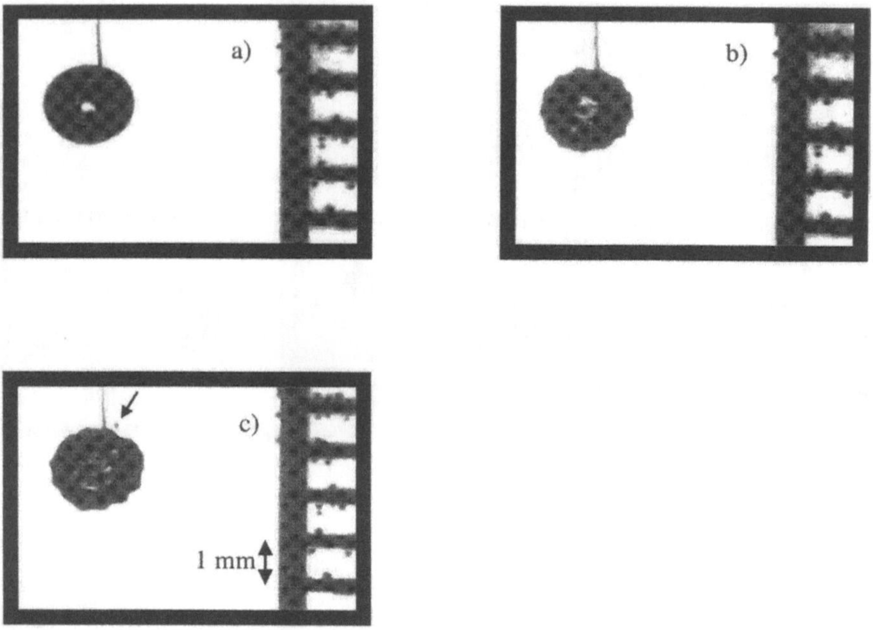

Figure 9.3 Video frames at 30 frames per second of a tethered bubble driven at 4.4 kHz at resonance (a) below the amplitude necessary to generate surface waves; (b) above the threshold, showing surface waves around the bubble wall; and (c) showing the generation of a small bubble (arrowed) pinched off as a result of surface wave activity. To the right in the pictures is a mm scale (photograph: A.D. Phelps and T.G. Leighton).

lations can break a bubble up, usually generating a small number of fragments of roughly similar size, in contrast to the shedding of many microbubbles from a larger parent bubble, shown in Figure 9.1, row 8 (Longuet-Higgins, 1992b; Leighton, 1995a; Leighton *et al.*, 1995a). Shape oscillations are encouraged by anisotropies in the local environment. Common causes for these are the presence of other bubbles, particles or walls. If such are present during bubble growth, then during the subsequent collapse the bubble may involute, one wall passing through the bubble to form a high-speed liquid jet (Plesset and Chapman, 1971; Lauterborn and Bolle, 1975; Prosperetti, 1984). This is shown in row 10 of Figure 9.1. Though the jetting of a bubble may be quite complicated, with, for example, the formation of counterjets, the net effect is that the bubble will usually fragment (Neppiras, 1980). Jetting is a fourth mechanism by which cavitation can cause mechanical damage, usually manifesting as the formation of pits in a solid body, where the liquid impact occurs.

As can be seen, the model of an isolated spherical bubble in an infinite fluid is proving inadequate to describe the range of behaviours seen: departures from spherical symmetry in both the bubble wall and the environment must be incorporated. One such environmental feature which has proved to be important is the presence of solid bodies which surround the gas pocket to a greater or lesser extent. Whilst the jetting feature, described above, is usually thought of as a modification of the spherical bubble model, a new class of gas inclusion has been introduced in the last decade, the *stabilized gas body*. These are pockets of gas partially bounded by liquid, and partially by solid structures, which can stabilize them against dissolution. They might comprise, for example, approximately cylindrical pockets of gas contained within tubular vessels, such as are found in biological structures. Examples include structures in plants (Miller, 1979), insect tracheae (Child *et al.*, 1981), and, speculatively, mammalian blood vessels and ear canals (Leighton *et al.*, 1995b) (Figure 9.1, row 11). The structures may be rigid, in which case the curved interfaces between the solid and gas would not move. However the gas/liquid interfaces which comprise the endwalls of the cylinder might oscillate in a piston-like or a membrane-like manner (Miller and Nyborg, 1983; Miller, 1984; Miller and Neppiras, 1985). Oscillations of such gas pockets may cause the nucleation of free-floating bubbles into the liquid (Figure 9.1, row 12; Section 9.5.2).

Some clear regimes are apparent in the classification scheme illustrated in Figure 9.1. One is between those behaviours which exploit both the compressibility of the gas/vapour, and the inertia of the liquid (rows 3–12), and those which do not (rows 1 and 2). Whilst obvious, the distinction between the spherical (rows 3, 4 and 6) and the aspherical (rows 5 and 7–12) is somewhat artificial in that the forms are idealized. Rather more important perhaps is the role of asymmetries in the media, which in some circumstances must be solid (rows 11 and 12), and in others may be solid or not (rows 9 and 10). One of the most important distinctions is between those which for maximal effect rely on the resonance, and those which rely on inertial cavitation. The importance of this distinction can be illustrated through reference to the common paradigm that a bubble will undertake a steady increase in equilibrium size through a process called rectified diffusion (Section 9.3), to eventually reach the radius which is resonant with the sound field. At this point it will cause maximum mechanical, sonochemical or sonoluminescence effect. This scheme has been used in some theories of pulse enhancement (Section 9.6), or cyclic cavitation (Neppiras, 1980). Whilst sometimes valid, this scenario is often used inappropriately. Consider, for example, a situation where ultrasonically induced cavitation erosion is observed (Figure 9.4). It might arise through microstreaming or bubble-mediated radiation forces (Section 9.4), both of which are maximal when the bubble reaches resonance size (rows 6 and 7). However damage by jetting is more dependent on the local geometry than the resonance (row 10); and damage through rebound pressure is dependent instead on the threshold for inertial cavitation. The source of the erosion will depend on the mechanism and the prevailing conditions. The case is

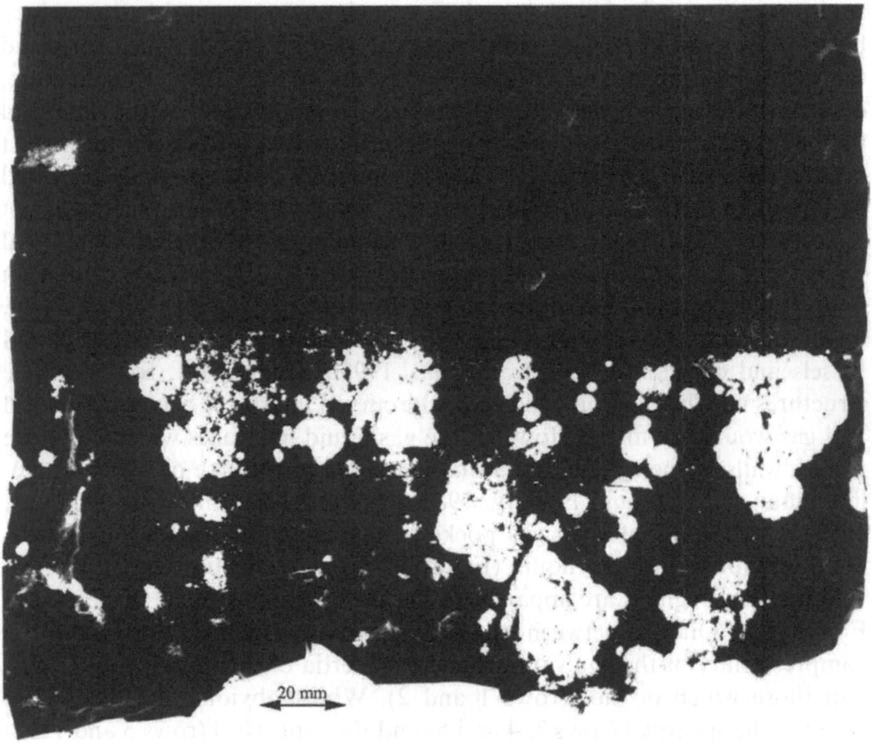

Figure 9.4 A back-lighted sample of aluminized mylar sheet, which had been subjected for around 1 minute to a cylindrical sound field in aerated water which, at the axial focus, had an acoustic pressure amplitude of 0.24 MPa. In the dark regions the aluminium has remained attached to the sheet. However in other regions it has been eroded away, such that the light shines through the transparent mylar in large, roughly circular regions (after Leighton 1994a). These can be seen to occur in the half of mylar which was submerged. The half which was not submerged is predominantly dark, the only surface removal there being a result of the mountings.

clearer for the generation of sonoluminescence and sonochemistry, which are dependent on the threshold for inertial cavitation. Therefore if the bubble is to grow by rectified diffusion to maximize a sonochemical or luminescent effect, it is likely to be the radius threshold for inertial cavitation which is critical, and not the resonance. Whether such schemes of bubble growth to critical sizes are likely when detailed bubble dynamics are incorporated into the theory is dependent on a number of parameters, not least the initial bubble size, the acoustic frequency, and the acoustic pressure amplitude (Church, 1988a). In the following sections the importance of these three parameters, and the characteristics of rectified diffusion and radiation forces on bubbles, will be expounded. Lastly, the issue of the measure of 'activity' of cavitation in an ultrasound field, and its suppression and enhancement, will be addressed.

9.3 Rectified diffusion

In many circumstances the size distribution of bubbles in a population is a key parameter in determining the acoustic effects. As introduced in the previous section, there is a critical size range in which a free-floating bubble must lie in order to undergo inertial cavitation. This will be examined further in Section 9.5.1. Similarly, the following section will detail the importance of the bubble size to the effect of acoustic radiation forces, and in turn how this can affect the cavitational activity in a population (Section 9.6). The interdependence and feedback between all these features is explored in Leighton (1995a).

The previous section also introduced the ability of a bubble to increase its equilibrium radius in an appropriate sound field through the process of *rectified* diffusion, which manifests as a net flux of dissolved gas out of solution and into the bubble. This flux is therefore counter to the direction that would be towards equilibrium were the sound field not present, since a gas bubble in a liquid will tend to dissolve owing to the excess internal gas pressure required to balance the pressure $2\sigma/R_0$ due to surface tension σ (Epstein and Plesset, 1950). Thus, in the absence of a sound field or any stabilizing mechanism, bubbles gradually dissolve (Gupta and Kumar, 1983). An appropriate stabilizing mechanism might be presented by the crevices shown in rows 11 and 12 of Figure 9.1: at equilibrium, the bubble wall is concave as seen from the liquid (row 11), as opposed to the convex meniscus of a spherical bubble, so that the pressure due to surface tension tends to stabilize the bubble against dissolution. Other mechanisms are discussed briefly in Section 9.5.1, and detailed in Chapter 2 of Leighton (1994).

In the presence of a sound field, the situation is quite different. During non-inertial cavitation, since evaporation and condensation take place so much more rapidly than the bubble dynamics, it is commonly assumed that, to first order, the vapour pressure within the bubble remains constant at the equilibrium value. However this is not so for the gas content of the bubble, a gas which will also be dissolved in the liquid. Harvey *et al.* (1944), studying the formation of bubbles in animals, suggested a mechanism by which bubbles undergoing stable cavitation in a sound field can experience a steady increase in their equilibrium radius, R_0. This inwardly directed *rectified diffusion* comes about through the active pumping of gas, initially dissolved in the liquid, into the bubble, using the energy of the sound field. There are two contributory elements to a full description of the processes, an 'area effect' and a 'shell effect'.

The area effect arises through the correlation between the direction of mass flux and the area of the bubble wall. While the bubble radius is less than equilibrium, the gas inside is at a greater pressure than the equilibrium value, and thus diffuses out into the liquid. Conversely, when the bubble radius is significantly greater than R_0, the internal gas pressure is less than the equilibrium value, and so gas diffuses from the liquid into the bubble interior. The net flow rate of the gas however, is not equal during the compressed and expanded phases of the bubble motion, because the area of the bubble wall (the transfer interface) is greater in the latter case than in the former. This process is illustrated over one acoustic cycle in Figure 9.5. Therefore, over a period of time, there will be a net influx of gas to the bubble interior.

The shell effect occurs because the diffusion rate of a gas in a liquid is proportional to the concentration gradient of the dissolved gas. As the bubble pulsates, a spherical shell of liquid surrounding the bubble will change volume, and so the concentration gradient will change. Consider the bubble shown at equilibrium in Figure 9.6(a). Two of the shells surrounding the bubble are illustrated schematically. When the bubble is expanded, each liquid shell contracts (Figure 9.6b). The concentration of dissolved gas in the liquid adjacent to the bubble wall is less than the equilibrium value (Henry's law), but the shell is thinner than when the bubble is at equilibrium radius, so that the gradient across the shell is higher. Therefore the rate of diffusion of gas towards and into the bubble is high. When the bubble is contracted, the liquid shells surrounding the bubble are expanded (Figure 9.6c). Although the concentration of gas near the bubble wall is higher than when the bubble is expanded (Henry's law), the increased thickness of the shell means that the concentration gradient is not as great as when the bubble is expanded. The two factors (gas concentration at the bubble wall, and shell thickness) work together when the bubble is expanded, but against one another when the bubble is contracted: On expansion there is a large concentration gradient driving gas a short distance, and in the second case a smaller gradient drives the gas a longer distance. The former effect is dominant.

The result is that both the bubble wall surface area and the dissolved gas diffusion rate are asymmetrical with respect to expansion and contractions:

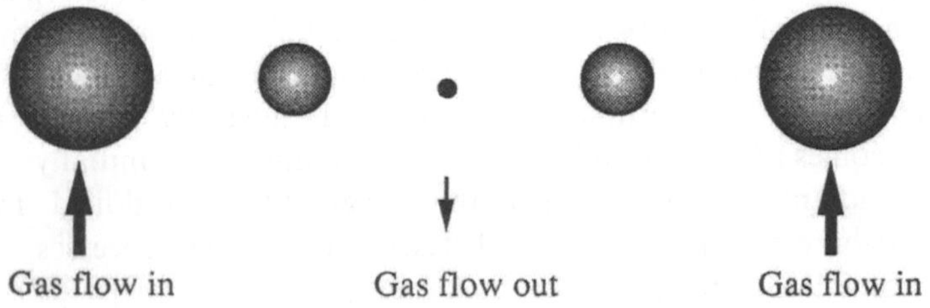

Figure 9.5 The 'area' effect. One oscillatory cycle of the bubble pulsation is illustrated, time increasing from left to right. The pressure within a pulsating bubble, and the surface area of the transfer (the bubble wall), both oscillate about an equilibrium value, causing a mass flux imbalance (after Leighton 1994a).

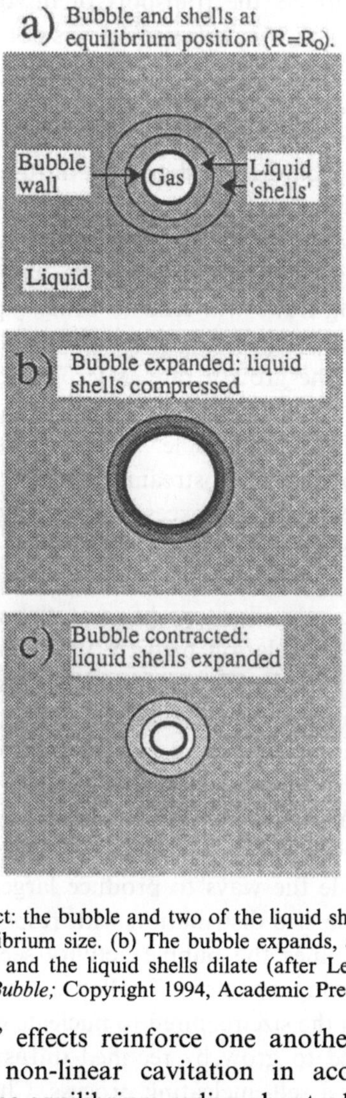

Figure 9.6 The 'shell' effect: the bubble and two of the liquid shells surrounding it are shown. (a) The bubble is at equilibrium size. (b) The bubble expands, and the liquid shells contract. (c) The bubble contracts, and the liquid shells dilate (after Leighton, 1994a). Reprinted by permission *The Acoustic Bubble;* Copyright 1994, Academic Press Ltd.

the 'area' and 'shell' effects reinforce one another. The combined effect means that during non-linear cavitation in acoustic fields which are sufficiently intense, the equilibrium radius about which the bubble pulsates will tend to increase (Eller and Flynn, 1965; Eller, 1972; Crum, 1984; Church, 1988a, b). Rectified diffusion of heat may also occur, whereby the bubble pulsation causes the establishment of thermal differentials against the gradients which would exist in the absence of a sound field (Wang, 1974).

The threshold acoustic pressure for growth is dependent on the bubble size and the acoustic frequency. Local minima in the acoustic pressure threshold are seen at harmonics and subharmonics of the resonance (Church, 1988a). The growth rate once the threshold has been exceeded is strongly influenced by the presence of surface-active agents, and by microstreaming, which affects the transport of dissolved gas beyond the bubble wall (Church, 1988b). Microstreaming was not incorporated into the above simple discussion, but its effects are qualitatively simple. As a bubble grows by rectified diffusion, the dissolved gas is taken from the liquid near the bubble. If there is no flow, then the rate at which the deficit is met depends on the rate at which dissolved gas can diffuse from regions farther out from the bubble. Since this is in general a slow process, the liquid outside the bubble wall will become depleted of dissolved gas. The resulting change in concentration gradient reduces the rate of further growth. However, microstreaming flows will tend to bring liquid from farther out close to the bubble wall. The convection of dissolved gas reduces the depletion, and increases the growth rate. Microstreaming will continually refresh the liquid at the bubble wall, giving it a dissolved gas concentration close to that found far from the bubble. The converse process is of course valid: if a bubble is dissolving, microstreaming will tend to remove from the region outside the bubble wall the excess dissolved gas concentration, so increasing the rate of dissolution.

Within the dynamics of a bubble population, rectified diffusion of gas may have a considerable role. The type of cavitational behaviour a bubble undertakes depends on the relation between its size and the other critical sizes (Figure 9.1). These include: the radius which is resonant with the sound field (governing, for example, radiation force effects, surface wave activity and microbubble shedding, etc.); the upper and lower limits of the radius range for the nucleation of inertial cavitation; and the threshold for rectified diffusion. Dissolution and fragmentation provide mechanisms by which bubble size reductions in the population can occur. Coalescence and rectified diffusion provide the ways to produce larger bubbles. How such changes affect the bubble size distribution with respect to the critical sizes mentioned above determines the bubble activity seen. For example, if a bubble fragments following an inertial collapse, and the resulting small bubbles are smaller than the size required to nucleate inertial cavitation, but larger than that required to grow by rectified diffusion, such growth can make inertial cavitation a self-nucleating process. Church (1988a) outlines some of these possible scenarios, and assesses their likelihood.

The two phenomena mentioned above, coalescence and bubble fragmentation, are dependent on inter-bubble spacing. For coalescence, the reasons are obvious. In fragmentation, the presence of a neighbouring bubble can excite the shape oscillations shown in row 9 of Figure 9.1 which, if sufficiently pronounced, can lead to bubble fragmentation (Leighton 1995a; Leighton *et al.*, 1995a). In an intense, continuous-wave or tone-burst acoustic field, among the most potent factors determining the inter-bubble spacing are the acoustic radiation forces: these are discussed in the following section.

9.4 Acoustic radiation forces on bubbles

In an accelerating liquid a gas bubble, being less dense than the surrounding fluid, will accelerate in the same direction as the surrounding liquid, but to a greater degree. That degree is governed by the magnitude of the difference in density between gas and liquid or, equivalently (assuming the bubble contains a fixed mass of gas), the instantaneous bubble volume. As an acoustic field passes through a liquid, the liquid particle acceleration oscillates, reversing direction periodically such that it is aligned with the direction of propagation for half the acoustic cycle, and contrary to it for the remainder. The bubble, being less dense than the liquid, will follow suit, accelerating in the same direction as the continually alternating liquid particle acceleration. However, if the bubble is pulsating with the same periodicity as the driving field (which a linear oscillator will do), then the phase relation is such that the bubble will be in the expansion phase (i.e. with its radius greater than equilibrium) as it travels in one direction, and in the compression phase as it travels in the other direction (Leighton *et al.*, 1990b). The bubble accelerates more during the expansion phase than the compression, since its volume is clearly greater, and its density less, during expansion than during the compression. Therefore such a bubble in a sound field will experience forces which reverse direction two times per acoustic cycle. However the net effect will be that it travels in the direction taken by the liquid acceleration when the bubble is in the expansion half-cycle. If the bubble is assumed to be an oscillator with a single degree of freedom, bubbles of less than resonance size will pulsate in antiphase to those of greater than resonance size. Therefore if one type of bubble is accelerated in one direction by the sound field, the other type will be forced in the opposite direction.

This behaviour is most readily observed in a standing wave field, where the radiation forces are commonly called ‘primary Bjerknes forces’. They cause bubbles of less than resonance size to travel up pressure gradients to collect at acoustic pressure antinodes, whilst bubbles of larger than resonance size migrate down the gradients to the nodes.

Similar comments apply in focused acoustic fields, where bubbles of less than resonance size migrate up pressure gradients towards the focus. To produce a bubble which is larger than resonance, it must either pre-exist (and continue to persist during insonation despite buoyancy and possible fragmentation); form through coalescence; or grow to a size larger than resonance by rectified diffusion. Acting against the latter scenario is the fact that it is on reaching resonance size that the bubble is most likely to lose gas through microbubble emission from surface waves (Figure 9.1, row 8). Also, if the bubble does pass through resonance intact, then rectified diffusion becomes far less efficient once a bubble exceeds resonance size, the pressure threshold increasing and the growth rate decreasing. There is also the general trend that the larger the bubble, the easier it is to excite shape oscillations (Figure 9.1, row 9) which can lead to fragmentation. Therefore in most fields, it is the aggregation of the smaller bubbles at the antinodes or the focus which is more commonly observed (Figure 9.7).

Formulations also exist for the radiation force on a bubble in travelling-wave conditions, where the force is greatest on resonant bubbles; and for the force on a particle suspended in a liquid close to a bubble (Coakley and Nyborg, 1978), an example of the latter being the aggregation of platelets in blood (Miller *et al.*, 1979). The so-called 'secondary' or 'mutual' Bjerknes force is exerted between two pulsating bubbles (Bjerknes, 1906, 1909, 1930; Prandtl, 1954; Batchelor, 1967; Leighton 1994a). There is a general rule that bubbles which are both less than, or both greater than, resonance size attract; but that if one bubble is greater than, and the other less than, resonance size, they repel. However, this is a simplification, and the results may be more complicated if the bubble population density is high or the incident sound field is strong. Such attractions can influence the size of individual bubbles by causing coalescence or, through the surface waves induced on one bubble by the proximity of its neighbour, fragmentation. If the bubble is oscillating close to a boundary (e.g. solid or free surface), migrations may be influenced through a secondary Bjerknes interaction between the bubble and its 'acoustic image', the wall acting as a form of

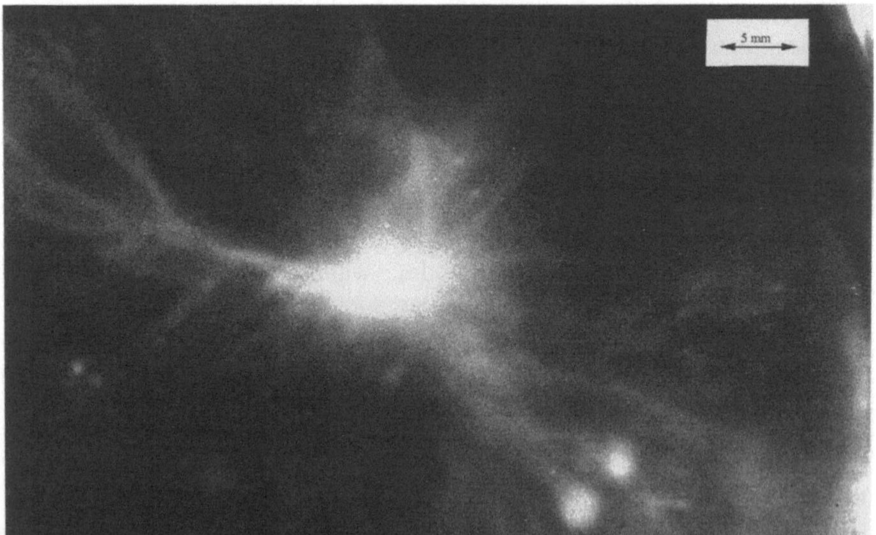

Figure 9.7 A photograph of aerated water cavitating in a cylindrically focused 10 kHz sound field, viewed along the line of the axial focus, where the acoustic pressure amplitude is 0.24 MPa. Exposure time 1/30th second. Streamers are clearly visible, comprising bubbles moving rapidly towards the focus, driven by radiation forces.

acoustic 'mirror' (Leighton, 1994a). Crevices in solids may attract bubbles, the subsequent expansion of which can extend the crevice (Brunton and Camus, 1970; Leighton, 1994a).

9.5 The thresholds between inertial and non-inertial cavitation, and the importance of stable gas bodies

Inertial cavitation is a threshold phenomenon, and whether or not it occurs depends critically on a number of parameters. The acoustic frequency is one, and this can be simply controlled [although the frequency content of an acoustic wave may change during propagation as a consequence of nonlinear effects (associated with, for example, the propagation of finite amplitude waveforms or parametric effects involving bubbles)]. The acoustic pressure amplitude of the sound field *at the bubble* is another critical threshold parameter. While the operator can readily determine the vibration of the transducer faceplate, and the acoustic field at the position of a monitor hydrophone, it is not so simple to control the field amplitude at any individual bubble. The third key parameter, the initial bubble size, is often the least controlled.

The bubble size and acoustic pressure amplitude are able to influence one another. Examples of ways in which the field can affect the size distribution were discussed in the previous sections (rectified diffusion; coalescence and fragmentation, possibly influenced by radiation forces). Other mechanisms for this, and for the effect of the population on the field, are key to the suppression or enhancement of cavitational activity.

This section therefore details the cavitation threshold, which determines the number of inertial cavitation events. Section 9.6 then describes how the 'activity' of the cavitation is a function of both the number of collapses and the ability of each to generate the effect in question; and how the suppression or enhancement of cavitational activity can be governed by the influence of the factors discussed in the preceding sections on the collapse dynamics outlined in this section.

9.5.1 The nucleation of inertial cavitation

The parameter which is often subject to least control is the initial size of the bubble. If it is too small, then surface tension forces prevent the initial sudden growth, and inertial cavitation does not occur. If it is too large, then the bubble may grow, but be too 'sluggish' to concentrate the energy sufficiently on collapse to generate free radicals, etc. There is therefore a critical size range in which, for a given sound field, the initial size of the bubble must fall if it is to nucleate inertial cavitation (Flynn, 1964; Flynn and Church, 1984; Holland and Apfel, 1989; Leighton, 1994a). The lower the frequency, the wider this range.

This is clearly shown in Figure 9.8, where the transition threshold between inertial and non-inertial cavitation is plotted, based on calculations by Apfel and Holland. They assumed that, in response to a single cycle of ultrasound, a bubble which is spherical at all times would grow and, on subsequent adiabatic collapse, the gas within the bubble should attain a temperature of at least 5000 K if the collapse is to be 'inertial' (Apfel, 1981; Apfel and Holland, 1991). Though there are clear approximations and the choice of such a criterion for defining inertial cavitation is not fundamental, this is nevertheless an extremely useful calculation, and illustrates that the acoustic pressure amplitude required to cause a bubble to undergo inertial cavitation is dependent upon the initial radius of that bubble. Since in most applications the frequency is the easiest of these three parameters to control, followed by the acoustic pressure amplitude at *the bubble,* with the radii of the nuclei present being the least accessible, then the graph can be interpreted in another manner. At a fixed frequency, say 10 MHz, an ultrasonic cycle with a peak negative pressure of 1.5 MPa (assumed to be constant throughout the field) will only generate inertial cavitation within a water sample if, according to this model, it contains bubbles having radii between 0.03 and 0.77 μm. As the pressure amplitude decreases, so does the range of bubble sizes which can nucleate inertial cavitation. The very lowest peak negative pressure which could give rise to inertial cavitation, according to this model, is P_{opt} (around 0.84 MPa at 10 MHz): and at this pressure

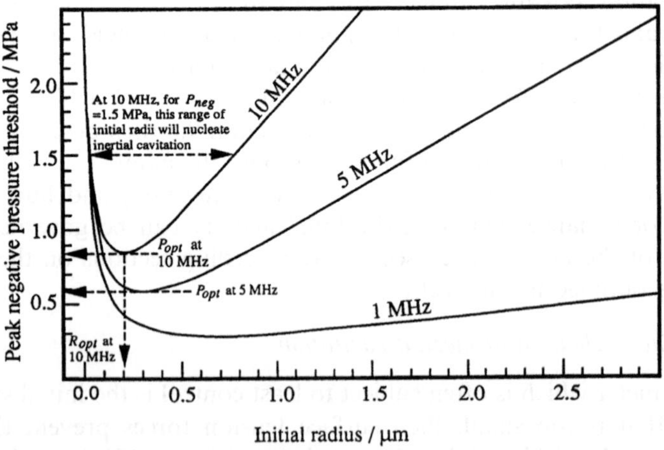

Figure 9.8 The threshold for inertial cavitation, as predicted by the theory of Apfel and Holland. For each frequency a line can be plotted: if the conditions of peak negative pressure and of the initial bubble radius are such that the point of interest on the graph lies below the line, non-inertial cavitation will occur. If the point of interest is above the line, inertial cavitation will occur (after Apfel and Holland, 1991). Reprinted by permission of Elsevier Science, Inc. from Gauging the liklehood of cavitation from short-pulse, low-duty cycle diagnostic ultrasound by Apfel R.E. and Holland C.K. *Ultrasound in Medicine and Biology,* Vol. 17, No. 2, pp. 179–185. Copyright 1991 by World Federation of Ultrasound in Medicine and Biology.

only bubble of a radius R_{opt} (0.2 μm at 10 MHz) could possibly nucleate inertial cavitation.

The theory assumes nucleation of inertial cavitation within the first acoustic cycle, so-called 'prompt cavitation', from a free-floating spherical bubble nucleus. In practice of course it is not a requirement that such nuclei be present before the start of insonation. Not only may bubbles which are initially too large to nucleate inertial cavitation enter the critical range (through, for example, dissolution, or fragmentation through a shape oscillation or microbubble shedding, as discussed in Section 9.3): bubbles too small to nucleate cavitation may enter it through rectified diffusion or coalescence. As discussed in Section 9.4, radiation forces can affect both size increases and size reductions. In Section 9.6 it will be shown how radiation forces in focused fields can convect suitable nuclei into the focus to nucleate inertial cavitation there.

The model employed to produce Figure 9.8 is based upon the dynamics of isolated, spherical bubbles, which *de facto* must be free floating. Other possible nuclei can be found naturally as gas pockets, stabilized against dissolution in crevices and cracks in the container wall or within free-floating particles within the liquid (Harvey *et al.*, 1944; Trevena, 1987; Atchley and Prosperetti, 1989; Leighton, 1994a). If inertial cavitation is undesirable, then such particles can be removed by filtering. However even this will not completely remove all suitable nuclei for inertial cavitation, which may be generated, for example, by the passage of cosmic rays through the sample (Greenspan and Tscheigg, 1967).

Crevice nucleation is illustrated in row 12 of Figure 9.1. High amplitude ultrasonic waves cause the gas pockets to either expand out of their crevice, or conceivably shed microbubbles through surface waves, to generate free-floating nuclei for cavitation. Such crevices are not the only mechanism through which a gas nucleus may be stabilized against dissolution: a hydrophobic 'skin' can collect on the bubble wall and cause free-floating nuclei to persist over long periods (Akulichev, 1966; Sirotyuk, 1970; Yount, 1979, 1982; Yount *et al.*, 1984).

The theory of Apfel and Holland has given rise to a measure called the 'mechanical index', which may be used to indicate the likelihood of exceeding the threshold required to nucleate cavitation.

9.5.2 *The mechanical index*

The model of Holland and Apfel investigates the threshold condition required to generate, within one acoustic cycle from the start of insonation, a collapse temperature of 5000 K. The threshold it predicts is relatively insensitive to the precise temperature criterion, and Holland and Apfel were able to compare their threshold predictions with those of Flynn and Church (1988), who used a different definition (based on the bubble achieving on expansion a maximum radius of at least twice its initial radius) for the onset of inertial cavitation, equivalent to the bubble achieving a maximum temperature of 960 K. A threshold for sonoluminescence of 1550 K has also been proposed (Sponer, 1990, 1991; Sponer *et al.*, 1990).

Apfel and Holland (1991) derive a *mechanical index*, which represents the likelihood that inertial cavitation will be nucleated, applicable (because of the assumption of 'prompt' cavitation) to microbubble growth in the limit of short-pulse, low duty cycle insonation. Clearly effects relating to longer insonation periods, such as growth by rectified diffusion, are not covered. However the index can be used to gauge the probability of inertial cavitation resulting from diagnostic ultrasound fields.

As described for Figure 9.8, the model demonstrates that for a given acoustic frequency there is a minimum in the curve. If the whole range of bubble size classes are present, then there is a specific initial bubble size for which the threshold acoustic pressure required to nucleate inertial cavitation is a minimum. As the insonation frequency increases, the bubble radius which requires minimum pressure to nucleate inertial cavitation decreases, since inertial and viscous forces increase with increasing frequency, and there is insufficient time to bring about the required amount of bubble growth. For the same reason the acoustic pressure required to nucleate inertial cavitation in all but the smallest bubbles increases with increasing frequency. Surface tension dominates the response of the smallest bubbles.

If one is interested in a worst-case assessment of the likelihood that inertial cavitation will occur when a liquid is insonated, clearly one must assume that the bubble population contains bubbles at the radius corresponding to the minimum in the threshold curve. At a given frequency, it is bubbles of this radius, R_{opt}, which the analysis predicts will require the smallest peak negative pressure, P_{opt}, to undergo prompt inertial cavitation in response to a single acoustic cycle. Apfel and Holland (1991) generate a plot of P_{opt} against frequency for water and whole blood, using pure fluid bulk property values for the σ, ρ and viscosity η relevant to the two fluids. The liquids are assumed to contain the relevant nuclei at size R_{opt}. Apfel and Holland employ a two-parameter least-squares fit to these plots in order to obtain a relationship between P_{opt} and the insonation frequency $f = \omega/2\pi$. They find a least-squares fit of:

$$\frac{(P_{opt})^a}{f} = b \tag{9.2}$$

where if P_{opt} is measured in MPa and f in MHz, the constant a takes values of 2.10 for water and 1.67 for blood, and b has values 0.06 for water and 0.13 for blood. For a given sound field with a maximum negative pressure of

P_{neg}, then by taking a value of $a \approx 2$ to approximate the appropriate physiologically relevant liquid, a mechanical index MI can be defined for the sound in that liquid

$$MI = \frac{(P_{neg}/\mathrm{MPa})}{\sqrt{f/\mathrm{MHz}}} \tag{9.3}$$

The mechanical index, *MI*, for prompt cavitation represents an approximate measure of the worst-case likelihood of nucleating inertial cavitation. As such it can be used to estimate the potential for nucleating inertial cavitation resulting from insonation by diagnostic ultrasound. Clearly it would be less appropriate to apply this index to tone-burst or continuous-wave insonations, during which a range of complicated processes, including rectified diffusion and enhancement of the cavitation by ultrasonic pulsing, can occur (Leighton, 1994a). Holland and Apfel recommend that the pulse length should not exceed 10 cycles, nor the duty cycle 1:100. The peak negative pressure output from the device, as measured in water, must be deaerated to give the appropriate peak negative pressure that would be attained *in vivo* at the location of the maximum pulse intensity integral (AIUM/NEMA, 1992). The centre frequency is used for f which, for accuracy, is expected to be of the order MHz. Apfel and Holland (1991) suggest that a mechanical index value below $\sqrt{0.5} \approx 0.7$ would indicate that, even in the presence of a broad size distribution of nuclei, the conditions are not sufficient to allow significant bubble expansion. If $MI \geq \sqrt{0.5}$, Apfel and Holland suggest that 'the user should be advised of the potential for bubble activity'. The AIUM, NEMA and FDA have adopted the mechanical index as a real-time output display to estimate the potential for cavitation *in vivo* during diagnostic ultrasound scanning (AIUM/NEMA, 1992).

Key points should be noted. First, the mechanical index gauges the likelihood of prompt cavitation, and nothing more: the effect of interest (e.g. a bioeffect) may be related to some other mechanism. Second, the model for the index is based on the assumption of a free-floating spherical nucleus of optimum size. In certain circumstances it may be that the nucleus is of a different type. Third, the effects of non-linear propagation are not included, and hence the index might underestimate conditions *in situ*. Fourth, when applied to diagnostic ultrasound instruments, the mechanical index describes conditions only at the focus, which is not necessarily the point of interest. Fifth, the mechanical index has arisen from a theory which gives smooth curves of the form shown in Figure 9.8: such curves may in fact show peaks when other effects are incorporated (Roy, 1996).

Lastly, the underlying theory is applied to derive the mechanical index in a way intended to elucidate the conditions which attain the threshold for nucleating inertial cavitation. The amount by which the mechanical index is exceeded is therefore only a guide to the degree of cavitational activity, and by no means an exact predictor. Consider the MHz range illustrated in Figure 9.8, where the range of nuclei size which can seed inertial cavitation is relatively narrow. In such a sound field of fixed frequency and increasing acoustic pressure, as one exceeds the threshold in Figure 9.8, the range of nuclei which may nucleate inertial cavitation increases. In a field containing a broad range of bubble sizes, with a uniform number of bubbles in each size class, the total number of nucleated events would be expected to increase. However if there is only a narrow distribution of bubble sizes, exceeding the threshold by increasing amounts in this manner would, to first order, have little effect on the *number* of inertial events which are nucleated. What might be expected to increase is the energy associated with each collapse. Similarly as, say, the acoustic frequency changes, it is no simple matter to predict how the mechanical index might correlate with the amount of cavitational 'activity' observed.

9.6 The 'activity' of cavitation

Whether a particular type of cavitation is desirable or undesirable in a given circumstance, or whether or not it is amenable to control, if cavitation does occur then the question of degree becomes important.

Consider inertial cavitation in a sound field which may show significant spatial variations. The relevant acoustic conditions, usually acoustic pressure and frequency, may be such that at any point within the sample there is theoretically no possibility of inertial cavitation occurring. To put it another way, at the frequency in question, the magnitude of the peak negative acoustic pressure is everywhere less than the P_{opt} calculated from the limits of the model described in the preceding section (Apfel and Holland, 1991). If however the peak negative acoustic pressure were to be greater or equal to P_{opt} at some point in the field, then inertial cavitation will only occur if suitable nuclei (i.e. having radii within the appropriate range) were present at that location. If such nuclei were not initially present, they may migrate to the point in question, perhaps through Bjerknes forces (Section 9.4) or convection by streaming, etc. (Madanshetty *et al.*, 1991). Alternatively, nuclei of a suitable size might be generated by the ultrasound in the region of interest, through for example, fragmentation of larger bubbles or the presence of suitable suspended agents (Madanshetty, 1995). Such considerations clearly become more important as the frequency becomes greater, because not only does the acoustic pressure threshold for inertial cavitation increase, but the range of allowable radii which a bubble must have to nucleate an inertial event becomes narrower (as illustrated in Figure 9.8 for the low MHz range).

Therefore knowledge of the sound field and the nuclei distribution can enable predictions of whether or not cavitation will occur, to within the accuracy of the model. Such threshold predictions are based on whether a given bubble will undergo inertial cavitation. However what is more difficult to predict is the 'amount of cavitation' that will occur if the threshold is exceeded, since this is clearly not amenable to models of single bubbles. This is partly because in referring to 'the amount of cavitation' what is actually observed is the magnitude of some effect which is brought about through cavitation, inertial or otherwise. The relationship between the 'amount of cavitation' and the ambient pressure is a case in point.

Inertial cavitation, as described in Section 9.5.2, involves two key phases: extensive initial growth, followed by a sudden, rapid collapse. Assuming that a bubble is to undergo inertial cavitation, increasing the static pressure will tend to affect a single bubble's inertial cavitation event in two ways (Walton and Reynolds, 1984): First, it makes it more difficult for the bubble to expand; and second, it increases the compressive forces that drive the bubble to collapse. The first would tend to decrease, and the second to increase, the energetics of the cavitation undergone by a single bubble. Whether the magnitude of the result (erosion, sonoluminescence, etc.) generated by the inertial cavitation of this bubble, and subsequently interpreted in terms of the energetics of the cavitation, increases or decreases depends on which effect has the greater influence. However there is a second factor to consider. The preceding discussion has been limited to the effect on a single bubble, which is in practice rarely the issue. In addition to altering the energy of each individual collapse, changes in the ambient pressure will alter the range of bubble nuclei that may seed inertial cavitation, and so affect the total number of collapses. It does this, not only through physically changing the size distribution of all bubbles present in the liquid, but also by altering the forms of the curves shown in Figure 9.8: in general an increase in static pressure would make the minima in the curves sharper, reducing the range of bubbles in the population that can nucleate inertial cavitation, primarily through inhibiting the growth phase.

Figure 9.9 shows the sonoluminescence measurements of Chendke and Fogler (1983) at $f = 20\,\text{kHz}$ for (a) nitrogen-saturated water, and (b) a saturated solution of water in carbon tetrachloride, containing dissolved nitrogen, for hydraulically increased pressures. The general trend is that small increases in static pressure (up to 7 bar) increase the violence of each collapse, which offsets the reduction in their number: as the pressure is increased further, the latter effect dominates, until the cavitation is suppressed. It has, as an aside, long been known by submariners that submerging the vessel will tend to reduce the noise emitted by flow-induced inertial cavitation at the propellers. However, when the cavitation is strong and the vessel is at high speed, increasing the depth of the vessel will initially cause an increase in the cavitation noise, before suppression occurs (Urick, 1983). This so-called 'anomalous depth effect' is due to the fact that, before the increasing pressure suppresses the growth phase, it first increases the violence of each collapse.

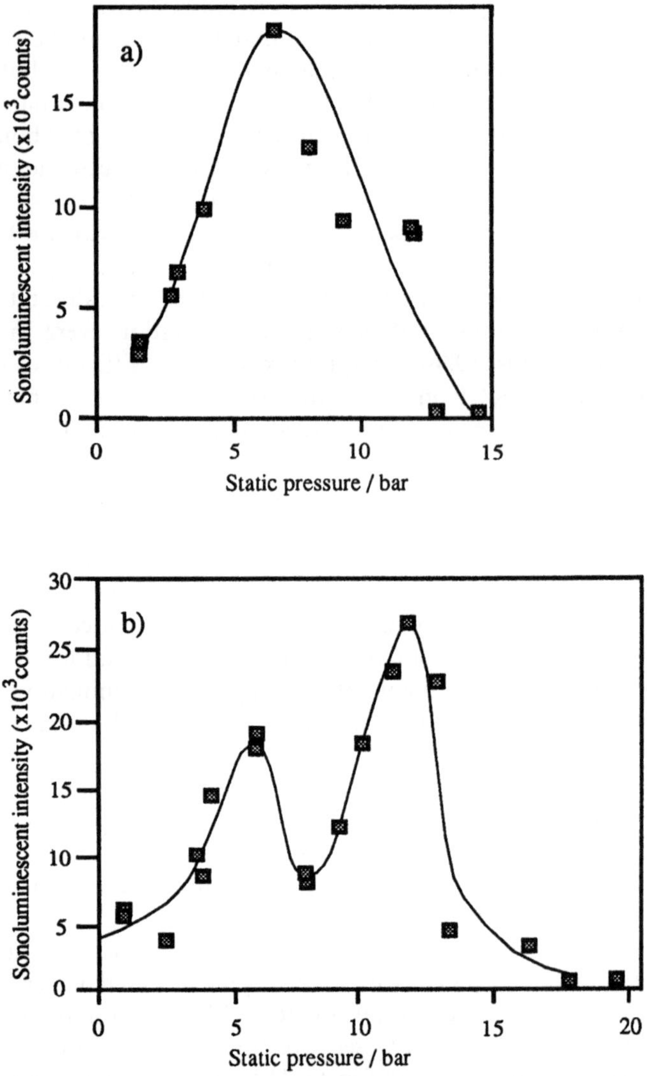

Figure 9.9 Variation of sonoluminescence with static pressure (hydraulically increased) at acoustic frequency of 20 kHz for (a) nitrogen-saturated water, and (b) a saturated solution of water in carbon tetrachloride, containing dissolved nitrogen (after Chendke and Fogler, 1983). Reprinted by permission *The Journal of Physical Chemistry,* 1983, vol. 87, pp. 1362; Copyright 1983 by the American Chemical Society and reprinted by permission of the copyright owner.

In Figure 9.9 the static pressure was increased hydraulically. If the pressure is increased by increasing a gas head, it will tend to increase the concentration of dissolved gas in the liquid. This will affect the cavitation further, as can be seen when the effect of increasing the gas concentration dissolved in the liquid (which can, of course, be achieved without increasing the ambient pressure) is considered in isolation. When the bubble expands during the growth phase of inertial cavitation, gas previously dissolved in the liquid will exsolve into the bubble, in a way similar to the first half-cycle of rectified diffusion in a bubble undergoing non-inertial cavitation. The greater the concentration of dissolved gas, the greater the exsolution for a given degree of expansion. This exsolution decreases the pressure reduction within the bubble which occurs just prior to collapse. The bubble volume change during the inertial collapse is so rapid that little gas re-dissolves, and the presence of a permanent gas component within the bubble will tend to 'cushion' the collapse. Therefore dissolving a gas such as carbon dioxide, which will readily come out of solution during the growth phase, will again cause a reduction in the violence of individual collapses. This is a technique often used to reduce the effects of inertial cavitation. (Clearly changing the gas content of the bubble will have other effects too, changing the acoustic, thermal and chemical properties of the gas content, although such effects will not be considered here.) However, although each individual collapse may be cushioned, the concentration of available nuclei is likely to have increased, increasing the total number of collapses. Similarly, though degassing the liquid will tend to increase the violence of each individual collapse, there will be fewer appropriate bubble nuclei, and therefore probably fewer inertial collapses (Flynn, 1964). The activity of inertial cavitation can also be manipulated through a variety of techniques which remedy the self-suppression effect that cavitation may generate in some modes of operation, as will now be discussed.

There is a range of mechanisms by which the suppression of cavitation is a by-product of the 'normal' method of insonating (e.g. fixed frequency, stationary transducer, etc.). In such circumstances it is therefore possible to enhance the cavitational activity by 'tweaking' the usual conditions. A simple example can be seen from extrapolation of the discussion of the primary Bjerknes forces on bubbles in standing wave fields (Section 9.4), where bubbles of smaller than resonance size migrate to the acoustic pressure antinode; and those of larger than resonance, to the node. The pattern of sonoluminescence from such a standing-wave field is shown in Figure 9.10. Hydrophone measurements in such fields show that the luminescence is clustered at the pressure antinodes, whilst the nodes are dark (Leighton *et al.*, 1988). In an almost identical field, Pickworth *et al.* (1989) demonstrated that the pressure antinodes were the dominant locations for cell lysis. This spatial distribution for cavitational effects is not surprising, since the larger bubbles are less common (Section 9.4), and tend to lie above the upper radius limit in the allowable size range for the nucleation of inertial cavitation (Section 9.5). The existence of regions of high, and of low, activity may be useful, for example with focused ultrasonic therapy (ter Haar *et al.*, 1991). However if the requirement is for uniform treatment by cavitation of a sample, then gradients in the acoustic pressure amplitude

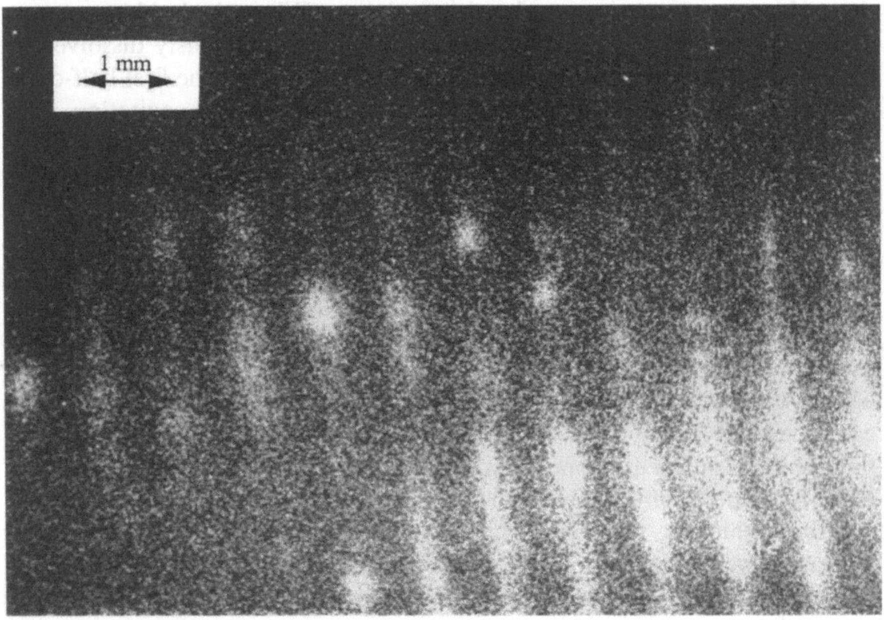

Figure 9.10 Bands of sonoluminescence generated by bubbles aggregating at the pressure antinodes in a standing wave field generated in water from a 1 MHz physiotherapeutic transducer generating a spatial-average acoustic intensity of $3\,W/cm^2$. The distance between centres of the bands of light is 0.75 mm, i.e. half an acoustic wavelength at 1 MHz in water (Image intensified photograph: T.G. Leighton and M.J.W. Pickworth).

(caused by reflections from the container walls, near-field effects, etc.) can lead to unwanted regions of inactivity (Walton and Reynolds, 1984). This problem can be reduced through a number of techniques which involve 'tweaking' the sound field by, for example, changing the ultrasonic frequency (Leighton, 1995a) or rocking the transducer (Leighton *et al.*, 1988).

Aggregations of bubbles may hinder sample treatment in other ways. If the transducer power is too great intense cavitation close to the transducer can prevent sound from reaching the bulk of the sample (Walton and Reynolds, 1984). The bubble clusters act as acoustic shields. This is the reason for some observations that an increase in power supplied to a continuous-wave transducer can result in a decrease in yield.

Acoustic shielding by bubble aggregations can also occur in continuous-wave standing-wave fields such as those described above (Leighton *et al.*, 1989a). Such aggregations may disperse if the sound is applied in tone-burst pulses. Though delivery of the acoustic energy in such pulses gives a lower time-average acoustic intensity than if a continuous-wave sound field of the same geometry and acoustic pressure amplitude were used, nevertheless there have been numerous reports of such 'pulse enhancement' of cavitation. These, and the mechanisms proposed, are reviewed by Leighton (1994a), as is the related phenomenon of the enhancement of a cavitational effect when the liquid sample is rotated in the ultrasound beam. In addition to shielding, other mechanisms by which sample rotation and acoustic pulsing may cause more thorough treatment of the sample may involve: the rise-time, saturation, and persistence of the 'activity' of the system (Henglein and Gutierrez, 1986; Leighton *et al.*, 1989a); the repeated cycling of a bubble through resonance size as it grows by rectified diffusion and subsequently dissolves when the sound is pulsed (Hill *et al.*, 1969; Clarke and Hill, 1970); the survival of unstabilized nuclei to seed further cavitation events (Ciaravino *et al.*, 1981; Flynn and Church, 1984); excitations associated with the start of insonation (Leighton *et al.*, 1989b); and with the local degassing of liquid (Pickworth *et al.*, 1988; Leighton *et al.*, 1989a). The role of the high-speed translations of the bubble through a sample was introduced in Section 9.2. Rotation of the sample will allow these bubbles to pass, not once, but repeatedly through the sample, enhancing their effectiveness (Miller and Williams, 1989; Miller *et al.*, 1991).

Finally, the relative importance of nucleation and activity to inertial cavitation needs clarifying. If, for example, no free-floating spherical nuclei exist, then one must look to stabilized gas bodies to seed cavitation (Section 9.5.1). If the threshold for the seeding from gas bodies is greater than the mechanical index for the free-floating bubbles generated from such gas bodies, then the key threshold to be exceeded to generate inertial cavitation is that seeding threshold, not the mechanical index. Once seeding has occurred, cavitation will occur; and once cavitation has occurred, it will through bubble fragmentation generate the free-floating nuclei for the next generation of inertial cavitation (providing off-times in the sound field are not too long). Indeed some workers have described the observation of two distinct thresholds: a 'nucleation' threshold, which indicates when cavitation first starts; and an 'activity' threshold, which indicates the onset of enhanced cavitational activity (Calabrese, 1996). For many industrial applications of power ultrasound, it is likely to be the latter, and not the former, which is critical.

9.7 Conclusions

There is a wide range of possible types of cavitation, many of which may occur simultaneously and interact in a power ultrasound field. Even though the effect of primary concern may be brought about through one form of cavitation (e.g. inertial), it is important to appreciate the role of other types since these may critically influence the main event by affecting either the sound field or the bubble size distribution.

When the effects of ultrasonically induced inertial cavitation are important, not only must the energetics of the individual collapses be considered, but also the number of collapses, features which may be interdependent.

Formulations are available to enable approximate calculation of the threshold conditions, particularly acoustic pressure amplitude, nuclei availability, and acoustic frequency. Once cavitation has commenced, the latter two may influence one another through feedback.

The amount of inertial cavitation 'activity' depends on both the energetics of each collapse and on the number of collapses. However it is far easier to monitor the *effects* of inertial cavitation (mechanical, chemical, biological), and these are in most practical situations the issue of prime importance.

9.8 References

Ahmad M., Pitt Ford T.R. and Crum L.A. (1987) Ultrasonic debridement of root canals: Acoustic streaming and its possible role. *J. Endodont.*, **13**, 490–9.

Akulichev V.A. (1966) Hydration of ions and the cavitation resistance of water. *Sov. Phys. Acoust.*, **12**, 144–9.

American Institute of Ultrasound in Medicine (1992) *Standards for Real-time Display of Thermal and Mechanical Acoustic Indices on Diagnostic Ultrasound Equipment*, Laurel, MN: AIUM.

Apfel R.E. (1981) Acoustic cavitation. In *Methods in Experimental Physics*, Editor P.D. Edmonds, Academic Press, New York, Vol. 19, pp. 355–413.

Apfel R.E. and Holland C.K. (1991) Gauging the likelihood of cavitation from short-pulse, low-duty cycle diagnostic ultrasound. *Ultrasound Med. Biol.*, **17**, 179–85.

Atchley A.A. and Prosperetti A. (1989) The crevice model of bubble nucleation. *J. Acoust. Soc. Am.*, **86**, 1065–84.

Barber B.P., Hiller R., Arisaka K., Fetterman H. and Putterman S. (1992) Resolving the picosecond characteristics of synchronous sonoluminescence, *J. Acoust. Soc. Am.*, **91**, 3061–3.

Batchelor G.K. (1967) *An Introduction to Fluid Dynamics*, Cambridge University Press, Cambridge, pp. 452–55.

Bjerknes V.F.J. (1906) *Fields of Force*, Columbia University Press, New York.

Bjerknes V.F.J. (1909) *Die Kraftfelder* (Brunswick), No. 28 of the series Die Wissenschaft.

Bjerknes (1930), Zeitschr. f. d. phys. u. chem. *Unterricht*, **43**, 1.

Blake F.G. (1949) Technical Memo No. 12, Acoustics Research Laboratory, Harvard University, Cambridge, MA.

Brunton J.H. and Camus J.J. (1970) The application of high speed photography to the analysis of flow in cavitation and drop impact studies, in *Proceedings of the 9th International Congress on High Speed Photography*, Denver, CO, 2–7 August 1970.

Calabrese A.M. (1996) Threshold measurements and production rates for inertial cavitation due to pulsed, megahertz-frequency ultrasound. PhD dissertation, University of Mississippi.

Chendke P.K. and Fogler H.S. (1983) Sonoluminescence and sonochemical reactions of aqueous carbon-tetrachloride solutions. *J. Phys. Chem.*, **87**, 1362.

Child S.Z., Carstensen E.L. and Lam S.K. (1981) Effects of ultrasound on *Drosophila* – III. Exposure of larvae to low-temporal-average-intensity, pulsed irradiation. *Ultrasound Med. Biol.*, **7**, 167–73.

Church C.C. (1988a) Prediction of rectified diffusion during nonlinear bubble pulsations at biomedical frequencies. *J. Acoust. Soc. Am.*, **83**, 2210–17.

Church C.C. (1988b) A method to account for acoustic microstreaming when predicting bubble growth rates produced by rectified diffusion. *J. Acoust. Soc. Am.*, **84**, 1758–64.

Ciaravino V., Flynn H.G, Miller M.W. and Carstensen E.L. (1981) Pulsed enhancement of acoustic cavitation: a postulated model. *Ultrasound Med. Biol.*, **7**, 159–66.

Clarke P.R. and Hill C.R. (1970) Physical and chemical aspects of ultrasonic disruption of cells. *J. Acoust. Soc. Am.*, **50**, 649–53.

Coakley W.T. and Nyborg W.L. (1978) Chapter II: Cavitation; dynamics of gas bubbles; applications, in *Ultrasound: Its Applications on Medicine and Biology*, Editor F. Fry, Elsevier, New York, Part 1, pp. 77–159.

Crum L.A. (1984) Rectified diffusion. *Ultrasonics,* **22L**, 215–23.

Crum L.A. and Cordry S. (1993) Single-bubble sonoluminescence, in *Bubble Dynamics and Interface Phenomena* (*Proc. IUTAM Symposium*, Birmingham, 6–9 September 1993), Editors J.R. Blake, J.M. Boulton-Stone and N.H. Thomas, Kluwer, pp. 287–97.

Crum L.A. and Eller A.I. (1969) Motion of bubbles in a stationary sound field. *J. Acoust. Soc. Am.*, **48**, 181–9.

Devin C. Jr (1959) Survey of thermal, radiation, and viscous damping of pulsating air bubbles in water. *J. Acoust. Soc. Am.*, **31**, 1654.

Eberlein C. (1996) Sonoluminescence as quantum vacuum radiation. *Phys. Rev. Lett.*, **76**, 3842.

Eller A.I. (1970) Damping constants of pulsating bubbles. *J. Acoust. Soc. Am.*, **47**, 1469–70.

Eller A.I. (1972) Bubble growth by diffusion in an 11-kHz sound field. *J. Acoust. Soc. Am.*, **52**, 1447–9.

Eller A.I. and Flynn H.G. (1965) Rectified diffusion through nonlinear pulsations of cavitation bubbles. *J. Acoust. Soc. Am.*, **37**, 493–503.

Epstein P.S. and Plesset M.S. (1950) On the stability of gas bubbles in liquid–gas solutions. *J. Chem. Phys.*, **18**, 1505–9.

Flynn H.G. (1964) Physics of acoustic cavitation in liquids, in *Physical Acoustics*, Editor W.P. Mason, Academic Press, New York, vol. 1, Part B, pp. 57–172.

Flynn H.G. and Church C.C. (1984) A mechanism for the generation of cavitation maxima by pulsed ultrasound. *J. Acoust. Soc. Am.*, **76**, 505–12.

Flynn H.G. and Church C.C. (1988) Transient pulsations of small gas bubbles in water. *J. Acoust. Soc. Am.*, **84**, 985–98.

Gaitan D.F. and Crum L.A. (1990) Observation of sonoluminescence from a single cavitation bubble in a water/glycerine mixture. In *Frontiers of Nonlinear Acoustics, 12th ISNA*, Editors M.F. Hamilton and D.T. Blackstock, Elsevier, New York, p. 459.

Greenspan M. and Tschiegg C.E. (1967) Radiation-induced acoustic cavitation; apparatus and some results. *J. Res. Natl. Bur. Stand., Section C*, **71**, 299–311.

Gupta R.S. and Kumar D. (1983) Variable time step methods for the dissolution of a gas bubble in a liquid. *Comput. Fluids,* **11**, 341–9.

Harvey E.N., Barnes D.K., McElroy W.D., Whiteley A.H., Pease D.C. and Cooper K.W. (1944) Bubble formation in animals. *J. Cell Comp. Physiol.*, **24**, 1–22.

Henglein A. and Gutierrez M. (1986) Chemical reactions by pulsed ultrasound: memory effects in the formation of NO_3^- and NO_2^- in aerated water. *Int. J. Radiat. Biol.*, **50**, 527–33.

Hill C.R., Clarke P.R., Crowe M.R. and Hannick J.W. (1969) Ultrasonics for Industry Conference Papers, pp. 26–30.

Holland C.K. and Apfel R.E. (1989) An improved theory for the prediction of microcavitation thresholds. *IEEE Trans Ultrasonics, Ferroelectrics, and Frequency Control,* **36**, 204–8.

Kamath V., Prosperetti A. and Egolfopoulos F.N. (1993) A theoretical study of sonoluminescence. *J. Acoust. Soc. Am.*, **94**, 248–60.

Kornfeld M. and Suvorov L. (1944) On the destructive action of cavitation. *J. Appl. Phys.*, **15**, 495–506.

Lauterborn W. and Bolle H. (1975) Experimental investigations of cavitation-bubble collapse in the neighbourhood of a solid boundary. *J. Fluid Mech.*, **72**, 391–9.

Leighton T.G. (1994a) *The Acoustic Bubble,* Academic Press, London.

Leighton T.G. (1994b) Acoustic bubble detection. I. The detection of stable gas bodies. *Environ. Engng,* **7**, 9–16.

Leighton T.G. (1995a) Bubble population phenomena in acoustic cavitation. *Ultrasonics Sonochemistry,* **2**, S123–36.

Leighton T.G. (1995b) Acoustic bubble detection. II. The detection of transient cavitation. *Environ. Engng,* **8**, 16–25.

Leighton T.G. and Walton A.J. (1987) An experimental study of the sound emitted from gas bubbles in a liquid. *Eur. J. Phys.*, **8**, 98–104.

Leighton T.G., Pickworth M.J.W., Walton A.J. and Dendy P.P. (1988) Studies of the cavitational effects of clinical ultrasound by sonoluminescence: 1. Correlation of sonoluminescence with the standing-wave pattern in an acoustic field produced by a therapeutic unit. *Phys. Med Biol.*, **33**, 1239–48.

Leighton T.G., Pickworth M.J.W., Walton A.J. and Dendy P.P. (1989a) The pulse enhancement of unstable cavitation by mechanisms of bubble migration. *Proc. Inst. Acoust.*, **11**, 461–9.

Leighton T.G., Walton A.J. and Field J.E. (1989b) High-speed photography of transient excitation. *Ultrasonics*, **27**, 370–3.

Leighton T.G., Wilkinson M., Walton A.J. and Field J.E. (1990a) The forced oscillations of bubbles in a simulated acoustic field. *Eur. J. Phys.*, **11**, 352–358.

Leighton T.G., Walton A.J. and Pickworth M.J.W. (1990b) Primary Bjerknes forces. *Eur. J. Phys.*, **11**, 47–50.

Leighton T.G., Fagan K.J. and Field J.E. (1991) Acoustic and photographic studies of injected bubbles. *Eur. J. Phys.*, **12**, 77–85.

Leighton T.G., Schneider M.F. and White P.R. (1995a) Study of bubble fragmentation using optical and acoustic techniques, in *Sea Surface Sound '94* (*Proc. 3rd Meeting on Natural Physical Processes related to Sea Surface Sound*), Editors M.J. Buckingham and J.R. Potter, World Scientific, Singapore, pp. 414–28.

Leighton T.G., White P.R. and Marsden M.A. (1995b) Applications of one-dimensional bubbles to lithotripsy, and to diver response to low frequency sound. *Acta Acustica*, **3**, 517–29.

Lepoint T., Voglet N., Faille L. and Mullie F. (1993) Bubbles deformation and interface distortion as a source of sonochemical and sonoluminescent activity, in *Bubble Dynamics and Interface Phenomena* (*Proc. IUTAM Symposium*, Birmingham, 6–9 September 1993), Editors J.R. Blake, J.M. Boulton-Stone and N.H. Thomas, Kluwer, pp. 321–33.

Longuet-Higgins M.S. (1989) Monopole emission of sound by asymmetric bubble oscillations. Part 1. Normal modes. *J. Fluid Mech.*, **201**, 525–41.

Longuet-Higgins M.S. (1992a) Nonlinear damping of bubble oscillations by resonant interaction. *J. Acoust. Soc. Am.*, **91**, 1414–22.

Longuet-Higgins M.S. (1992b) The crushing of air cavities in a liquid. *Proc. R. Soc. Lond.*, **439**, 611–26.

Madanshetty S.I. (1995) A conceptual model for acoustic microcavitation. *J. Acoust. Soc. Am.*, **98**, 2681–9.

Madanshetty S.I., Roy R.A. and Apfel R.E. (1991) Acoustic microcavitation: its active and passive detection. *J. Acoust. Soc. Am.*, **90**, 1515–26.

Margulis M.A. (1985) Sonoluminescence and sonochemical reactions in cavitation fields. A review. *Ultrasonics*, **23**, 157–69.

Medwin H. and Beaky M.W. (1989) Bubble sources of the Knudsen sea noise spectra. *J. Acoust. Soc. Am.*, **86**, 1124–30.

Miller D.L. (1979) A cylindrical bubble model for the response of plant-tissue gas-bodies to ultrasound. *J. Acoust. Soc. Am.*, **65**, 1313–1321.

Miller D.L. (1984) Gas body activation. *Ultrasonics*, **22**, 261–9.

Miller D.L. and Neppiras E.A. (1985) On the oscillation mode of gas-filled micropores. *J. Acoust. Soc. Am.*, **77**, 946–953.

Miller D.L. and Nyborg W.L. (1983) Theoretical explanation of the response of gas-filled micropores and cavitation nuclei to ultrasound. *J. Acoust. Soc. Am.*, **73**, 1537–44.

Miller D.L. and Thomas R.M. (1993a) A comparison of hemolytic and sonochemical activity of ultrasonic cavitation in a rotating tube. *Ultrasound Med. Biol.*, **19**, 83–90.

Miller D.L. and Thomas R.M. (1993b) Frequency dependence of cavitation activity in a rotating tube exposure system compared to the mechanical index. *J. Acoust. Soc. Am.*, **93**, 3475–80.

Miller D.L. and Williams A.R. (1989) Bubble cycling as the explanation of the promotion of ultrasonic cavitation in a rotating tube exposure system. *Ultrasound Med. Biol.*, **15**, 641–8.

Miller D.L., Nyborg W.L. and Whitcomb C.C. (1979) Platelet aggregation induced by ultrasound under specialized condition *in vitro*. *Science*, **205**, 505.

Miller D.L., Thomas R.M. and Williams A.R. (1991) Mechanisms for hemolysis by ultrasonic cavitation in the rotating exposure system. *Ultrasound Med. Biol.*, **17**, 171–8.

Miller D.L., Thomas R.M. and Buschbom R.L. (1995) Comet assay reveals DNA strand breaks induced by ultrasound cavitation *in vitro*. *Ultrasound Med. Biol.*, **21**, 841–8.

Minnaert M. (1933) On musical air-bubbles and sounds of running water. *Phil. Mag.*, **16**, 235–48.

Neppiras E.A. (1980) Acoustic cavitation. *Phys. Rep.*, **61**, 159–251.

Neppiras E.A. and Noltingk B.E. (1951) Cavitation produced by ultrasonics: theoretical conditions for the onset of cavitation. *Proc. Phys. Soc.*, B, **64**, 1032–8.

Noltingk B.E. and Neppiras E.A. (1950) Cavitation produced by ultrasonics. *Proc. Phys. Soc.*, B, **63**, 674–85.

Phelps A. and Leighton T.G. (1996) High resolution bubble sizing through detection of the subharmonic response with a two frequency excitation technique. *J. Acoust. Soc. Am.*, **99**, 1985–92.

Phelps A. and Leighton T.G. (1997) The subharmonic oscillations and combination frequency emissions from a resonant bubble. *Acta Acustica*. **83**, 59–66.

Pickworth M.J.W., Dendy P.P., Leighton T.G. and Walton A.J. (1988) Studies of the cavitational effects of clinical ultrasound by sonoluminescence: 2. Thresholds for sonoluminescence from a therapeutic ultrasound beam and the effect of temperature and duty cycle. *Phys. Med. Biol.*, **33**, 1249–60.

Pickworth M.J.W., Dendy P.P., Twentyman P.R. and Leighton T.G. (1989) Studies of the cavitational effects of clinical ultrasound by sonoluminescence: 4. The effect of therapeutic ultrasound on cells in monolayer culture in a standing wave field. *Phys. Med Biol.*, **34**, 1553–60.

Plesset M.S. and Chapman R.B. (1971) Collapse of an initially spherical vapour cavity in the neighbourhood of a solid boundary. *J. Fluid Mech.*, **47**, 283–90.

Plesset M.S. and Prosperetti A. (1977) Bubble dynamics and cavitation. *Ann. Rev. Fluid Mech.*, **9**, 145–85.

Prandtl L. (1954) *Essentials of Fluid Dynamics*, Blackie, London, pp. 180 and 342–5.

Prosperetti A. (1984) Bubble phenomena in sound fields: part 2. *Ultrasonics*, **22**, 115–24.

Pumphrey H.C. and Crum L.A. (1990) Free oscillations of near-surface bubbles as a source of the underwater noise of rain. *J. Acoust. Soc. Am.*, **87**, 142–8.

Rayleigh, Lord (1917) On the pressure developed in a liquid during the collapse of a spherical cavity. *Phil. Mag.*, **34**, 94–8.

Roy R.A. (1996) The demographics of cavitation produced by medical ultrasound. *J. Acoust. Soc. Am.*, **99**, 2485 (abstract).

Sirotyuk M.G. (1970) Stabilisation of gas bubbles in water. *Sov. Phys. Acoustics*, **16**, 237–40.

Sponer J. (1990) Dependence of ultrasonic cavitation threshold on the ultrasonic frequency. *Czech J. Phys., B*, **40**, 1123–32.

Sponer J. (1991) Theoretical estimation of the cavitation threshold for very short pulses of ultrasound. *Ultrasonics*, **29**, 376–80.

Sponer J., Davadorzh C. and Mornstein V. (1990) The influence of viscosity on ultrasonic cavitation threshold for sonoluminescence at low megahertz region. *Studia Biophys.*, **137**, 81–9.

Strasberg M. (1956) Gas bubbles as sources of sound in liquids. *J. Acoust. Soc. Am.*, **28**, 20–6.

Suslick K.S., Hammerton D.A. and Cline R.E. Jr (1986) The sonochemical hot-spot. *J. Am. Chem. Soc.*, **108**, 5641–2.

ter Haar G., Rivens I., Chen L. and Riddler S. (1991) High intensity focused ultrasound for the treatment of rat tumours. *Phys. Med. Biol.*, **36**, 1495–501.

Trevena D.H. (1987) *Cavitation and Tension in Liquids*, Adam Hilger, Bristol.

Urick, R.J. (1983) *Principles of Underwater Sound*, 3rd edn, McGraw-Hill, New York, p. 337.

Vaughan P.W. and Leeman S. (1986) Some comments on mechanisms of sonoluminescence. *Acustica*, **59**, 279–81.

Vyas B. and Preece C.M. (1976) Stress produced in a solid by cavitation. *J. Appl. Phys.*, **47**, 5133–8.

Walton A.J. and Reynolds G.T. (1984) Sonoluminescence. *Advances in Physics*, **33**, 595–660.

Wang T. (1974) Rectified heat transfer. *J. Acoust. Soc. Am.*, **56**, 1131–43.

Wu C.C. and Roberts P.H. (1993) Shock-wave propagation in a sonoluminescing gas bubble. *Phys. Rev. Lett.*, **70**, 3424–7.

Yount D.E. (1979) Skins of varying permeability: a stabilisation mechanism for gas cavitation nuclei. *J. Acoust. Soc. Am.*, **65**, 1429–39.

Yount D.E. (1982) On the evolution, generation, and regeneration of gas cavitation nuclei. *J. Acoust. Soc. Am.*, **71**, 1473–81.

Yount D.E., Gillary E.W. and Hoffman D.C. (1984) A microscopic investigation of bubble formation nuclei. *J. Acoust. Soc. Am.*, **76**, 1511–21.

10 Ultrasound: a new opportunity for food preservation

R.G. EARNSHAW

10.1 Introduction

There is presently much industrial interest in developing mild food preservation procedures which could replace the severe heat-based methods which are currently in common use. Often termed minimal processing, the benefits of these approaches are an important aspect of current and future commercial product development. Quality attributes which can be protected by application of minimal process technologies are:

- flavour and odour
- visual appearance – colour and texture
- nutritional value
- absence of additives.

Minimal processing can be applied to a wide variety of foods including short shelf-life products such as fresh fruit and vegetables, chilled ingredients and convenience dishes through to long-life ambient stable foods such as cooked meats and vegetables. This commercial challenge has opened up new opportunities for combined preservation systems incorporating mild heat treatments or low frequency use of food additives and provided impetus for the introduction of new variables for microbial control.

Ultrasound is a new tool in the food preservation technologist's armoury and appears to provide significant opportunity to develop high-quality, minimally processed foods. This chapter considers the important aspects of the antimicrobial effects of ultrasound treatments in combination with elevated temperature and antimicrobial chemicals and explores the opportunities available to devise effective combined preservation systems based on a sound understanding of the *modus operandi* of ultrasound treatment.

10.2 Traditional preservation using elevated temperature

Food preservation by elevation of temperature for short periods of time is still the most common form of food preservation process. In most cases the process variables and controls have been derived by empirical investigation of the effect of temperature and time of exposure on microbial survival kinetics with little regard given to food quality in relation to thermal effects on food composition and structure. The loss of quality is brought about by deformation of plant and animal structures, modification of macromolecules and the production of new substances from heat-catalysed reactions. Ultrasound offers some exciting opportunities to reduce these undesirable effects.

The high temperatures to which microbial cells are exposed will not affect one specific target, as the thermal energy in the cell is an integral part of an entire complex system. The heat energy affects a wide variety of cellular constituents, including structures, molecules and the reactions in the cell. These targets provide opportunities to harness the cell damaging ability of ultrasound for combined preservation systems.

10.3 Ultrasound: a new method for the magnification of heat damage and catalysis of cell death

Biological structures and systems can be perturbed by ultrasound and much work has been done to investigate the mechanism by which this occurs. When applied with sufficient intensity, ultrasound has the potential to cause cell death. The killing potential of ultrasound was discovered when sonar was being investigated for use in anti-submarine warfare, when it was noticed that the sound waves used were killing fish. Thereafter, research into ultrasound as a method for destroying or inactivating cells flourished. In the 1960s research concentrated on understanding the mechanism of ultrasound interaction with microbial cells (Hughes and Nyborg, 1962). Cavitation phenomenon and associated shear disruption, localized heating and free radical formation were found to be primary effectors. By 1975, it was shown that brief exposure to ultrasound caused a thinning of cell walls attributed to the freeing of the cytoplasmic membrane from the cell wall (Alliger, 1975). Following these findings, research continued to look at the effect of ultrasound in combination with other treatments such as heat (Ordonez *et al.*, 1984, 1987; Garcia *et al.*, 1989; Wrigley and Llorca, 1992) and chemicals (Lillard, 1993). The problem with these high-power systems is that they have limited feasibility by virtue of impractical power requirements and the obvious damage which such high levels of ultrasound would impart in foods. Cells contain several targets for the action of heat, and so it can be proposed that the basal heat resistance of microorganisms can be reduced if the intrinsic stability of macromolecules, i.e. RNA, ribosomes, nucleic acids, enzymes, proteins and the membrane, can be affected. These targets provide opportunities to harness the cell damaging ability of ultrasound for combined preservation systems.

10.4 Mechanisms of ultrasound-induced cell damage

Several theories have been proposed to describe the mechanism of ultrasound cell damage. Ultrasonic waves passing through a liquid consist of alternate rarefactions and compressions. If the waves have a sufficiently high amplitude, bubbles or cavities are produced, and this phenomenon is known as cavitation (Scherba *et al.*, 1991).

Cavitation may be stable or transient and this produces two different effects (Scherba *et al.*, 1991). Stable cavitation arises from small bubbles which oscillate during compression and rarefaction. It takes many thousands of cycles during which interchanges of gas take place and the bubbles begin to grow in size (Atchley and Crump, 1988). Stable cavitation is associated with small bubbles dissolved in a liquid. When the ultrasound passes through the liquid, the bubbles vibrate causing strong eddies to be produced in the surrounding liquid (Scherba *et al.*, 1991). Other small bubbles are attracted into the sonic field and these too can create micro-currents around themselves which spread into the liquid (Hughes and Nyborg, 1962). This effect is known as microstreaming. The microstreaming effect provides a large force, without the bubbles having to burst (Williams *et al.*, 1970). This force effectively abrades the membrane surfaces of cells, causing them to fracture and leak (Kinsloe *et al.*, 1954). It is likely that cells can repair these leaky patches but the kinetic aspects of this have yet to be defined. It is probable that the energetic cost of a leaky membrane works in concert with sublethal heat and antimicrobial chemicals to bring about subsequent cell death.

Transient cavitation occurs when the bubble size changes much more quickly, i.e. within a few oscillatory cycles, and the bubbles finally collapse with different intensities (Sala *et al.*, 1995). When bubbles collapse, high pressures (up to 100 MPa) and high temperatures (up to 5000 K) are momentarily produced (Mason *et al.*, 1994). The localized high temperatures produced by transient cavitation bombard the cell membranes and are strong enough to damage cell wall structures or to remove particles from surfaces (Schuett-Abraham *et al.*, 1992). A possible mechanism for the cell's energetic cost of ultrasound treatment could be the displacement of weakly bound ATPase moieties from the cell membrane.

Application of ultrasound to a liquid can lead to the formation of free radicals, which may or may not be beneficial. Water can be broken down by ultrasound (sonolysis) and OH^- and H^+ species and hydrogen peroxide can be produced (Mason *et al.*, 1994; Suslick, 1988). These species have important bactericidal properties. The primary target site of free radicals in the cell is the DNA. Sonication of DNA molecules causes breakages in the chain at various points, causing small fragments to be formed. These fragments of DNA are highly sensitive to attack by free radicals where, for example, hydroxyl radicals attack the hydrogen bridges which hold the polymer together (Hughes and Nyborg, 1962). The chemical environment in which microbes are exposed to ultrasound thus has critical implications in terms of process efficacy. Oxy species are likely to be important and simple variables such as oxygen partial pressure could be manipulated in order to achieve optimized cell killing.

The extent to which cavitation occurs is influenced by a number of parameters. For instance, cavitation bubbles are produced more rapidly at higher temperatures. This is because the vapour pressure is increased and the tensile strength decreases. More bubbles are produced at higher temperatures; however, this is offset by the fact that the high vapour tension cushions the intensity of bubble collapse (Alliger, 1975). Cell damaging effects can be maintained by reducing the cushioning effect by the application of pressure to the process system.

The frequency of ultrasound is also an important parameter, and determines the maximum bubble size (Suslick, 1988). At lower frequencies (e.g. 20 kHz) the bubbles produced are bigger in size and when they collapse higher energies are produced. At higher frequencies, bubble formation becomes increasingly difficult and at frequencies above 2.5 MHz, cavitation does not occur (Alliger, 1975). The intensity of cavitation is influenced by the amplitude of the ultrasound wave: if high intensity cavitation is required then higher amplitude ultrasound is necessary. Another important factor which influences the effectiveness of cavitation is the viscosity of the liquid environment. In highly viscous products, ultrasound diffusion is easily disrupted and this reduces the degree to which cavitation occurs. Low-frequency, high-power ultrasound will be better at penetrating viscous products than higher frequency ultrasound, which is more easily dispersed within a viscous liquid, and effectively reduces the level of ultrasound received. Ways of overcoming the viscosity effects are to use higher intensity, lower frequency ultrasound. Some liquids could undergo viscosity reduction during heating to overcome apparent viscosity-related hurdles at room temperature.

It is widely believed that cellular stress is caused by the cavitation effect. When the bubbles produced collapse, the high pressures are thought to be responsible for cell disruption and leakage (Scherba *et al.*, 1991). The high temperatures produced may also have some effect, but as these temperature changes occur momentarily during cavitation, only the liquid in the immediate surroundings is heated and therefore only a small number of cells would be affected. The effect of the pressures produced on the cell membrane disrupts its structure and causes the cell wall to break down. If ultrasound is combined with other preservation techniques such as heat or chemicals, then the bacterial cells undergo a synergistic attack on their vital processes and structures.

10.5 Practical aspects of application for food preservation

It has been found that resistance of microorganisms to ultrasound in a food product is different when compared to that obtained from experiments carried out in microbiological broths (Lee *et al.*, 1989). For example, foods that have high fat content reduce the effect of ultrasound killing. Differences in survival could be due to the intrinsic effect of the environment on the ultrasound action or a reflection of changes in ultrasound penetration and energy distribution.

Different species of microorganism may be more susceptible to ultrasound treatment than others (Alliger, 1975). It has been shown that in general, larger cells are more sensitive to ultrasound (Ahmed and Russell, 1975). With the increased surface area, larger sized cells are increasingly bombarded with the pressure produced from cavitation, making them more vulnerable to sonication treatment. *Staphylococcus aureus* and *Bacillus subtilis,* and the Gram-negative *Pseudomonas aeruginosa* and *Escherichia coli* have been studied in detail (Scherba *et al.,* 1991). Gram-positive cells are more resistant to ultrasound than Gram-negative cells (Ahmed and Russell, 1975). This may be because the cell wall in Gram-positive cells is thicker than in Gram-negative cells and contains a more tightly adherent layer of peptidoglycans. Scherba *et al.* (1991) found that the effect of differing cell wall morphology contradicted earlier findings and there was no significant difference between the percentage of Gram-negative and Gram-positive microorganisms killed by ultrasound treatment. It was suggested therefore that the target of ultrasonic damage may be the inner cytoplasmic membrane, although the exact nature of the damage caused could not be postulated. It has been found that coccus-shaped cells are more resistant than rod-shaped bacteria (Alliger, 1975) and that spore-formers are more resistant than vegetative bacteria (Sanz *et al.*, 1985). The ability of certain species of microorganisms to produce spores provides protection and it has been found that those spore-formers, such as *Bacillus,* which are resistant to heat treatments are similarly resistant to ultrasound.

10.6 Combination of ultrasound and increased temperature – thermosonication

In general, bacterial cells become more sensitive to heat treatment if they have undergone ultrasound treatment (Ordonez *et al.* 1984, 1987; Garcia *et al*, 1989). Increased cell death has been demonstrated in cells that are subjected to a combined ultrasound and heat treatment compared with cells subjected to ultrasound treatment only (Hurst *et al.*, 1995) – see Figures 10.1 and 10.2. This increased heat sensitivity caused by sonication can be quantified in terms of D value (Tables 10.1 and 10.2). In addition, similar

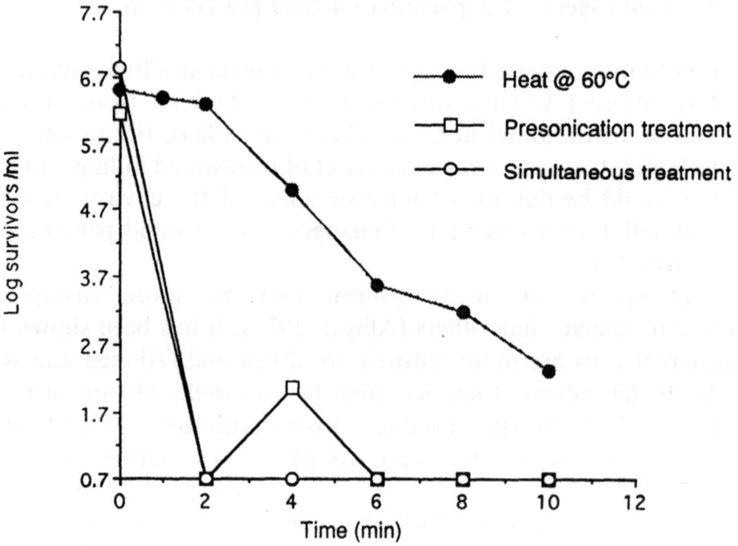

Figure 10.1 Effect of ultrasound (20 kHz) and heat (60°C) on the survival of *L. monocytogenes* in UHT whole milk.

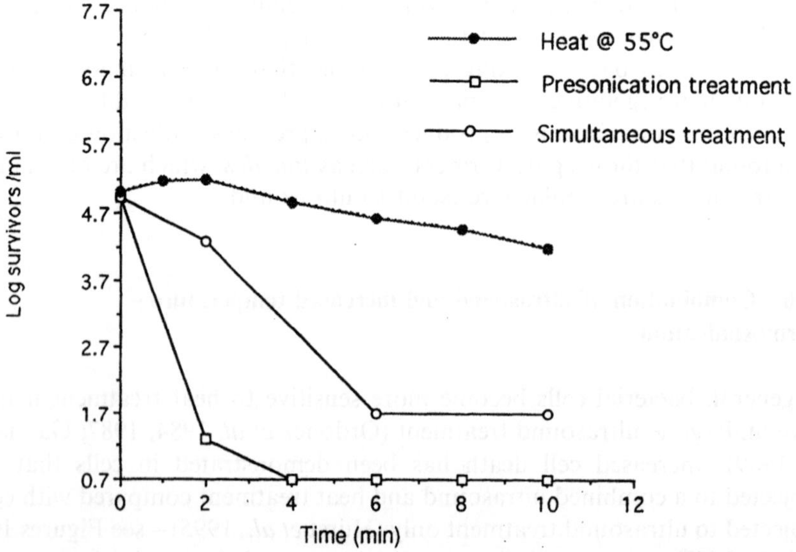

Figure 10.2 Effect of ultrasound (20 kHz) and heat (55°C) on the survival of *Z. bailii* in orange juice.

Table 10.1 *D*-values (in min) for *L. monocytogenes*

Product	Heat only (60°C)	Presonication (kHz) 20	38	800	Simultaneous (kHz) 20	38	800
UHT milk	2.1	0.4	0.3	> 10	0.3	1.3	1.4
Rice pudding	2.4	NT	NT	0.3–5.9	NT	NT	4.5–3.4

NT, not tested.

(Reprinted from *International Journal of Food Microbiology*, **28**, 197–219, Earnshaw *et al.*, (1995). Understanding physical inactivation processes: combined preservation opportunities using heat, ultrasound and pressure. With kind permission of Elsevier Science – NL, Sara Burgerhartstraat 25, 1055 KV, Amsterdam, The Netherlands.)

Table 10.2 *D*-values (in min) for *Z. bailii*

Product	Heat only (55°C)	Presonication (kHz) 20	38	800	Simultaneous (kHz) 20	38	800
Orange juice	10.5	2.4	0.9	1.4	3.9	1.8	> 10
Rice pudding	11.0	2.3	0.5	NT	1.0	NT	> 10
Orange juice + 5% starch	15.4	NT	NT	3.4	0.5–2.1	NT	NT

NT, not tested.

(Reprinted from *International Journal of Food Microbiology*, **28**, 197–219, Earnshaw *et al.*, (1995). Understanding physical inactivation processes: combined preservation opportunities using heat, ultrasound and pressure. With kind permission of Elsevier Science – NL, Sara Burgerhartstraat 25, 1055 KV, Amsterdam, The Netherlands.)

cell death is seen in cells subjected to high temperature treatment, compared with those that are given the combined treatment using ultrasound and a significantly reduced process temperature (Raso *et al.*, 1994).

Sanz *et al.* (1985) suggested that spore-formers cannot be affected by sonication on the basis of the limited perturbation which would occur within the spore; however, other researchers have reported a reduction of 43% in the heat resistance of *B. subtilis* when sonicated in water within the temperature range 70–95°C (Garcia *et al.*, 1989), and similar reductions in heat resistance of *B. cereus* and *B. licheniformis* have been reported after treatment at 20 kHz (Burgos *et al.*, 1972). The combined application of ultrasound with heat has been termed thermosonication (Hurst *et al.*, 1995).

If temperatures exceed the boiling point of the liquid there is a loss of cavitation effect due to the high water vapour tension at these temperatures (Garcia *et al.*, 1989). This can be overcome by combining pressure treatment with the heat and ultrasound treatment (mano-thermosonication). Pressurizing the liquid during thermosonication treatment enables cavitation to be maintained at temperatures above boiling point. This treatment has been used effectively against spore-forming bacteria such as *B. cereus, B. stearothermophilus* and *B. coagulans* (Raso *et al.* 1994), using pressures of 300 kPa in combination with a temperature of 112°C. This increased the lethalities by 10-fold compared with non-pressurized samples.

During treatment with a combination of pressure and thermosonication, bacterial spores are subjected to intense vibration and cavitation, and undergo a violent motion (Raso *et al.*, 1994). As a result, chemicals such as dipicolinic acid and low molecular weight polypeptides are released from spores (Palacios *et al.*, 1991). Degradation of the cortex is reported to cause the protoplast to undergo rehydration and this results in loss of heat resistance (Gould and Dring, 1975), making the cells more susceptible to the heat and ultrasound treatment. Presumably, ultrasound treatment is able to cause such cortex rehydration.

10.7 Engineering considerations of thermosonication

Ultrasound processing in combination with lower heat process temperatures has many advantages. It can lead to better quality products, with improvements in taste, texture and appearance. It also could result in reduced energy requirements and therefore reduced cost. The application of ultrasound with heat will require the design of new types of processing equipment. The design of heat exchangers and heating systems which incorporate powerful ultrasonic transducers will not be easy. The fact that heat can be applied after ultrasound treatment with useful effects is encouraging if one considers the engineering aspects. The compatibility of ultrasound processing with existing material handling and packaging systems is encouraging. Plant modification will be needed rather than complete equipment replacement: this has significance in terms of commercial realization and the demands on capital investment needed to put the technology to work.

The likely applications are difficult to predict at this stage but thermosonication is particularly suited to pumpable liquids and liquids containing solids which could be processed in a continuous flow arrangement with clean or aseptic filling. The prospect of applying ultrasound to foods already packed into containers is conceptually possible but the practical constraints need to be considered in detail, particularly in terms of ultrasound penetration.

10.8 Increasing the effectiveness of antimicrobial chemicals

Not only has heat and/or pressure in combination with ultrasound proved to be effective against microorganisms, so too has ultrasound in combination with chemical treatment. Chemicals such as chlorine are often used to clean and decontaminate food products and food processing surfaces. Lillard (1993) demonstrated the increased effectiveness of ultrasound in combination with chlorine in decontaminating broiler breast skin inoculated with *Salmonella*. The *Salmonella* cells were released from the skin by bombardment with ultrasound and it is possible that the chlorine was able to penetrate the cells more easily, causing enhanced inactivation.

Ultrasound is likely to cause bacterial cells to be dispersed in suspensions and systems containing complex solid–liquid interfaces, making them more susceptible to treatment with sanitizing agents. An advantage of combined treatment could be large reductions in the concentrations of chemicals required for sanitation and disinfection. This too has additional advantages, with less likelihood of residual cleaning agents contaminating equipment after cleaning and a reduced environmental cost.

10.9 Combination with pH control

Microorganisms have been observed to vary in their response to ultrasound depending on the surrounding pH (Kinsloe *et al.*, 1954; Utsunomyia and Kosaka, 1979). In particular, acidic pH reduces the resistance of microorganisms to sonication treatment (Utsunomyia and Kosaka, 1979).

10.10 The future

Ultrasound is being used in food processing for mixing and blending, emulsification of fats and oils, to accelerate freezing and to speed up ageing in meats and wines (Mason *et al.*, 1994; Sala *et al.*, 1995). Many other applications are described in this volume including cleaning, homogenization, foam breaking and liquid clarification.

The use of ultrasound for food preservation needs to be considered very carefully for reasons of ethics and safety. A number of important questions relating to the reliability of the process and its suitability for application to mass production of food remain to be answered. We have yet to see commercial development of the technology for food production but the research hurdles do not seem to be insurmountable and the benefits to manufacturers and consumers are large. Crystal-ball gazing in food technology is a risky business but it is not unreasonable to suppose that ultrasound could be applied to the pasteurization or sterilization of fruit juice, milk or other liquid products within the next 5–20 years.

10.11 References

Ahmed F.I.K. and Russell C. (1975) Synergism between ultrasonic waves and hydrogen peroxide in the killing of microorganisms. *J. Appl. Bacteriology*, **39**, 31–40.
Alliger H. (1975) Ultrasonic disruption. *American Laboratory*, **10**, 75–85.

Atchley A.A. and Crump L.A. (1988) Acoustic cavitation and bubble dynamics, in *Ultrasound. Its Chemical, Physical and Biological Effect*, Editor K.S. Suslick, VCH, New York, pp. 1–64.
Burgos J., Ordonez J.A. and Sala F.J. (1972) Effect of ultrasonic waves on the heat resistance of *Bacillus cereus* and *Bacillus licheniformis* spores. *Appl. Microbiol.*, **24**, 497–8.
Garcia M.L., Burgos J., Sanz B. and Ordonez J.A (1989) Effect of heat and ultrasonic waves on the survival of two strains of *Bacillus subtilis. J. Appl. Bacteriol.*, **67**, 619–28.
Gould G.W. and Dring G.J. (1975) Heat resistance of bacterial endospores and concept of an expanded osmoregulatory cortex. *Nature*, **258**, 402–5.
Hughes D.E. and Nyborg W.L. (1962) Cell disruption by ultrasound. *Science*, **138**, 108–14.
Hurst R.M., Betts G.D., and Earnshaw R.G. (1995) The antimicrobial effect of power ultrasound. CCFRA R&D Report No. 4, Chipping Campden.
Kinsloe H., Ackerman E. and Reid J.J. (1954) Exposure of microorganisms to measured sound fields. *J. Bacteriol.*, **68**, 373–80.
Lee B.H., Kermasha S., and Baker B.E. (1989) Thermal ultrasonic and ultraviolet inactivation of *Salmonella* in thin films of aqueous media and chocolate. *Food Microbiol.*, **6**, 143–52.
Lillard H. S. (1993) Bactericidal effect of chlorine on attached salmonellac with and without sonification. *J. Food Protection*, **56**(8), 716–17.
Mason T., Newman A., Sukhvinder P. and Charter C. (1994) Sound solution. *World Water and Environ. Engng*, April, p. 16.
Ordonez J.A., Sann B., Hermandez P.E. and Lopez-Lorenzo P. (1984) A note on the effect of combined ultrasonic and heat treatments on the survival of thermodurie streptococci. *J. Appl. Bacteriol.*, **56**, 175–7.
Ordonez J.A., Aguilera M.A., Garcia M.L. and Sanz B. (1987) Effect of combined ultrasonic and heat treatment (thermoultrasonication) on the survival of a strain of *Staphylococcus aureus. J. Dairy Research*, **54**(1), 61–7.
Palacios P., Burgas J., Hoz L., Sanz B. and Ordonez J.A. (1991) Study of substances released by ultrasonic treatment from *Bacillus stearothermophilus* spores. *J. Appl. Bacteriol.*, **71**, 445–51.
Raso J., Condon S. and Sala Trepat F.J. (1994) Mano-thermosonication: a new method of food preservation, in *Food Preservation by Combined Processes.* Final report for FLAIR Concerted Action No.7, Subgroup B.
Sala F.J., Burgos J., Condon S., Lopez P. and Raso J. (1995) Manothermosonication, in *New Methods of Food Preservation*, Editor G.W. Gould, Blackie, London.
Sanz P., Palacios P., Lopez P. and Ordonez J.A. (1985) Effect of ultrasonic waves on the heat resistance of *Bacillus stearothermophilus* spores, in *Fundamental and Applied Aspects of Bacterial Spores*, Editors G.J. Dring, D.J. Ellars and G.W. Gould, Academic Press, New York, pp. 251–9.
Scherba G., Weigel R.M. and O'Brien J.R. (1991) Quantitative assessment of the germicidal efficiency of ultrasonic energy. *Appl. Environ. Microbiol.*, **57**, 2079–84.
Schuet-Abraham I., Trommer E. and Levetzow R. (1992) Ultrasehall im SteriBeken. Zum einsatz von ultrasehall in eihrichtung zur reinigung und desinfektion von messern am arbeitsplatz in schlact und zerlegebetrieben. *Fleischwirtschaft*, **72**, 864–7.
Suslick K.S. (1988) Homogenous sonochemistry, in *Ultrasound. Its Chemical, Physical and Biological Effects*, Editor K.S. Suslick, VCH, New York.
Utsunomyia Y. and Kosaka Y. (1979) Application of supersonic waves to foods. *J. Faculty Appl. Biol. Sci.*, **18**, 225–31.
Williams A.R., Stafford D.A., Callely A.G. and Hughes D.E. (1970) Ultrasonic dispersal of activated sludge floes. *J. Appl. Bacteriol.*, **33**, 656–63.
Wrigley D.M. and Llorca N.G. (1992) Decrease *of Salmonella typhimurium* in skim milk and egg by heat and ultrasonic wave treatment. *J. Food Protection*, **55**, 678–80.

11 Ultrasonically assisted separation processes

E.S. TARLETON AND R.J. WAKEMAN

11.1 Introduction

The use of additional force fields in filtration processes has been attracting an increasing amount of attention in recent years. Although current applications of these new technologies at anything above laboratory scale are relatively scarce, there is a growing acceptance by industry for their potential uses. Magnetically assisted filters are now accepted as technical alternatives in the choice of solid–liquid separation equipment, commercial electrically assisted filters exist but they have not been adopted widely, and ultrasonically assisted filters are found almost exclusively in research laboratories. While there is a general understanding for the role of magnetic and electric fields in the separations to which they have been applied, the same cannot be claimed for many ultrasonically assisted separations. Here postulations abound and considerable difficulties exist in understanding the many complex phenomena which are thought to contribute.

This chapter presents an overview of how field-assisted separation processes have progressed to their current state, and provides particular emphasis on filtration and ultrasonics. While much of the authors' work in these areas has concentrated on improving the separation of aqueous colloidal and near-colloidal mineral suspensions at the laboratory scale, the general principles employed are applicable to a wide range of feeds, including many food and biological products. The concepts of filtration both with and without field assistance are detailed with reference to previous work and some data are presented to highlight the improvements in separation rates which can be expected using imposed force fields. The factors which limit the improvements are discussed and some conclusions are drawn regarding the future of field-assisted separations.

11.2 The need for field-assisted separations

The use of cake filtration is widespread in many of the chemical processing industries. Referring to Figure 11.1, a suspension under a pressure gradient is forced directly toward a semi-permeable medium where the solids are

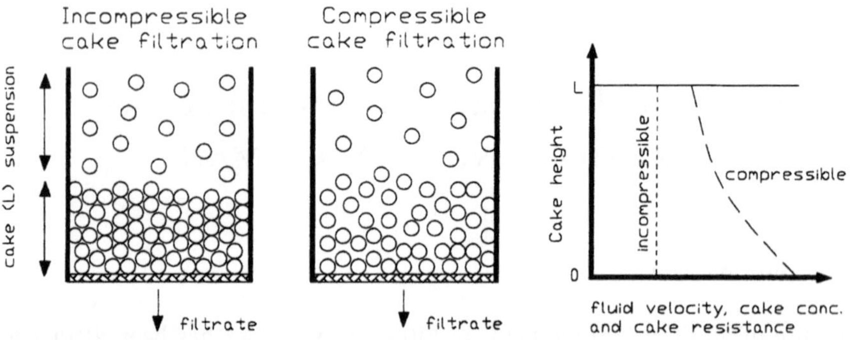

Figure 11.1 Schematic representations of cake filtration processes.

retained as a cake and the clear filtrate passes through. If the cake structure is able to withstand the cumulative drag induced by the flow of filtrate then the cake is regarded as incompressible, otherwise, particle rearrangement occurs with time and the cake is considered to be compressible. With finer particulates and colloids the influence of surface phenomena and compressibility is generally increased to the extent that separations become progressively more difficult due to the formation of filter cakes with low permeabilities. In an attempt to overcome these difficulties in so-called 'dead-end' cake filtration the concept of crossflow filtration was introduced. Here the feed suspension is introduced tangential to, rather than directly toward, the filter medium (usually a membrane), and the shear forces generated at the filtering surface limit particulate deposition there and lead to a pseudo-equilibrium flux as time progresses.

The development of pressure-driven membrane processes such as ultrafiltration and microfiltration promised to solve some of the problems inherent with other, more conventional, technologies in the processing of finer particle feeds. Although membranes have found many applications in this important area fouling problems continue to limit their more widespread use. The problems are perhaps typified by crossflow microfiltration and the dewatering of aqueous suspensions with particulates in the range 0.05–30 μm. Whilst crossflow velocities up to 3 $m\,s^{-1}$ were quite normal in the earlier days of crossflow filtration, the failure to reliably promote sufficiently high fluxes (say $>1\,m^3\,m^{-2}\,h^{-1}$) led industry to use increasingly higher filtration pressures and crossflow velocities up to 8 $m\,s^{-1}$. Frequently such combinations did not yield the desired improvements in flux rates (see Tarleton and Wakeman, 1993) and a rapid decline from the relatively high membrane clean water flux and subsequently low quasi-steady flux levels were the undesirable consequences.

Over the years many attempts have been made to improve our fundamental understanding of membrane fouling (e.g. Fane, 1986; Rautenbach and Albrecht, 1989; Lu and Ju, 1989). In parallel research programmes various other techniques, including the use of abrasives and filter aids, backwashing, backpulsing, flow reversal and pulsatile crossflows, have also been devised to reduce the influence of fouling by finer particulates. All of these generally lead to reduced membrane lifetimes or to undesirable complexities in the filter operating cycle and researchers have been moved to find alternative ways of improving filtration rates. One of the most promising avenues of research has been directed towards using imposed force fields. Both electric and, more recently, ultrasonic fields have been shown to improve the separation of a range of particulate suspensions through a number of different filtration routes. As an important synergy exists when these two types of force field are applied simultaneously, it is necessary to highlight the principles by which both can enhance filtration.

11.3 The concepts of field-assisted separations

The effects of electric fields in solid–liquid separation were first observed in the early part of the nineteenth century (see Creighton, 1943). At this time studies showed how a direct current (dc) electric field could influence the filtration of aqueous clay suspensions whereby water migrated through a porous diaphragm (electro-osmosis) whilst the clay particles moved in the opposite direction toward an electrode suspended above the filtering surface (electrophoresis). From this early discovery a number of commercial processes and patents evolved over the early part of the twentieth century with the most viable employing dc electric fields to promote the dewatering of fine particle clay beds at rates up to 15 tons per day of dry clay and power consumption equivalent to 20 kWh ton^{-1}. Of more interest here, however, is the recent use of electrokinetic effects to reduce membrane fouling in both micro- and ultrafiltration (Wakeman and Tarleton, 1987; Tarleton and Wakeman, 1988; Visvanathan and Ben Aim, 1989).

Figure 11.2 illustrates a typical laboratory crossflow electrofilter comprising a planar semi-permeable membrane sandwiched between two electrodes enclosed within a filter chamber. During filtration the feed suspension is transported under pressure via a flow circuit through the annular space in the electrofilter where a suitably polarized dc constant voltage potential gradient (E) in the range 0–100 V cm^{-1} is applied between the two electrodes. Charged particulates are caused to move away from the filtering surface by inducing an electrophoretic velocity (ν) according to equation (11.1), thereby reducing membrane fouling and improving filtrate flux such that

$$\nu \approx \frac{2\varepsilon_0 D\zeta E}{3\mu} \tag{11.1}$$

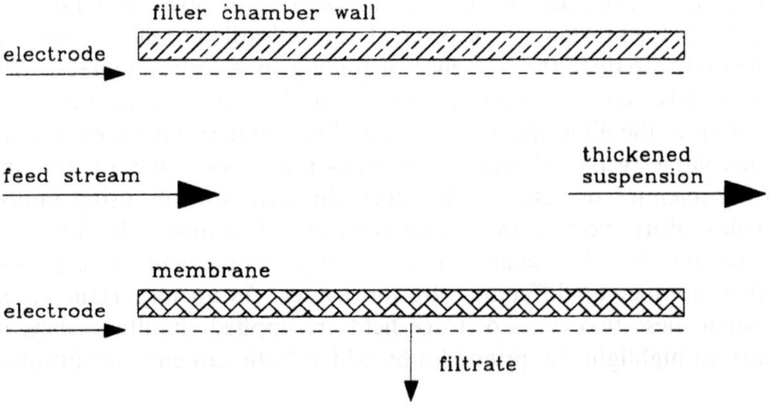

Figure 11.2 Schematic diagram of a cross-section through a crossflow electrofilter.

where ε_0 is the permittivity of a vacuum, D is the fluid medium dielectric constant, ζ is the particle zeta (ζ-) potential and μ is the fluid medium dynamic viscosity. Equation (11.1) suggests that the field strength is increased so that less fouling by particulates occurs and this is confirmed by the electrofiltration data in Figure 11.3 for anatose suspensions. The ζ-potential, which is a measure of particle surface charge and a property of the electrical double layer which surrounds particles in aqueous suspension, also

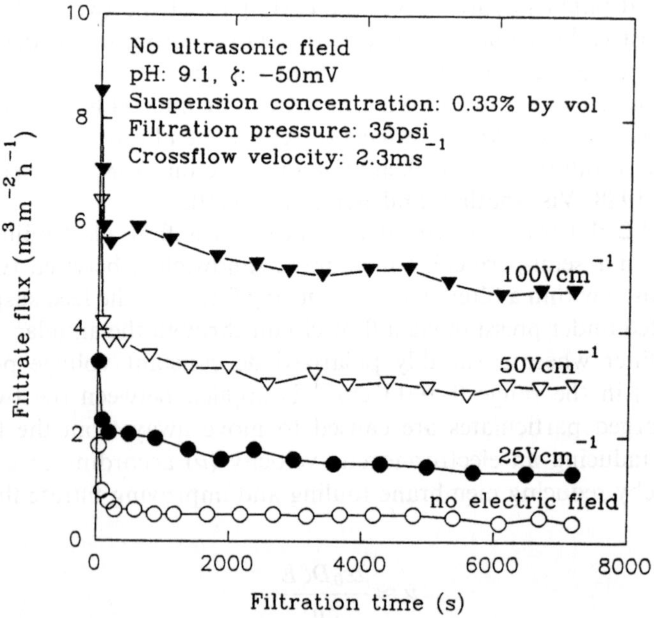

Figure 11.3 Typical effect of an imposed electric field in crossflow microfiltration.

influences the degree of fouling. The effects shown in Figure 11.3 can be repeated with a range of sub-5 μm aqueous suspensions provided the ζ-potential of the dispersed phase is greater than $|20|$ mV.

The extent of the increase in filtration rates by electric fields is primarily due to the magnitude of particle size, ζ-potential and the applied field gradient with the electrical power consumed dependent on the conductivity of the feed stream and the electrode separation. Flux increases of at least an order of magnitude are not uncommon in either dead-end or crossflow electrofiltration when even moderate field gradients are applied. Thus, in principle electric fields can be used to enhance the membrane filtration of a wide range of suspensions, including foods and biological cells, provided prerequisite criteria are satisfied. With the latter, however, careful electrode design would be required to prevent potential contamination of the sensitive feeds.

In contrast to other areas of macrosonic processes engineering such as cell disruption, atomization and emulsification, reports on the uses of ultrasonic fields in solid–liquid separation were until recently restricted to a few, largely qualitative, papers that illustrated some basic applications and physico-chemical effects of ultrasound waves (e.g. Weissler, 1950; Sollner, 1950). Fairbanks (1973) and Semmelink (1973) both showed increased liquid flow rates through porous media by the application of 20 kHz ultrasound waves at intensities between 0.6 and 6.0 W cm^{-2}. Several researchers, including Anada *et al.* (1974), demonstrated the agglomeration of particulates in high-frequency standing waves and thereby improved settling rates. The latter concept has since been adapted by Dobhlhoff-Dier *et al.* (1994) and Hawkes *et al.* (1997) to produce what they term laboratory scale continuous filtration devices.

In the mid-1980s interest in ultrasonically assisted separations began to increase when there were several contributions concerning the batch dewatering of coal, sewage, mineral sludge and food product filter cakes (Muralidhara *et al.*, 1985, 1986, 1988a, b). Here ultrasonic transducers operating at 20 kHz and intensities up to 2.5 W cm^{-2} were placed in contact with the top of a cake and the combination of pressure, ultrasonic, and sometimes electric fields, produced synergistic effects and drier cakes. A commercially available continuous particle classifier has also been described by Miyazaki *et al.* (1988) where a 19 kHz ultrasonic source is used to help remove the cake from a rotating metallic filter candle.

The authors have considerable experience using power ultrasonics in both cake filtration (Tarleton, 1988) and crossflow filtration to reduce fouling by finer particulates (Tarleton and Wakeman, 1990; Wakeman and Tarleton, 1991; Tarleton, 1992). A schematic representation of their laboratory scale crossflow acoustic filter is shown in Figure 11.4. The pressure-driven filter cell contains a microfiltration membrane and ultrasonic transducers mounted on the upstream side of the membrane such that the ultrasound

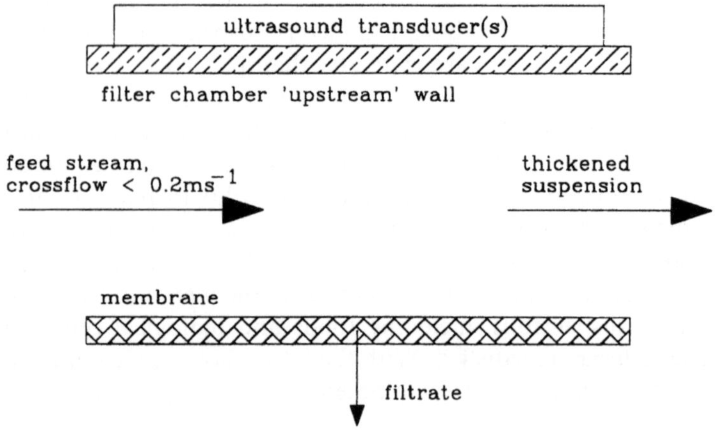

Figure 11.4 Schematic diagram of a laboratory scale crossflow acoustic filter.

waves transmitted through the flowing suspension impinge on the surface of any deposits formed at the membrane surfaces and dislodge them. Figure 11.5 shows data obtained for the filtration of a dilute aqueous china clay suspension both with and without ultrasonic assistance. The increase in flux seen when the ultrasound waves are applied is indicative of a change in both the structure and the amount of the deposit fouling the membrane surfaces.

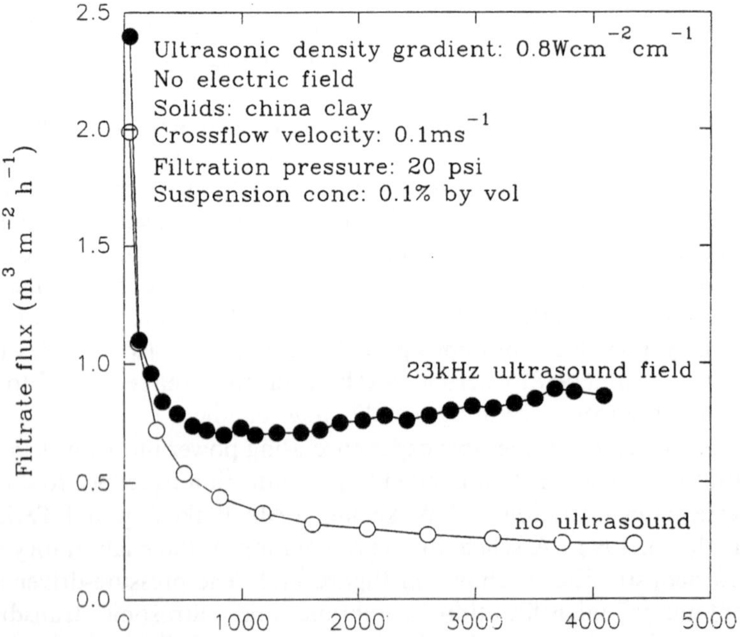

Figure 11.5 Typical effect of an imposed ultrasonic field in crossflow microfiltration.

The enhancement of membrane filtration by ultrasonics is considered to be primarily a consequence of cavitation and thus directly related to the intensity (I) and frequency (f) of the applied ultrasound such that

$$I_x = I_0 \exp(-\alpha x) = I_0 \exp(-kf^2 x) \tag{11.2}$$

where x is the distance from the ultrasound source, α is the intensity absorption coefficient and k is a constant of proportionality. Experimental data presented below indicate that the degree of enhancement is also dependent on a number of other factors including particle type, particle size. suspension viscosity and solids concentration. Sufficient data have been acquired to suggest that ultrasonic enhancement of filtration is applicable to a wide range of finer particulate suspensions.

11.4 The enhancement of flux in filtration processes

While a number of researchers have improved 'filtration' by using standing ultrasonic waves to enhance particle agglomeration (Dobhlhoff-Dier *et al.*, 1994; Hawkes; 1997), very few have attempted to use ultrasound to enhance more conventional filtration processes involving separation of particulates with a semi-permeable medium. The typical laboratory equipment required is shown in Figure 11.6. The apparatus comprises a flow loop incorporating an electroacoustic filter around which the suspension flows at constant

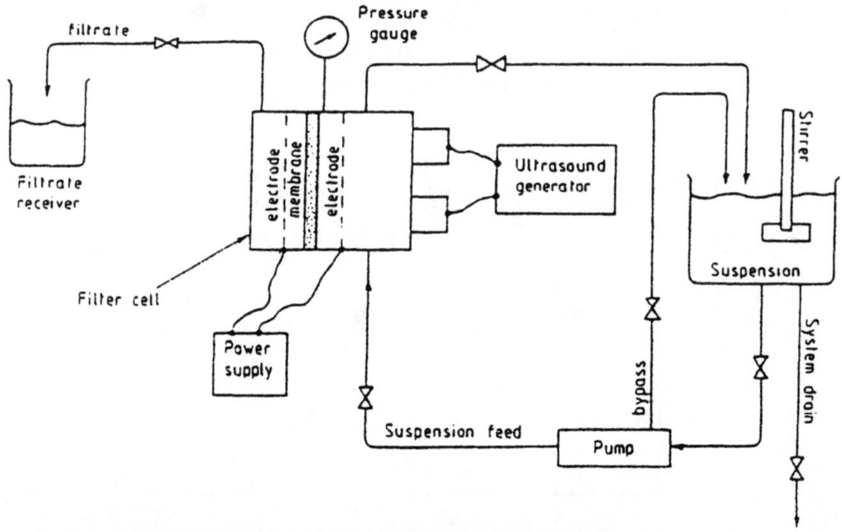

Figure 11.6 Schematic diagram of typical electroacoustic apparatus.

temperature and trans-membrane pressure and, when operated in the crossflow mode, at constant crossflow velocity. The electroacoustic filter assembly is a combination of the two representations shown in Figures 11.2 and 11.4. Suitable electrodes sandwich the planar membrane (of an area up to 240 cm^2) and ultrasound generators are placed in contact with the suspension on the upstream side of the membrane such that the generated ultrasound waves travel through the feed suspension to impinge on membrane deposits. Thus, the apparatus can be operated in the dead-end (cake) or crossflow filtration modes, with individual or combined electric and ultrasonic force fields using a range of aqueous colloidal and near colloidal suspensions.

11.4.1 Influence of force fields in dead-end cake filtration

The influence of power ultrasound on filtration rate, and hence the cumulative volume of filtrate, in dead-end filtration is typically modest and dependent on a number of factors including solution environment and the frequency of the applied ultrasound. Figure 11.7 shows that for anatase, a form of titania, improvements in flux are seen at certain pH levels. The level of improvement varies to the extent that minima are observed when the feed suspension is close to the points of zero and maximum particle surface charge, respectively. Such general trends can be seen with a range of feeds, though for suspensions and cakes that exhibit a degree of structure the impingement of ultrasonic waves can produce a wholly detrimental effect on

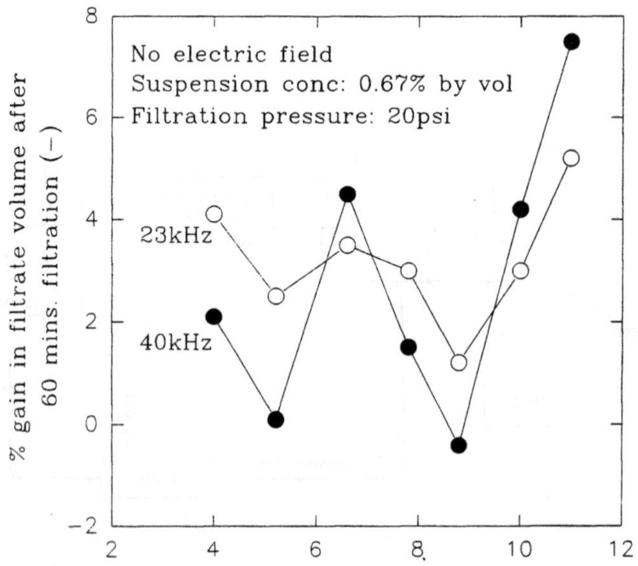

Figure 11.7 Effect of ultrasound and pH on the dead-end microfiltration of anatase suspensions.

flux levels (Tarleton, 1988). Here, the passage of ultrasound waves reduces cake porosity and the hydraulic pressure induces sufficient cake collapse to lower filtration rates. In dead-end filtration an electric field typically produces greater enhancements of flux and synergistic effects when combined with an ultrasonic field. Figure 11.8 shows that for anatase suspensions the use of combined fields provides an additional increase over that attained with an electric field alone and the increase gets progressively larger as the electric field strength is increased.

The synergy between the two force fields is particularly interesting and has also been reported for the enhanced dewatering of filter cakes (Muralidhara *et al.*, 1986). Although there is uncertainty about the exact coupling mechanisms, several hypotheses have been proposed. One theory supposes that the passage of the ultrasound waves through a suspension provides for better electrical continuity to induce improved electro-osmotic flow during filtration. Another postulates that the ultrasound initiates changes in the suspension properties at the microscopic level to promote more favourable conditions for electrofiltration. It is noted that ultrasound-induced particulate dispersion occurred during the experiments represented in Figures 11.7 and 11.8 and this could easily account for the synergy observed, as finer particles tend to be influenced to a greater extent by electric fields. Whatever the cause of the synergy the effects observed are repeatable with a number of different suspensions over a range of conditions.

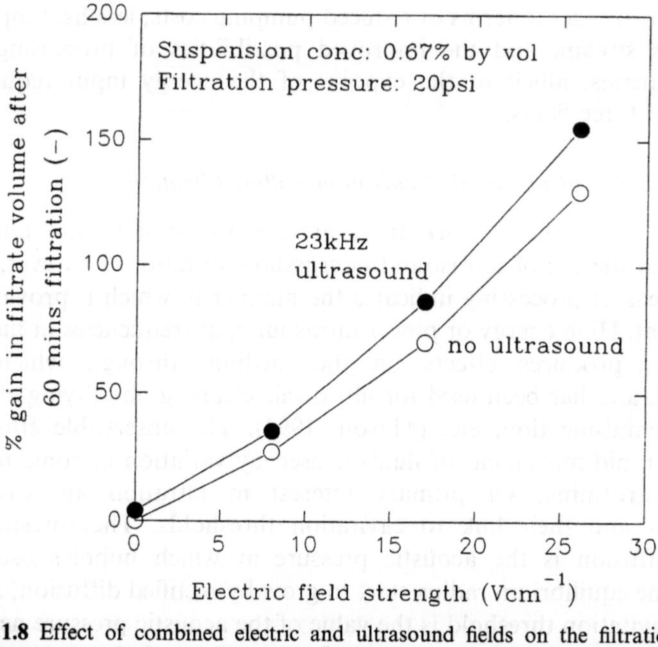

Figure 11.8 Effect of combined electric and ultrasound fields on the filtration of anatase suspensions.

11.4.2 Influence of electric fields in crossflow filtration

The influence of a dc electric field on filtrate flux during the crossflow filtration of anatase suspensions has been shown in Figure 11.3 where flux increases of four, nine, and 14 times were observed for electric field gradients of 25, 50 and 100 $V\,cm^{-1}$, respectively, with no loss of filtrate clarity. Greater flux enhancements are possible for finer particles carrying higher surface charges when using steeper field gradients. The field gradient is dependent on the applied voltage and the inter-electrode distance, and clearly electric power consumption can be reduced by designing the separation distance to be as small as possible. A further way of reducing power input to the filter is to reduce the crossflow velocity of the feed suspension. It has recently been proven that high crossflows do not guarantee a reduction of fouling (Tarleton and Wakeman, 1994a) and when an electric field is used a lower crossflow velocity can be a positive advantage, particularly when the particle charge is close to zero (Wakeman and Tarleton, 1991). Here a reduced crossflow velocity allows for an increased particle residence time in the filter and therefore a considerable flux improvement when an electric field is applied. In the industrial context the use of low crossflow is also significant and perhaps contrary to widespread opinion on the operation of conventional microfilters. Experimental investigation of this point shows that crossflow velocities as low as 0.1 $m\,s^{-1}$ can be used to advantage in microfiltration enhanced by either electric or ultrasonic force fields. Such low velocities also offer potential advantages in terms of reduced pumping costs, less heat input into the process stream, and the improved possibilities of processing shear sensitive streams, albeit at the expense of the energy input required to generate the force fields.

11.4.3 Influence of ultrasonic fields in crossflow filtration

Other than the authors' work there are almost no literature references pertaining to the use of ultrasound in crossflow filtration. However, its use in other areas of processing indicates the manner in which it provides flux enhancement. High energy or power ultrasound, at frequencies in the range 18–100 kHz, produces effects on the medium through which it is propagated, and has been used for ultrasonic cleaning, modifying chemical reactions, emulsification, etc. (Mason, 1993). The observable effects are cavitation, rapid movement of fluids caused by variation of sonic pressure and microstreaming. Of primary interest in filtration are cavitation mechanisms and their link to cavitation thresholds. The threshold for rectified diffusion is the acoustic pressure at which bubbles oscillating around some equilibrium radius start to grow by rectified diffusion, and the transient cavitation threshold is the value of the acoustic pressure necessary for the cavity to expand in an unstable fashion. General observations of the effects of liquid properties on the transient cavitation threshold are summarized in Table 11.1. If gaseous nuclei aid membrane filtration it should be recognized that they may be different in behaviour from those formed in a large stagnant volume, as the environment and process conditions under which the bubbles are formed can affect their behaviour. For example, a streaming behaviour with little coalescence of the bubbles may be observed in a large liquid volume but in a filter streaming is likely to be followed by rapid coalescence due to the constrained height in which the liquid volume is contained.

Figure 11.9 shows the typical effect of ultrasonic frequency on filtrate flux. The greater flux improvements with a 23 kHz ultrasound field are in accordance with equation (11.2) and suggest that cavitation is an important factor in the flux enhancement process. A more vigorous 'cleaning' action

Table 11.1 Parameters affecting cavitation threshold

Parameter	Influence of increasing parameter on threshold
Dissolve gas saturation	Decrease
Hydrostatic pressure	Increase
Surface tension	Decrease
Temperature	Decrease
Solids concentration	Decrease
Particle size	Decrease

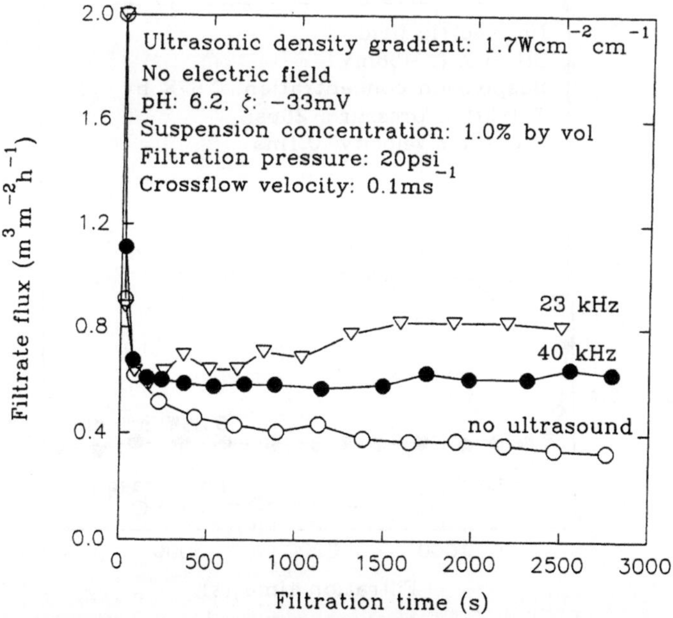

Figure 11.9 Effect of frequency on the microfiltration of china clay suspensions enhanced with ultrasound.

and less sound absorption by the particulate matter would be expected at the 23 kHz frequency compared with 40 kHz. The intensity of the ultrasound field in Figure 11.9 is expressed as a power density gradient with the units $W\,cm^{-2}\,cm^{-1}$. Equation (11.2) suggests that as the density gradient is increased so greater flux improvements should be seen. Figure 11.10 shows an example of such a result where the gradient was varied by using an ultrasonic source with a fixed power output ($3\,W\,cm^{-2}$) and the separation distance from the source to the membrane surface was altered over the range 11.5–3.8 cm. As the source is brought closer to the membrane the effect of the ultrasound on improving filtration rates increases considerably. In contrast to dead-end filtration, the use of power ultrasound in crossflow filtration is almost always beneficial to flux performance, despite the tendency to form higher resistance fouling layers on the membrane through particulate dispersion. It seems that the disruptions induced by cavitation at the membrane–particle surfaces dislodge the fouling layers sufficiently for even low crossflow velocities to partially re-entrain them into the flowing stream and facilitate improved flux.

Increasing suspension concentration generally reduces the flux enhancements possible with an ultrasonic field. Figure 11.11 shows this where the percentage gain in the filtrate flux as a result of using ultrasound is plotted as a function of the volume fraction of solids in the feed and the filtration time.

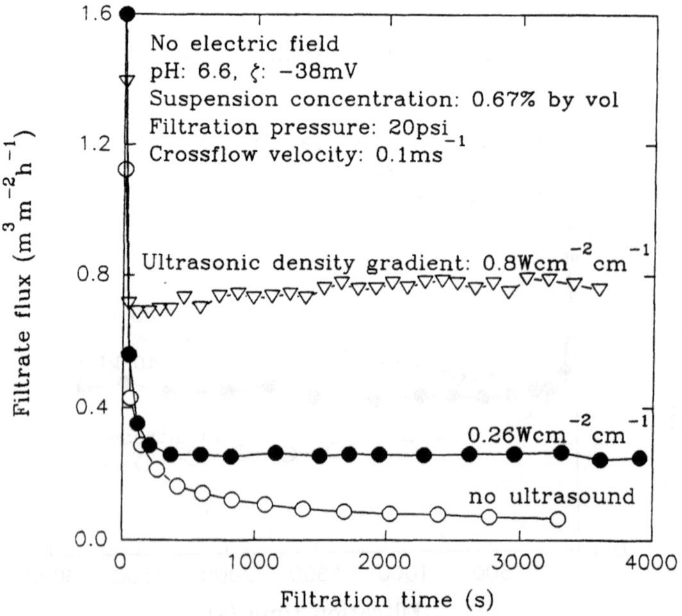

Figure 11.10 Effect of field gradient on the microfiltration of anatase suspensions enhanced with ultrasound.

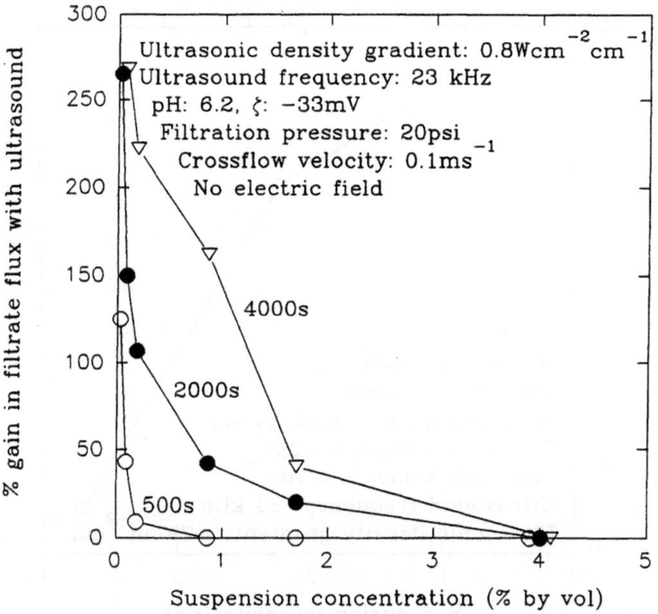

Figure 11.11 Effect of feed concentration on the microfiltration of china clay suspensions enhanced with ultrasound.

The data indicate the typically observed effect of flux improvement increasing at longer filtration times; the ultrasound causing the fouling layer thickness to equilibrate more rapidly during filtration, leading to the more rapid establishment of a constant flux than is the case when no ultrasound is used. However, it must also be noted that a higher particle concentration produces greater attenuation of the ultrasound waves as they pass through the crossflowing suspension due to increased acoustic impedance. The degree of attenuation varies with different feed solids and experimental conditions, and is considered to be an important parameter controlling the efficacy of an ultrasonic field in filtration.

Theoretical considerations suggest that an increased suspension viscosity will cause greater attenuation of an ultrasonic field (Suslick, 1986) and thus reduced flux enhancements. Tests performed with 1% v/v china clay dispersions in glycerol–water mixtures indicated that for the chosen experimental conditions attenuation of the sound was effectively total at a suspension viscosity of 4 cP (see Figure 11.12). At higher viscosities flux gains were constant throughout a test, but at lower viscosities the gain was time dependent. The 2.2 cP data indicated in Figure 11.12 showed a 23% gain after 1 min filtration, 80% after 30 min, and 105% after 60 min.

The typical effect of particle size is shown in Figure 11.13, where data for unground and wet ground calcite suspensions with mean sizes of 8 and 3 μm,

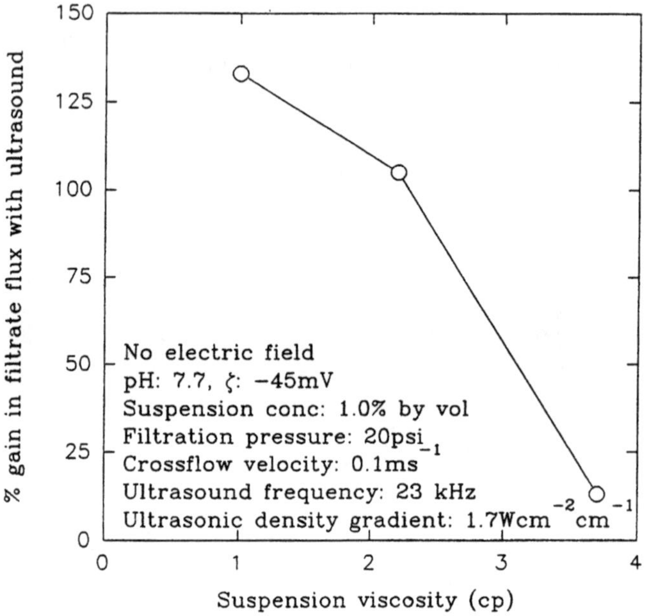

Figure 11.12 Effect of suspension viscosity on the microfiltration of china clay suspensions enhanced with ultrasound.

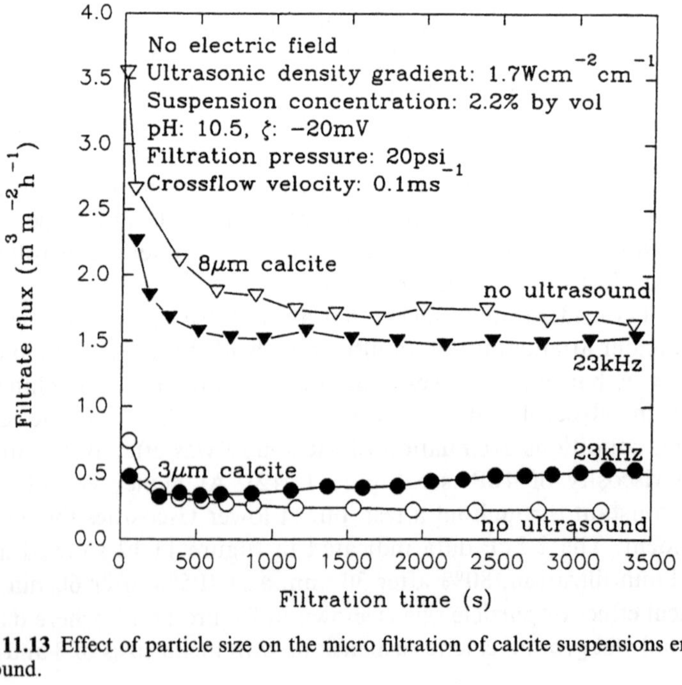

Figure 11.13 Effect of particle size on the micro filtration of calcite suspensions enhanced with ultrasound.

respectively, are compared. At the smaller particle size the filtration flux is enhanced by ultrasound, suggesting that less fouling had occurred at the membrane surface. With a smaller mean size the particle movement in the ultrasound field could reasonably be expected to follow more closely that of the suspending fluid, and the ultrasonic field could possibly promote sufficient motion at or near the fouling layer surface to cause the particles to stay in suspension or to re-suspend. Alternatively, smaller particles in suspension may cause less attenuation of the sound field. It has to be noted, however, that the situation may be complicated by the substantial increase in the number of cavitation nucleation sites generated on the new particle surfaces by the grinding process. Thus, although the influences of particle size are obviously important no clear operating mechanism can be identified. Figure 11.13 also shows that at the larger particle size (slightly) more rapid fouling occurs when ultrasound is applied. A possible explanation here is that the momentum needed to re-suspend the particles is more than that which can be supplied by the motion of the bubbles resulting from cavitation. The bubbles probably cause sufficient agitation of the foulant layer to help it pack more densely and subsequently yield a lower filtrate flux in a similar manner to that observed in dead-end filtration experiments.

11.4.4 Influence of combined electric and ultrasonic fields in crossflow filtration

The contributions of each field to combined field filtrations are shown in Figures 11.14 and 11.15. Both electrical and ultrasonic fields are seen to reduce fouling when applied individually, but the extent of improvement by the ultrasonic field can be minimal in these cases due to the relatively high suspension concentration and consequent ultrasound attenuation. In general an applied electric field provides a greater flux improvement than that due to an individual ultrasonic field, particularly when the particles are well dispersed at high ζ-potentials. Comparisons of the effects of the electric fields on Figures 11.14 and 11.15 show up marked differences in the early stages of filtration. In Figure 11.14 the approach to a pseudo-equilibrium flux after a few hundred seconds is quite orderly, with a progressive fouling of the membrane occurring until equilibrium is reached. In Figure 11.15 there is an initial sharp drop in the filtrate flux due to the high initial flux causing rapid deposition of 'cake' at the membrane surfaces. The large number of particles approaching the membrane surface causes most of the pores to be bridged and the flow resistance is simply due to cake thickness. The cake is subsequently largely removed by the electric field, and the filtrate flux rises. However, once most of the cake has been removed finer particles start to penetrate the membrane surface and cause the flux to decline once more in a manner typical of the behaviour of a higher concentration feed. When the two fields are applied simultaneously the filtrate flux is significantly better than the additive effects of the individual fields. The synergy

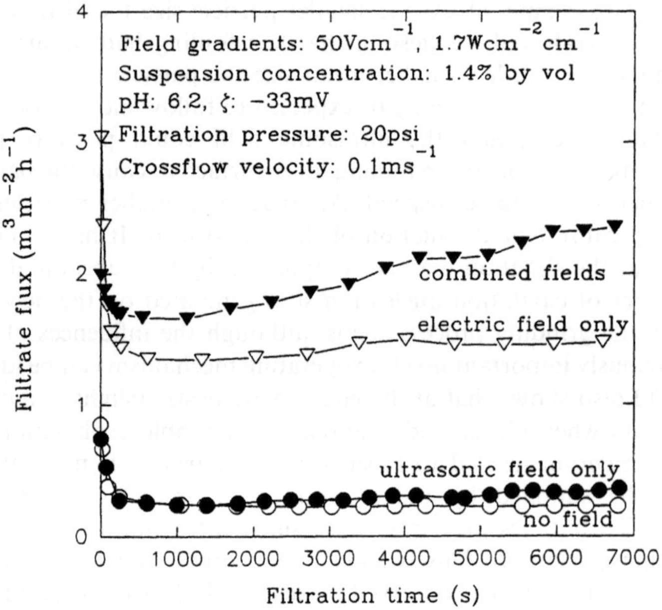

Figure 11.14 Synergy between electric and ultrasonic fields during the microfiltration of china clay suspensions.

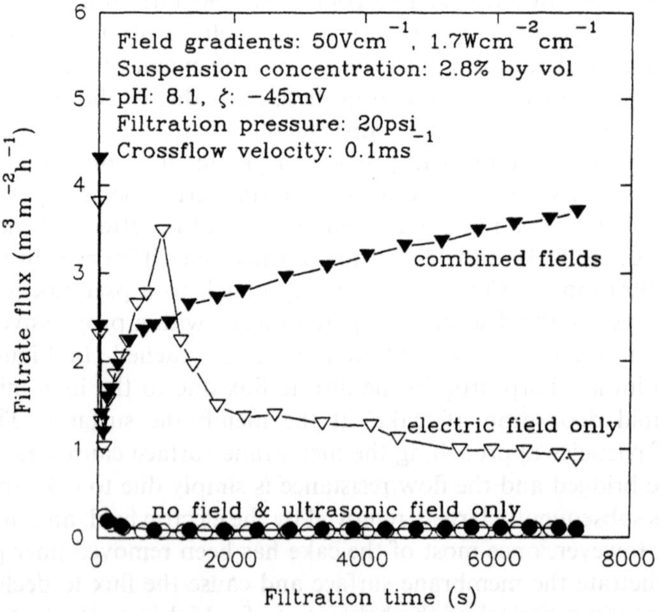

Figure 11.15 Synergy between electric and ultrasonic fields during the microfiltration of anatase suspensions.

between electric and ultrasonic fields in crossflow filtration can be repeated for a range of suspensions with, perhaps surprisingly, a greater synergism being observed with the more problematic suspensions and in particular at higher feed concentrations.

In crossflow microfiltration without field assistance the effect of membrane type on filtrate flux is minimal in the absence of fouling by adsorption at both higher feed concentrations and when the particle size is larger than the pore size of the membrane (Tarleton and Wakeman, 1994b). Membrane type also appears to have a negligible effect when simultaneous electric and ultrasound fields are applied to microfiltration and there is currently no evidence that membranes of differing structure respond differently in either electric, ultrasonic or combined fields. However, Figure 11.16 indicates that the surface charge carried by the dispersed phases is important to the flux enhancements by combined fields. When no fields are used the flux is similar at both surface charge states. With the addition of the two force fields substantial increases in filtration rates are observed. At a higher ζ-potential the particles are better dispersed and a greater improvement in filtration rate is measured. At the pH corresponding to the point of zero mean surface charge a range of ζ-potentials actually exist and some of the potential foulants are removed from the membrane surface by the combined action of the fields.

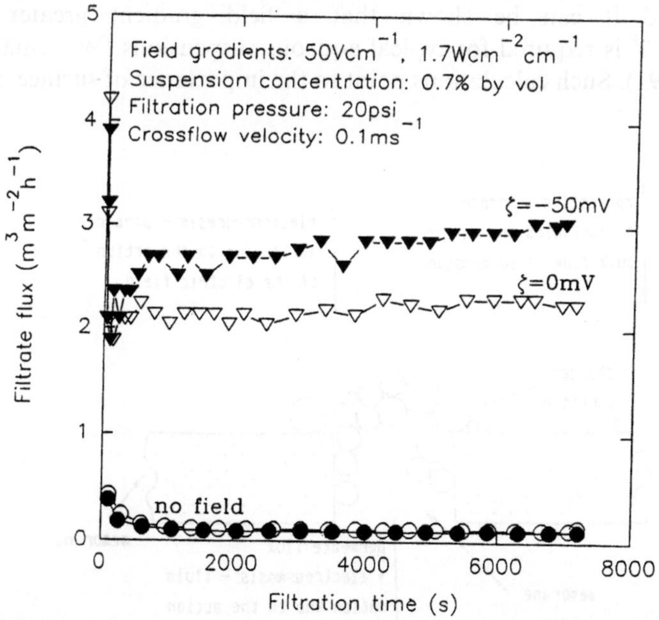

Figure 11.16 Effect of ζ-potential on the electro acoustically assisted filtration of china clay suspensions.

11.5 Enhancement mechanisms and limiting parameters in field-assisted filtration

The parameters which dictate the levels of flux enhancement with imposed force fields are highlighted by the data in Figures 11.3–11.16. They indicate that finer particle suspensions are more likely to respond to the actions of both individual and combined electric and ultrasonic fields, but do not highlight the factors likely to limit the potential enhancements. The data do, however, point to the possibility of intensive processing for suspensions where the benefits of lower overall power consumptions, reduced pumping requirements and smaller equipment/filtration areas potentially exist.

In crossflow microfiltration the bulk suspension flow is tangential to the membrane. The movement of filtrate is transposed on this and a convective flow causes particles to be transported laterally toward the membrane and the pore entrances. The convective flow of particles can be counterbalanced to a greater or lesser extent by giving an electrophoretic velocity to the particles through the application of a dc electric field gradient (see Figure 11.17). In accordance with equation (11.1) the contribution of fine particles and colloids to membrane fouling can be reduced by this means. An acceptable filtrate flux level from a microfilter is about $1\,m^3\,m^{-2}\,h^{-1}$, although instances of filters operating at lower fluxes are not infrequent. The successful application of electrical assistance requires that an electrophoretic velocity greater than $1\,m\,h^{-1}$ ($\sim 1\,mm\,s^{-1}$) be imparted to the suspended particles and it can be shown that a field gradient greater than $\sim 590/\zeta\,V\,m^{-1}$ is required for typical aqueous suspensions (Wakeman and Tarleton, 1991). Such calculations point to the importance of surface charge

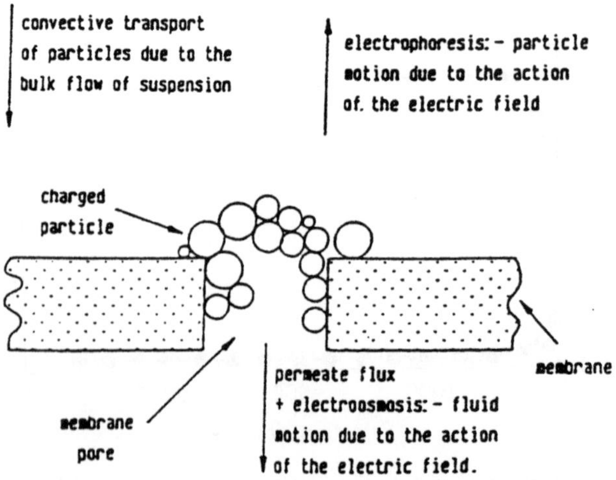

Figure 11.17 Schematic diagram showing some of the mechanisms involved in electrofiltration.

in electrically assisted microfiltration and give an idea of the order of magnitude for the parameters involved. In the practical context, however, it should be realized that this relatively simple analysis assumes that particles are flowing normal to the membrane surface as they approach it and that all particles in suspension have identical surface properties. Neither is true in reality, and the field gradient actually required may be of the order of only 10–20% of the calculated value.

When ultrasonic sound waves are applied to a suspension of particulates the waves propagate through the mixture in the form of mechanical vibratory energy. At the solid–liquid phase boundaries high inertial forces are generated that may be of sufficient magnitude to cause effects such as particulate dispersion, viscosity reduction, changes in the surface properties of the particles in suspension and perhaps, most pertinently, cavitation. The latter is manifested as the rapid formation, oscillation, growth and subsequent violent destruction of gaseous microbubbles that initiate from nucleation sites. In a filtration system these sites occur on vessel walls, suspended particulates, existing air bubbles in the fluid medium, particles forming fouling layers and membrane pores. Although particulate dispersion is typically observed when power ultrasonics are applied to filtration, cavitation bubbles are thought to dislodge fouling layers sufficiently during their growth and destruction to allow even low crossflows to re-entrain constituent particulates and produce flux improvements.

Power ultrasound is characterized by its ability to transmit substantial amounts of mechanical power at small mechanical movements, and primary vibratory inputs from the ultrasonic transducer lead to the occurrence of compound acoustic phenomena. For instance, it is known that acoustic pressure causes cavitation and microstreaming in liquids and ultrasonic acceleration is responsible for surface instabilities at fluid–fluid interfaces. Vapour bubbles driven by sound fields cool slightly and take in heat from the surrounding fluid on the expansion part of the cycle, and on compression the reverse occurs. Since the expanding area is slightly greater than the contracting area, the bubble gains some heat over a complete cycle. Not only will the increased temperature of the bubble mean that its saturation vapour pressure is increased so that evaporation of liquid to vapour must occur to restore equilibrium, but if the bubble is attached to or very close to the surface of a particle a local change of temperature and ionic composition will occur around the ‘double layer’ at the particle surface. Therefore it might be assumed that ultrasonic acceleration causes property changes at fluid–solid interfaces. A 23 kHz transducer operating at a peak displacement amplitude of 50 μm has a peak velocity of $\sim$6 m s^{-1} and a peak acceleration of $\sim$90 000 g. The peak velocity is thus $\sim 10^4$ greater than a typical filtrate velocity of $\sim$0.1 mm s^{-1} and both peak velocity and peak acceleration at the transducer surface are very large compared with the fluid and particle velocities and accelerations at the membrane surface.

The difference in the acoustic impedance of the solid and liquid phases in the suspension causes high inertial and elastic forces to be generated at their interfaces and hence induces phenomena such as particulate dispersion and cavitation. The latter phenomenon is recognized as being initiated from the nucleation sites present as 'free' microbubbles in the carrier liquid, gas pockets on/within suspended particle surfaces and gas pockets at vessel walls or, in the case of a microfiltration system, membrane pores. The rapid growth, oscillation and destruction of these gaseous areas, particularly at or near the membrane pore throats and within fouling layers, are likely to loosen some of the constituent particles. It is estimated that a displacement of up to 4 μm is needed to move particles to more favourable positions where a crossflowing stream can re-entrain them. Experimental data obtained with ultrasound suggest that the mechanisms of ultrasound transport through the feed, the interactions between the ultrasound, feed constituents and membrane and the ultrasound field strength at the filtering surface are of paramount importance. Although a rigorous theoretical analysis of the processes occurring is not warranted here, some explanation can be gained by considering the motion of a two-dimensional sound wave through a particulate suspension. Equation (11.2) indicates that when viscosity and/or particulate matter is present in a suspension any travelling ultrasound waves are attenuated as they pass through, with the intensity being related to the depth of penetration. The parameter which characterizes the attenuation is the absorption coefficient (α) which for china clay suspensions in the absence of cavitation has been shown (Urick, 1948) to be

$$\alpha = C\left(\frac{k^4a^3}{6} + \frac{sk(\sigma - 1)^2}{s^2 + (\sigma + \tau)^2}\right) \qquad (11.3)$$

where a is particle radius, C is volume fraction of solids in the suspension, σ is the ratio of particle to fluid densities and k, s and τ are parameters related to the ultrasound frequency and suspension properties. Values of α calculated from equation (11.3) are raised by increasing the solids concentration in the suspension or by increasing the size of the particles. For instance, α varies between 0.001 and 0.5 m^{-1} over the solids concentration range 0.01–5% v/v for china clay and, assuming equation (11.3) is valid for other solids, α changes from 0.55 to 1.2 m^{-1} as calcite particle size is changed from 3 to 8 μm. The ultrasound intensities at any distance from the source are correspondingly reduced and this confirms the observed experimental trends. The values of α given by equation (11.3) are strictly only valid for spherical, non-interacting particles in the presence of a non-cavitating sound field. Whilst the former assumptions are reasonable approximations, it appears that both 'stable' and 'transient' cavitation bubbles can form in a typical filtration apparatus enhanced with ultrasound. Further analysis to evaluate a coefficient which accounts for attenuation through both particle absorption and cavitation yields significantly higher values for α than that given by equation (11.3) (see Wakeman and Tarleton, 1991). This suggests that cavitation is typically present in a system during filtration, and that cavitation substantially contributes to the processes occurring.

To provide some insight into the synergy between electric and ultrasonic fields in filtration it is necessary to consider some details of the likely mechanistic phenomena occurring close to the pore entrances in membranes. The form of the electric field gradients at the pore entrances is complex, particularly when the pores are non-cylindrical and possess a surface conductance. Preliminary analysis of the field gradients using finite difference techniques suggests that, under these conditions, the field gradient is lower at the pore entrance than it is above the areas between the pores. The differences between gradients can be substantial, potentially leading to the slower removal of particles from these regions. Moreover, an ultrasound field is likely to propagate bubbles, whether they originate from sites at the membrane or particle surfaces or from gas in solution. At pressures of 100 kPa streams of bubbles are formed by rectified diffusion and the equilibrium bubble diameter is $\sim 20\,\mu m$ if the bubble cycle is isothermal or $80\,\mu m$ if the cycle is adiabatic (Suslick, 1986). A smaller radius is associated with a lower acoustic pressure threshold, largely because as the bubble grows toward resonance size its pulsation amplitude increases which results in more rectified mass transfer per cycle. Hindrance of bubble motion close to the entrance of a pore by the presence of large numbers of particles at a fairly high concentration may cause rapid coalescence of the bubbles, thus increasing their buoyancy and their ability to disturb particles deposited at the membrane surface. Flow oscillations at the membrane surface, together with acoustic streaming, will further enhance rates of bubble growth and coalescence. For the bubble to be buoyant at the surface it has to move against the filtrate flux. An analysis using Stokes' law suggests that the unbounded velocity of a $22\,\mu m$ bubble will rise against a permeate flux $1\,m^3\,m^{-2}\,h^{-1}$, and a $36\,\mu m$ bubble against a flux of $2.5\,m^3\,m^{-2}\,h^{-1}$. These factors, together with the (typical) results shown in Figures 11.3–11.16, suggest that bubble streamers formed by rectified diffusion provide for filtration enhancement by ultrasonics.

When an ultrasonic field is used in combination with an electric field bubble streamers have the ability to displace the particles far enough, between 1 and $4\,\mu m$, toward the mainstream flow. Referring to Figure 11.18, with no imposed force fields, fouling of the membrane results from an accumulation of particulate matter at the separating surface (a). The deposited colloidal material is sufficiently well trapped both at, and within, the pore entrances to prevent re-entrainment by the crossflow stream. In the presence of an electric field particulates are caused to move away from the membrane by electrophoresis and facilitate an improved flux (b). However,

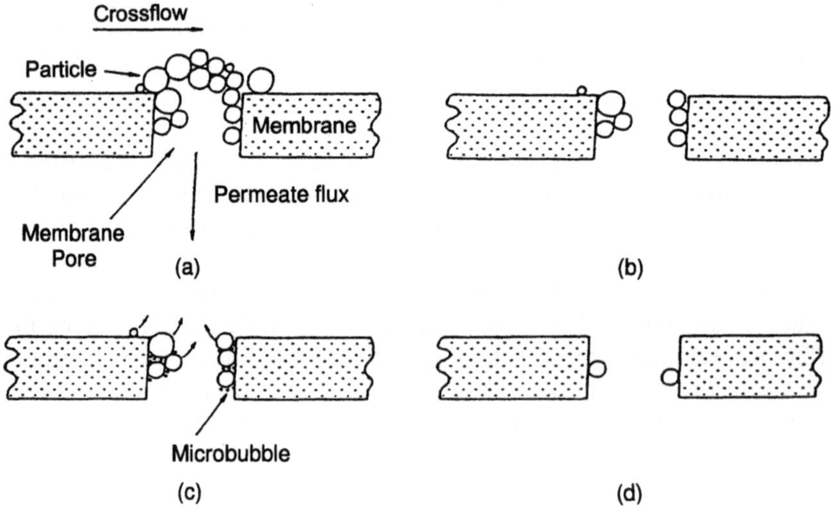

Figure 11.18 Possible mechanism for the synergistic coupling of electric and ultrasound force fields.

where the imposed electrical potential is below the threshold required to eliminate deposition, some particles can still be expected to foul the membrane. With the application of ultrasound it is postulated that the cavitational collapse of microbubble streamers in the pore entry regions can be sufficient to move the previously trapped particles to regions of more favourable field gradient. Here sufficient electrophoretic velocities are imparted to move the particles further into the bulk of the crossflowing stream (c and d).

11.6 The potential advantages of field-assisted filtration

To increase the filtration rate is not necessarily a sufficient criterion to assess filter performance as the energy consumed in achieving that rate can be equally as important. There is probably an optimum, which is specific to each filter application, between increasing the rate of separation and the added expenditure on energy. In conventional microfiltration a substantial amount of energy can be consumed by providing pumped crossflows at velocities up to $8\,m\,s^{-1}$. In combined field crossflow filtration additional contributions to energy consumption arise from the generation of the dc electric field and the ultrasonic field. Figure 11.19 gives an indication of the power requirements for the imposed field test data presented in Figures 11.14 and 11.15 and some comparison experiments using similar suspensions, no force fields and crossflow velocities of $2.3\,m\,s^{-1}$. Whilst it

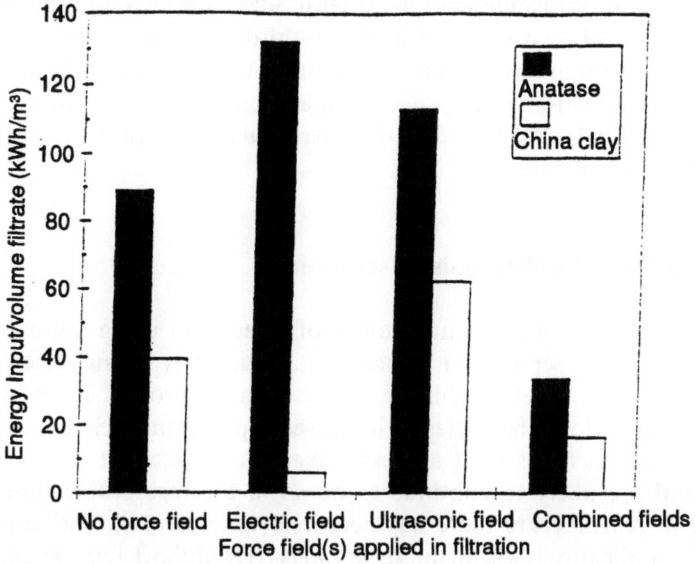

Figure 11.19 Energy inputs during no field and imposed field(s) microfiltration.

should be realized that overall power inputs with imposed fields are generally higher than corresponding arrangements with no fields, the energy required to produce a unit volume of filtrate can be decreased significantly for a range of suspensions. Moreover, for the data shown in Figures 11.14 and 11.15 the time taken to extract a unit volume of filtrate from each suspension was reduced by 18 and ten times, respectively when combined electric and ultrasonic fields were applied.

The low crossflow velocities employed in field-assisted crossflow filtration also offer several other potential advantages. A high proportion of the operating cost involved with conventional membrane separations is associated with the pumping of the feed suspension. The need to provide velocities up to $8\,m\,s^{-1}$ is removed in field-assisted membrane separations where velocities as low as $0.1\,m\,s^{-1}$ can be utilized profitably. A reduced pumping duty also implies that less heat is dissipated into a process stream and that the processing of shear sensitive streams can be improved as less denaturing is likely to take place. Finally, by providing for improved flux rates the possibility of using smaller equipment through process intensification becomes more of a reality.

Although the experimental data shown in this chapter are encouraging they should be viewed in the light that (to date) little attempt has been made to design a filter that uses a minimum of power, either by imposed electric or ultrasonic fields. In field enhanced filtration it is considered feasible to reduce the energy consumed by an electric field by at least 25%, and that consumed by an ultrasonic field by at least 50%. This would reduce power input levels to between one-half and two-thirds of those shown in Figure 11.19 whilst maintaining similar filtration rates. Should such a goal be achievable then field-assisted processes may compare very favourably with more conventional crossflow filtration, particularly for difficult-to-filter or 'high value' suspensions.

11.7 The future for field-assisted separations

There is currently a significant amount of interest in using imposed force fields to enhance separation processes, particularly using electric and ultrasound fields. Whether this be in the area of process intensification, improved separation efficiencies or increased separation rates it is clear that many potential benefits exist and new avenues of research are constantly arising, and are likely to continue to do so in the foreseeable future. For example, in recent years there has been relevant research, and sometimes industrial application, within biotechnology (Dobhlhoff-Dier *et al.*, 1994; Hawkes *et al.*, 1997; Miles *et al.*, 1995; Davies, 1994), food processing (Muralidhara *et al.*, 1988) and the enhanced separation of various particulates/molecules (Wakeman and Sabri, 1995; Akay and Wakeman, 1996; Jurado and Bellhouse, 1993; Jagannadh and Muralidhara, 1996). Each of these applications utilize force fields to either alter feed stream properties or promote beneficial particulate motions. The principles highlighted in this chapter are believed to be applicable to a range of membrane processes, feed suspensions and process conditions and the information in Table 11.2 provides a summary of the influence of combined electric and ultrasound fields in crossflow separation. With membranes likely to become an increasingly important feature of separations technology the need to provide systems which control fouling has never been greater. Moreover, the processing of finer particle suspensions is an increasingly important requirement for industrial applications. Field-

Table 11.2 Typical effect of parameters on combined field crossflow filtration

Parameter (increasing magnitude)	Influence on filtration rate
Electric field strength	Increase
Ultrasound frequency	Decrease
Ultrasonic field strength	Increase
Crossflow velocity	Increase (diameter $< 6\ \mu m$), decrease (diameter $> 6\ \mu m$)
Particle size (diameter)	Decrease (upper limit 8 μm)
Liquid viscosity	Decrease (upper limit 10 cP)
Feed concentration	Increase

assisted and combined field separations would appear to provide one technology by which our future needs can be realized.

11.8 References

Akay G. and Wakeman R.J. (1996) Electric field intensification of surfactant mediated separation processes. *Trans. IChemE*, **74**(A), 517.

Anada H.R., Shah Y.T. and KIinzing G.E. (1974) Methods of improving settling of industrial pigment slurries. *Can. J. Chem. Engng*, **52**, 715.

Creighton H.J. (1943) *Principles and Applications of Electrochemistry*, Vol. 1, pp. 143–65, Chapman & Hall, London.

Davies G.B. (1994) The use of ultrasonics in beer clarification. *Proc. IChemE Research Event*, pp. 541–3, IChemE, Edinburgh.

Dobhlhoff-Dier O., Gaida Th., Katinger H., Burger W., Groschl M. and Benes E. (1994) A novel ultrasonic resonance field device for the retention of animal cells. *Biotech. Prog.*, **10**, 428.

Fairbanks H.V. (1973) Use of ultrasound to increase filtration rate. *Proc. Ultrasonics Symposium*, pp. 11–15, IPC, Guildford.

Fane A.G. (1986) Ultrafiltration: factors influencing flux and rejection, in *Progress in Filtration and Separation,* Vol. 4, Editor R.J. Wakeman, Elsevier, Amsterdam, pp. 101–79.

Hawkes J.J., Limaye M.S. and Coakley W. T. (1997) Filtration of bacteria and yeast by ultrasound-enhanced sedimentation. *J. Appl. Microbiol.*, **82**, 39.

Jurado J. and Bellhouse B.J. (1993) The application of electric fields and vortex mixing for enhanced ultrafiltration. *Proc. Filtech Conference*, pp. 249–264, Filtration Society, Karlsruhe. Germany.

Jagannadh S.N. and Muralidhara H.S. (1996) Electrokinetic methods to control membrane fouling. *Ind. and Eng. Chem. Res.*, **35**(4), 1133.

Lu W.-M. and Ju S.-C. (1989) Selective particle deposition in crossflow filtration. *Sep. Sci. Tech.*, **24**(7/8), 517.

Mason T.J. (1993) Sonochemistry: a technology for tomorrow. *Chem. & Ind.*, 18 January, 47.

Miles C.A., Morley M.J., Hudson W.R. and Mackey B.M. (1995) Principles of separating micro-organisms from suspensions using ultrasound. *J. Appl. Bacteriol.*, **78**(1), 47.

Miyazaki K., Takahashi Y. and Shiomi K. (1988) Particle classification in a slurry with an ultrasonic filter. *Proc. Conference on Particle Technology in Relation to Filtration & Separation*, pp. 3.121–3.132, K.VIV, Antwerp.

Muralidhara H.S., Ensminger D. and Putnam A. (1985) Acoustic dewatering and drying (low and high frequency). *Drying Technol.*, **3**(4), 529.

Muralidhara H.S., Senapati N., Ensminger D. and Chauhan S.P. (1986) Electro-acoustic separation process for fine particle suspension. *Filtration & Separation,* **23**(6), 351.

Muralidhara H.S., Hampel H. and Kroebning H. (1988a) Dewatering of Hamburg dredged material by electroacoustic dewatering. *Drying Technol.*, **6**(3), 535.

Muralidhara H.S., Chauhan S.P., Senapati N., Beard R., Jirjis B. and Kum B.C. (1988b) Electroacoustic dewatering (EAD) a novel approach for food processing and recovery. *Sep. Sci. Tech.*, **33**, 2143.

Rautenbach R. and Albrecht R. (1989) *Membrane Processes*, John Wiley, Chichester.

Semmelink A. (1973) Ultrasonically enhanced liquid filtering. *Proc. Ultrasonics Symposium*, IPC, Guildford, pp. 7–10.

Sollner K. (1950) Colloidal effects of ultrasonics. *Ultrasonics Symposium Series*, **47**, 715.

Suslick K.S., Editor *Ultrasound*, VCH, New York.

Tarleton E.S. (1988) How electric and ultrasonic fields assist membrane filtration. *Filtration & Separation*, **25**(6), 402.

Tarleton E.S. (1992) The role of field assisted techniques in solid/liquid separation. *Filtration & Separation*, **29**(3), 246.

Tarleton E.S. and Wakeman R.J. (1988) Prevention of flux decline in electrical microfiltration. *Drying Technol.*, **6**(3), 547.

Tarleton E.S. and Wakeman R.J. (1990) Microfiltration enhancement by electrical and ultrasonic force fields. *Filtration & Separation*, **27**(3), 192.

Tarleton E.S. and Wakeman R.J. (1993) Understanding flux decline in crossflow microfiltration: Part 1 – Effects of particle and pore size. *Trans. IChemE*, **71**(A), 399.

Tarleton E.S. and Wakeman R.J. (1994a) Understanding flux decline in crossflow microfiltration: Part 2 – Effects of process parameters. *Trans. IChemE*, **72**(A), 431.

Tarleton E.S. and Wakeman R.J. (1994b) Understanding flux decline in crossflow microfiltration: Part 3 – Effects of membrane morphology. *Trans. IChemE*, **72**(A), 521.

Urick R.J. (1948) The absorption of sound in suspensions of irregular particles. *J. Acous. Soc. Am.*, **20**(3), 283.

Visvanathan C. and Ben Aim R. (1989) Application of an electric field for the reduction of particle and colloidal membrane fouling in crossflow microfiltration. *Sep. Sci. Tech.*, **24**(5/6), 383.

Wakeman R.J. and Sabri M.N. (1995) Utilising pulsed electric fields in crossflow microfiltration of titania suspensions. *Trans. IChemE*, **73**(A), 455.

Wakeman R.J. and Tarleton E.S. (1987) Membrane fouling prevention in crossflow microfiltration by the use of electric fields. *Chem. Engng Sci.*, **42**(4), 829.

Wakeman R.J. and Tarleton, E.S. (1991) An experimental study of electroacoustic crossflow microfiltration. *Trans. IChemE*, **69**(A), 386.

Weissler A. (1950) Physico-chemical effects of ultrasonics. *Ultrasonics Symposium Series*, **47**, 22.

12 Ultrasonic monitoring of shelf life in food emulsions

M.M. ROBINS

12.1 Introduction

Emulsions are physical mixtures of oil and water, one phase usually dispersed as droplets within the other, continuous, phase (Becher, 1983). The majority of food emulsions are oil in water, where the oil is dispersed as droplets typically 0.5–10 μm in diameter in a continuous aqueous phase that contains other flavour ingredients such as sugar, acid and salt (Dickinson and Stainsby, 1982). There are a number of aspects to the concept of shelf life in food emulsions. Here we are concerned only with the changes in physical stability, i.e. the integrity and uniformity of the emulsion structure.

Oil-in-water emulsions are intrinsically unstable, and eventually will separate into the two component phases via coalescence of the droplets. Other instabilities that contribute to the functionality of the product are flocculation, where the droplets remain for a significant time as loose aggregates before coalescing into larger droplets, and creaming which is the movement of the buoyant oil droplets to the top of the container. Figure 12.1 illustrates the principal instability mechanisms.

The emulsions contain various additives to reduce the rate of separation. The flocculation stage, which precedes coalescence, is controlled by the long-range interactions between the droplets (Russel *et al.*, 1989). The rate of flocculation is reduced by designing repulsive interparticle interactions, using emulsifier molecules that adsorb at the oil–water interface, thus providing a protective layer on the droplets (Hunter, 1992). These surface-active molecules may be charged, so that the droplets repel each other by electrostatic forces (charged surfactants), or possess a dominant, bulky, hydrophilic end-group which maintains an aqueous layer between two approaching droplets (steric surfactants). When droplets have overcome the long-range repulsions and are in close proximity, the rate of coalescence depends on the interfacial surfactant film.

Creaming is universal in emulsions where there is a significant density difference between the oil and aqueous phases, and the droplets are large enough for diffusion to be unimportant (typically over 1 μm). The rate of

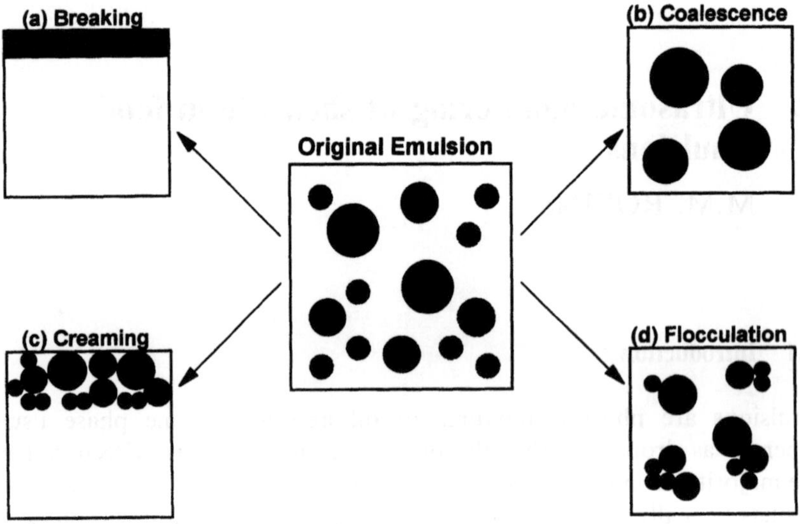

Figure 12.1 Emulsion instability mechanisms.

creaming increases with droplet size so it is aggravated by coalescence. The rate is often reduced by the addition of polysaccharide 'thickeners' which are conventionally thought to increase the viscosity of the continuous phase. In practice, as will be seen below, the polymer acts by causing the droplets to flocculate into a weak particulate network, which confers temporary stability to gravitational separation.

One of the problems of investigating emulsions is that most commercial emulsions are opaque to light and although considerable instability may be present, it is initially invisible. Attempts have been made to accelerate the structural collapse, for example by storage at elevated temperatures, but whether these give a reliable predictor of the long-term behaviour is highly dependent on the detailed instability mechanisms. The characteristic polydispersity of droplet size displayed by food emulsions is another complication, as a considerable amount of oil may have creamed to the top of a container, but the small droplets remaining in suspension maintain the product's opacity, for a while at least.

The requirement to detect separation occurring in emulsions at an early stage, for rapid product development, has led a number of workers to investigate the use of ultrasound as a probe of emulsion structure (Javanaud, 1988; Povey 1989). Although opaque to light, emulsions are transparent to ultrasound up to very high oil concentrations. The velocity of ultrasound is usually different in the oil and aqueous phases, and the velocity through the emulsion is an indicator of the relative amounts of oil and water. Since any movement of droplets due to creaming changes the vertical distribution of oil, measuring the ultrasonic velocity through the emulsion as a function of height provides a profile of oil concentration. This principle is the basis of the IFR ultrasonic creaming monitor (Howe *et al.*, 1986) described in the next section.

12.2 Ultrasonic creaming monitor

The instrument is shown in Figure 12.2. The sample cell, typically of horizontal dimensions $32 \times 32\,mm^2$ and 150 mm in height, is immersed in a water bath. The bath is for temperature control and to enable ultrasonic coupling between the transducers and the sample. The transducers are mounted either side of the cell, and can be moved vertically to access different sample heights. The instrument measures the time of flight of a pulse of ultrasound (typically with a centre frequency of 6.4 MHz) as it travels through the sample. The time of flight is converted to a group velocity of ultrasound by calibrating the cell dimensions using two known fluids. The measured velocity of ultrasound through an emulsion of known oil content is shown in Figure 12.3. In many cases it is possible to fit a theoretical relationship based on simple mixing (Urick, 1947), also shown in Figure 12.3. In other cases, more complicated analysis is required (Pinfield *et al.*, 1996). Thus the collection of time-of-flight data as a function of height enables the oil concentration as a function of height to be determined. The precision of the measurements approaches ± 0.1 m/s, which converts to

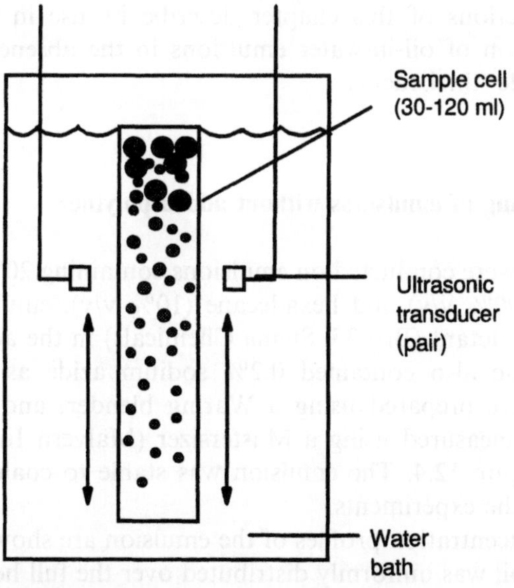

Figure 12.2 Ultrasonic creaming monitor.

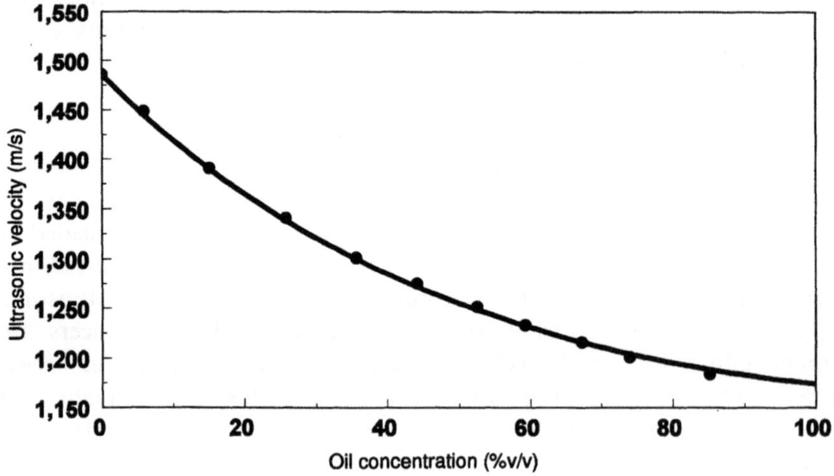

Figure 12.3 Ultrasonic velocity and oil concentration. The measured data (•) are compared with the prediction (——) of Urick (1947) which is based on the density and ultrasonic velocity of each phase alone.

±0.03% oil for alkane emulsions and typically ±0.2% oil for vegetable oil emulsions. The transducers are 10 mm in diameter, but in practice the vertical resolution is better than 2 mm. A detailed description of an automated ultrasonic creaming monitor, based on the IFR instrument, is given in Povey (1995, 1997).

The instrument may be used on a wide range of dispersions, and the remaining sections of this chapter describe its use in the detection and characterization of oil-in-water emulsions in the absence and presence of polysaccharide 'thickeners'.

12.3 Creaming of emulsions without added polymer

Experiments were conducted on emulsions containing 20% v/v of a mixture of heptane (90% v/v) and hexadecane (10% v/v), emulsified with 0.35% non-ionic surfactant (Brij 35, Sigma Chemicals) in the aqueous phase. The aqueous phase also contained 0.2% sodium azide as preservative. The emulsions were prepared using a Waring blender, and their droplet size distribution, measured using a Mastersizer (Malvern Instruments, UK) is shown in Figure 12.4. The emulsion was stable to coalescence during the timescale of the experiments.

The oil concentration profiles of the emulsion are shown in Figure 12.5a. Initially the oil was uniformly distributed over the full height of the cell, as shown by the initial horizontal line (shown in bold) measured after 0.092

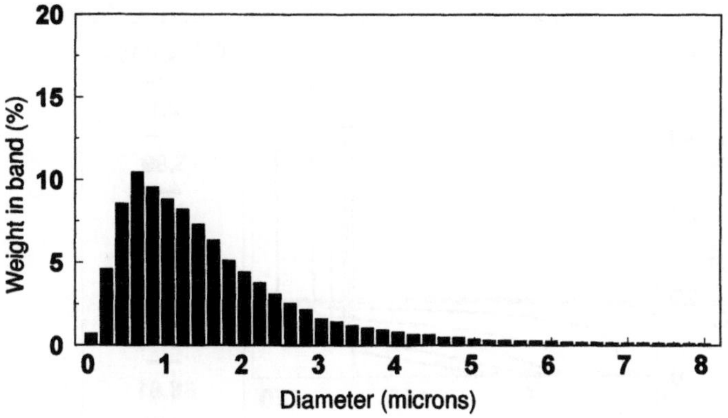

Figure 12.4 Droplet size distribution in emulsions, measured on diluted samples using a light diffraction particle sizer.

days. After a few days, however, the data show that the oil began to accumulate at the top of the cell, forming a concentrated cream layer of over 70% oil. The oil concentration gradually decreased at the base of the cell, until, after over two months, all the oil had reached the cream layer (dotted line in Figure 12.5a).

The creaming rate ν_s of particles in a liquid is given by Stokes' law (Batchelor, 1972)

$$\nu_s = \frac{\Delta\rho \cdot d^2 \cdot g}{18 \cdot \eta} \tag{12.1}$$

where d is the (spherical) particle diameter, $\Delta\rho$ the difference in density between the dispersed and continuous phases, g the acceleration due to gravity and η the viscosity of the continuous phase. Stokes' law is valid for very dilute dispersions after the particles have reached terminal velocity. It is obtained by equating the upthrust on the particles with the viscous drag imposed on them by the surrounding liquid. For dispersions of $>1\%$ concentration, the velocity is reduced significantly by the presence of neighbouring particles which serve to hinder the rise of any single particle. Particle interactions such as flocculation also alter the terminal velocity and these interactions can be included explicitly in a creaming model (Pinfield *et al.*, 1997). However, a simple way to incorporate the effect of particle concentration on the creaming rate is to treat it as an effective medium of increased viscosity (Ball and Richmond, 1980).

$$\eta = \eta_0 \left(1 - \frac{\phi}{\phi_m}\right)^{-2.5\phi_m} \tag{12.2}$$

(a)

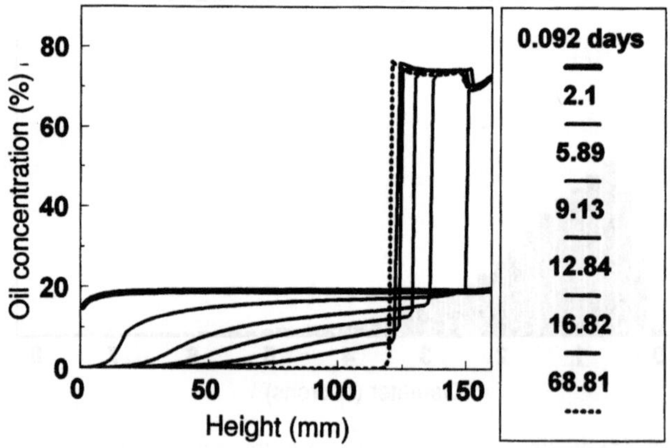

(b)

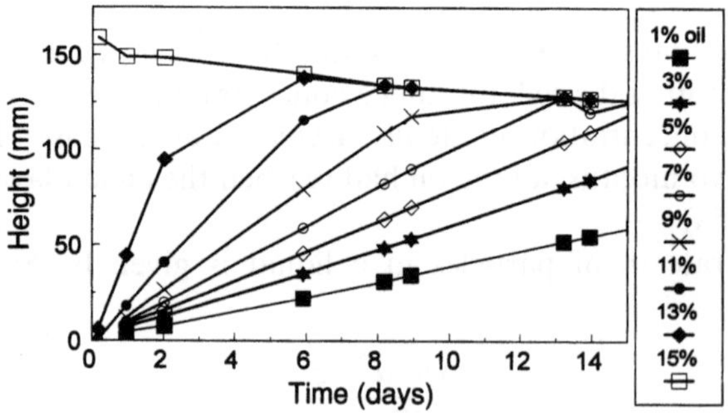

(c)

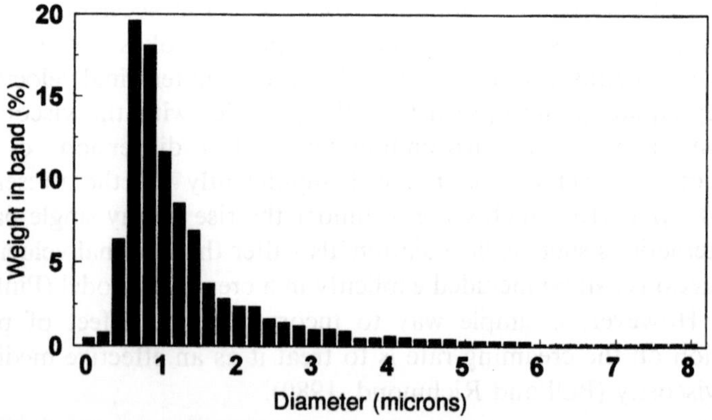

Figure 12.5 Creaming of emulsion without added polymer. (a) Concentration contours measured using the ultrasonic creaming monitor showing the progression from uniform (heavy line) to fully creamed state (dotted line). (b) Height of contours of constant oil concentration with time. The gradient of each contour is related to the Stokes' velocity of the size fraction represented by each concentration. (c) The droplet size distribution inferred from the creaming contour velocities.

where η_0 is the viscosity of the continuous phase alone, ϕ is the local oil volume fraction and ϕ_m is the volume fraction of oil at which no creaming can occur because the droplets are effectively close-packed.

Stokes' law thus enables prediction of the terminal creaming velocity of the droplets of known diameter. If the droplets are all the same diameter then we expect the creaming to be at a single rate, with a sharp boundary between the emulsion and the continuous phase layer left behind. The rate of rise of the boundary, ν, is thus governed by the droplets' Stokes velocity, modified by the hindrance factor

$$\nu = \frac{\Delta\rho \cdot d^2 \cdot g}{18 \cdot \eta_0}\left(1 - \frac{\phi}{\phi_m}\right)^{2.5\phi_m} \tag{12.3}$$

In the case of polydisperse droplets, if they move individually then there is a range of relevant Stokes' velocities. This leads to diffuse boundaries in the creaming emulsion, as the faster-moving (larger) droplets are depleted from the base at an early stage, leaving behind the smaller fraction. The diffuse boundaries seen in Figure 12.5a are indicative of a range of droplet velocities, due to a range of droplet diameters.

By the same reasoning, we can analyse the creaming rates observed and obtain the effective droplet size distribution in the creaming emulsion from equation (12.3). This enables us to check whether the droplets are indeed moving individually, or whether flocculation has occurred. The analysis of creaming rate is simple for monodisperse particles, as a clear boundary is visible and it is straightforward to measure its rate of rise up the cell. With the polydisperse droplets, exhibiting a range of creaming rates, our approach is to examine contours of constant oil concentration as these move with time. For example, consider the boundary corresponding to 10% oil (half the total oil concentration). The change in local concentration from the original 20% to 10% indicates that half the droplets have already cleared that height but the smaller half are moving more slowly. By measuring the rate of rise of the contour representing 10% oil we obtain the velocity of the 'average' (median) droplet fraction. Extending this to other oil concentrations we convert the data in Figure 12.5a to height/time graphs as in Figure 12.5b. It is clear that each fraction of oil represented by contours of different concentration rises with constant speed until it reaches the cream layer at a height of 120 mm. The speed of each contour is a measure of the Stokes' velocity (with the hindrance factor) of each size fraction. The linearity of the contours indicates that the creaming droplets are not changing size by flocculation or coalescence on the timescale of creaming.

The speed of each fraction of oil (contour) can thus be used to infer the effective hydrodynamic size distribution of the droplets using equation (12.3). Figure 12.5c shows the droplet diameters obtained from the creaming contours. The size distribution is consistent with that measured by the

Mastersizer (Figure 12.4), showing that the assumption of individual droplet movement according to Stokes' velocity is justified (Carter *et al.*, 1988).

12.4 Creaming of emulsions with low concentrations of non-adsorbing polymer

Figure 12.6a shows the creaming behaviour of an emulsion whose continuous phase contained 0.02% of a polysaccharide, hydroxyethylcellulose (HEC, Natrosol 250HR from Hercules, UK). This emulsion had the same droplet size distribution as the previous emulsion and was also

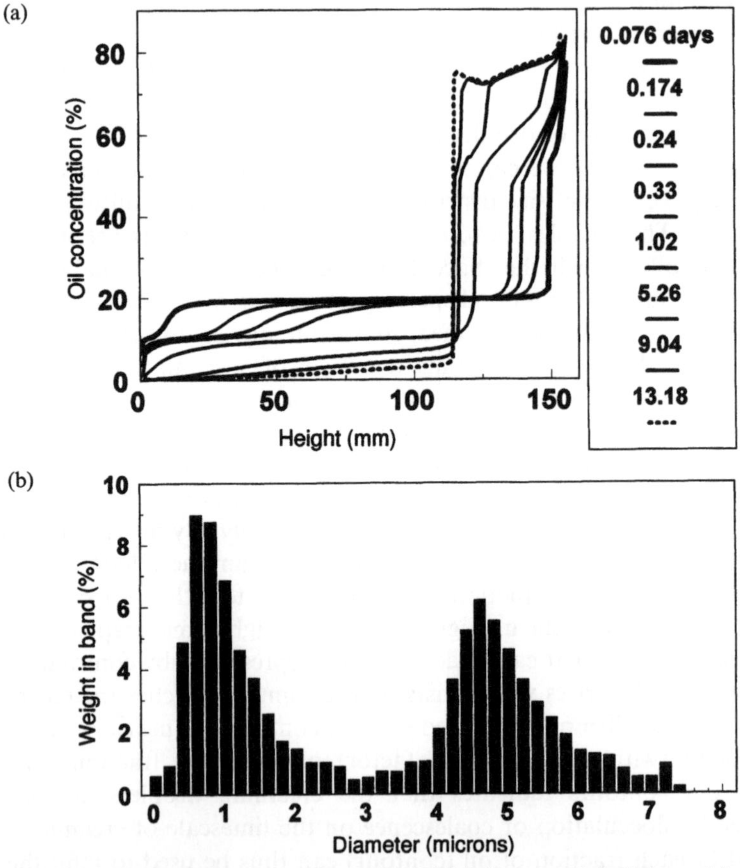

Figure 12.6 Creaming of emulsion with 0.02% hydroxyethylcellulose. (a) Concentration contours measured using the ultrasonic creaming monitor showing the progression from initial (heavy line) to fully creamed state (dotted line). (b) The droplet size distribution inferred from the creaming contour velocities. The larger fraction represents the equivalent hydrodynamic diameter of small flocs.

stable to coalescence. The creaming behaviour, however, showed two distinct populations of droplets, one fraction moving rapidly to the top of the container leaving behind about half the oil which creamed in a manner more similar to the polymer-free emulsion described above.

As before, the data can be analysed in terms of the rate of rise of contours of constant oil concentration. Analysis of the rates yielded the effective droplet size distribution in Figure 12.6b. In addition to the primary size distribution, there was a second fraction of droplets whose inferred diameters were much larger. These are droplets that flocculated and then creamed as small aggregates. The interesting point is that only part of the emulsion became flocculated, and that a flocculated fraction was in coexistence with the primary particles (Fillery-Travis *et al.*, 1993). This phenomenon, however, is predicted to occur when non-adsorbing polymer is added to an otherwise stable suspension or emulsion (Napper, 1983, Russel *et al.*, 1989).

The degree of flocculation varied with polymer concentration, in this case in the range 0.01–0.04% HEC. The creaming profiles enable the amount of flocculation to be determined accurately. Figure 12.7 shows that the ratio of flocculated to unflocculated droplets vary exponentially with the concentration of HEC. Data for similar emulsions of 20% alkane with the same droplet size distribution but containing xanthan (SBI Industries, Carentan, France) are also shown in Figure 12.7. The same dependence on polymer concentration is observed, although the large xanthan molecules act as an effective flocculant at lower concentration than HEC (see also Dickinson *et al.*, 1994a, b). Both are behaving as non-adsorbing (depletion) flocculants.

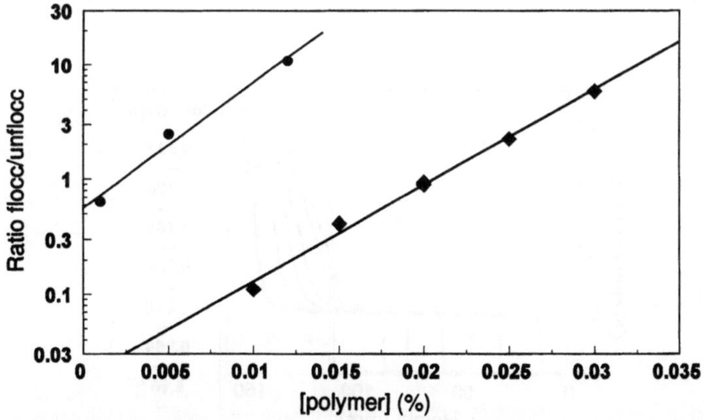

Figure 12.7 Degree of flocculation and polymer concentration in the coexistence regime for emulsions containing hydroxyethylcellulose (◆) or xanthan (•). The data are expressed as the ratio of the concentration of flocculated droplets to the concentration of individual droplets.

12.5 Creaming of emulsions with high concentrations of non-adsorbing polymer

When the concentration of HEC was increased to 0.04% the emulsion creamed as a single entity, with a sharp boundary that was clearly visible. Figure 12.8 shows the data from the ultrasonic monitor. The rapid creaming resulted in a cream layer that initially was only at a concentration of ~68%, but with time the cream layer compacted to an even higher concentration than in the polymer-free emulsion.

This behaviour may be interpreted as a flocculated network. Each droplet is presumed to be in 'contact' with two or more neighbouring droplets, and the network creamed by a compaction mechanism. The individual droplets did not coalesce; if a sample was removed from the emulsion and diluted for sizing the original droplet size distribution (Figure 12.4) was obtained.

As the polymer concentration is further increased, the creaming becomes very slow, and may be inhibited completely on the timescale of the experiments. Figure 12.9a shows the concentration profiles of an emulsion containing 0.95% HEC. The creaming is slow, but there also is a significant delay before creaming starts. This is shown most clearly by plotting the boundary height as a function of time, as in Figure 12.9b. The delay has also been observed for emulsions containing xanthan (Parker *et al.*, 1995) and casein (Dickinson *et al.*, 1997), and Figure 12.10 shows that as the concentration of polymer increases the delay period increases dramatically (up to three months in the case of 0.5% xanthan).

The reasons for the delay are not yet well established, but a possible explanation invokes a yield stress in the flocculated network (Parker *et al.*, 1995). The interdroplet attraction is presumed to be strong enough to hold the droplets in the structure. However, we speculate that the droplets in the flocculated network are continually re-arranging to form more compact

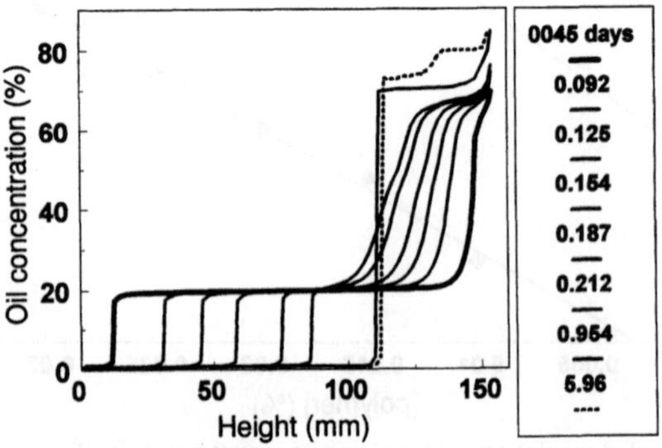

Figure 12.8 Oil concentration profiles during creaming of an emulsion with 0.04% hydroxyethylcellulose, showing a flocculated network.

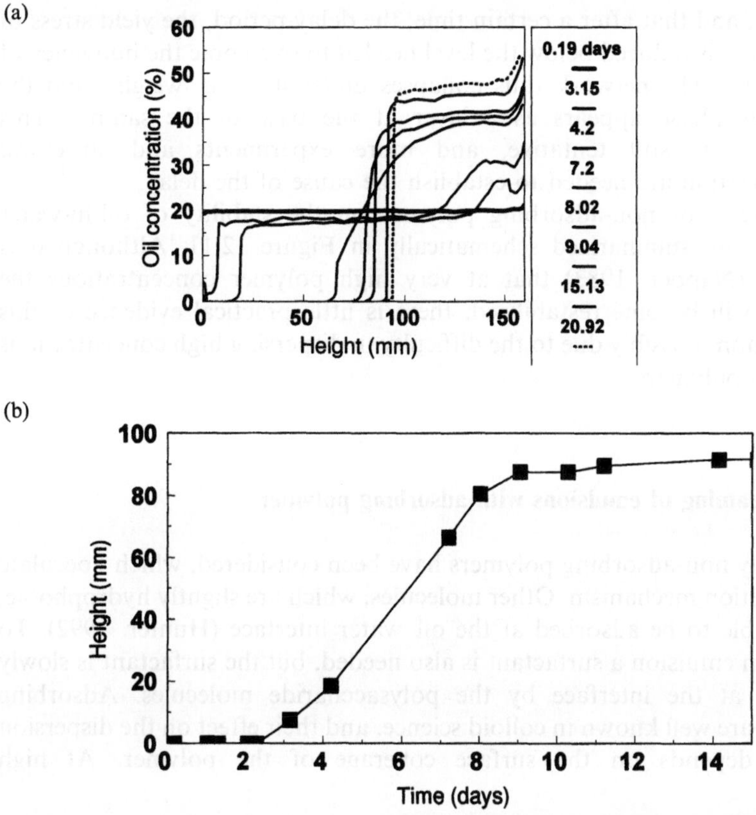

Figure 12.9 Creaming of emulsion with 0.95% hydroxyethylcellulose. (a) Oil concentration profiles showing a flocculated network. (b) Height of the base of the network with time, showing delay period.

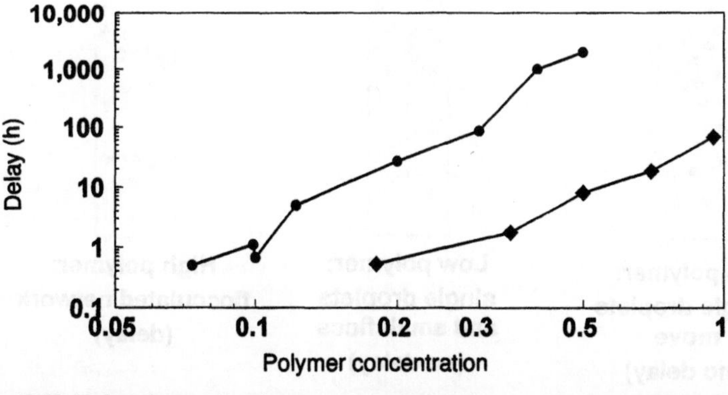

Figure 12.10 Delay time and polymer concentration for emulsions containing hydroxyethylcellulose (◆) or xanthan (●).

structures, and that after a certain time, the delay period, the yield stress of the structure is reduced below the level needed to overcome the buoyancy of the droplets. The network thus collapses under its own 'weight' and the continuous phase appears as a layer at the base of the sample. This explanation is still tentative, and more experiments and structural characterization are needed to establish the cause of the delay.

The effects of non-adsorbing polymer on the stability of oil-in-water emulsions are summarized schematically in Figure 12.11. Although it is predicted (Napper, 1983) that at very high polymer concentrations the emulsion will become restabilized, there is little practical evidence of this phenomenon, possibly due to the difficulty in dispersing high concentrations of viscous polymers.

12.6 Creaming of emulsions with adsorbing polymer

So far only non-adsorbing polymers have been considered, which flocculate by a depletion mechanism. Other molecules, which are slightly hydrophobic, may be able to be adsorbed at the oil–water interface (Hunter, 1992). To prepare an emulsion a surfactant is also needed, but the surfactant is slowly displaced at the interface by the polysaccharide molecules. Adsorbing polymers are well known in colloid science, and their effect on the dispersion stability depends on the surface coverage of the polymer. At high

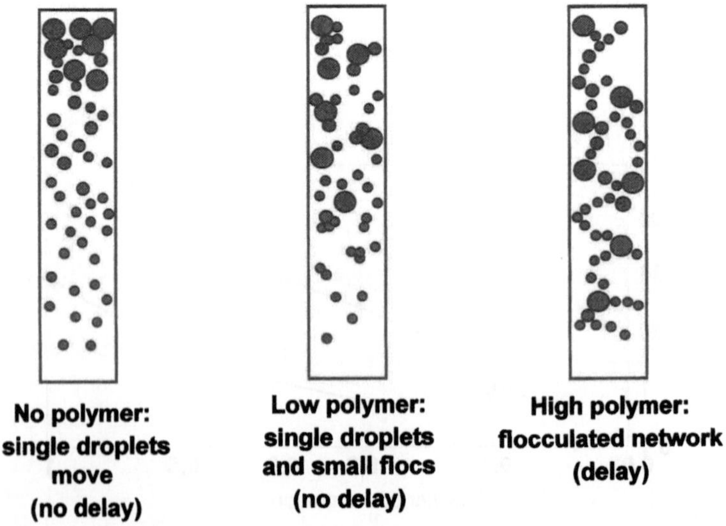

Figure 12.11 Schematic diagram of the effect of non-adsorbing polymer on the flocculation and creaming behaviour of emulsions.

concentrations, the particle surfaces are completely covered and flocculation is usually prevented. However, when the polymer is only able to cover part of the surface, then the polymer molecules may bridge between two droplets and cause persistent flocculation (Napper, 1983).

The adsorption process is likely to be slow compared with the timescale for depletion flocculation, where the depletion potential is present as soon as the polymer is added. Figure 12.12a shows the concentration profiles of an alkane-in-water emulsion containing a slightly hydrophobic polymer, hydroxypropyl cellulose (Klucel E, Aqualon, UK). The concentration profiles resembled those of the unflocculated system, with no sharp boundaries that might indicate a flocculated network. However, when the contours of constant oil concentration were derived from the creaming data, as in Figure 12.12b, they are distinct from those previously observed (Figure 12.5b). The contours did not rise linearly with a constant velocity showing

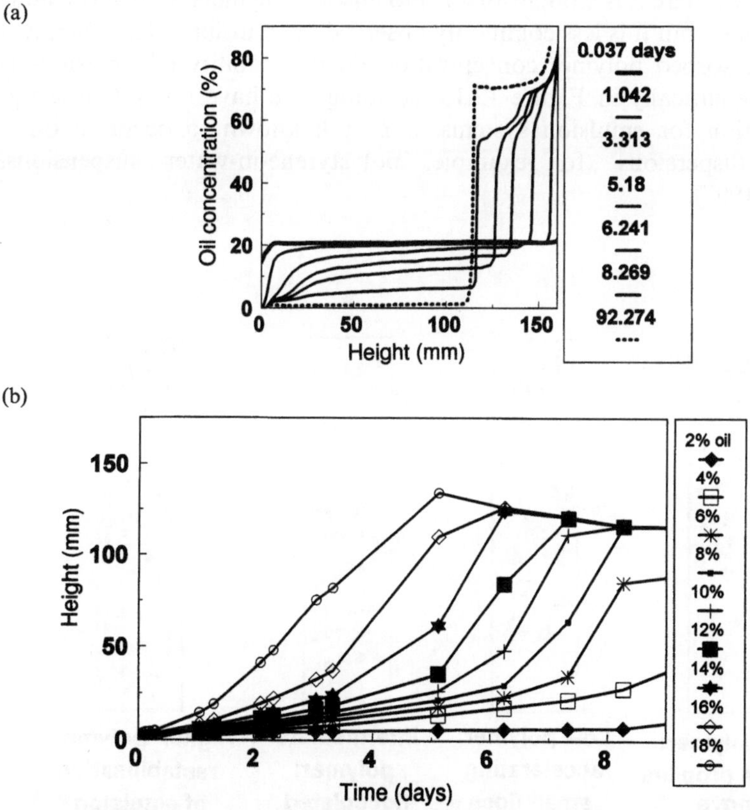

Figure 12.12 Creaming of emulsion with 0.9% hydroxypropylcellulose. (a) Concentration contours measured using the ultrasonic creaming monitor. (b) Height of contours of constant oil concentration with time showing increasing creaming velocities with time.

the steady terminal velocity of droplets or flocs of unchanging size. In this case the contours rose at increasing speed, which suggests that the creaming entities increased in size during the creaming process. The primary droplet distribution did not change, so there was no coalescence occurring, and the creaming was attributed to flocs that increased in size during the experiment. The creaming rates could not be used to derive a unique droplet size distribution, because the sizes were not constant. This behaviour is consistent with an emulsion that is simultaneously flocculating and creaming, due to slow adsorption of the polymer at the oil–water interface.

At higher polymer concentrations a network was formed, but usually it creamed a small amount first. It is not yet established whether there is a delay period in the case of bridging flocculation. The explanation offered above for the delay in depletion flocculation may also be relevant to bridging, but in the latter case it is likely that the droplets are less able to rearrange by rotating around each other.

Flocculation by a bridging mechanism is well known for particle suspensions, where it is straightforward to adsorb polymers onto surfactant-free interfaces, but it is less commonly observed in emulsions. The suggested effect of adsorbed polymer concentration on the stability of emulsions is shown schematically in Figure 12.13. Although we have not yet observed restabilization for emulsion systems, it is well known to occur in other colloidal dispersions (for example, polystyrene-in-water suspensions, Hibberd, 1997).

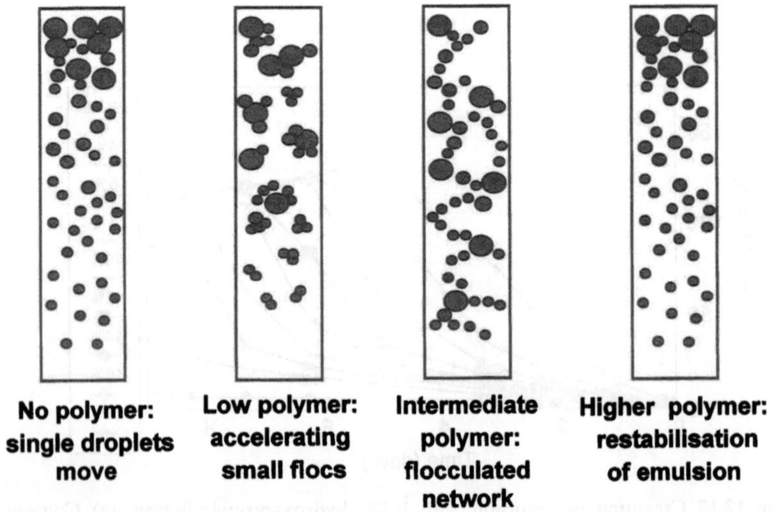

Figure 12.13 Schematic diagram of the effect of adsorbing polymer on the flocculation and creaming behaviour of emulsions.

12.7 Summary

The physical appearance of food emulsions is an important factor in their shelf life. Frequently the structure separates over a period of time to form an unsightly layer of oil at the top, and/or clear aqueous phase at the base. This separation, caused by the density difference between the oil and aqueous phases, may be detected at an early stage using the IFR ultrasonic creaming monitor.

The ultrasonic creaming instrument provides detailed profiles of oil concentration as a function of height and time. As well as early detection of separation, the concentration profiles enable us to estimate the particle size of the creaming droplets. This is an advantage when structures are sensitive to the intrusive sampling and dilution procedures needed for conventional particle sizing.

In the presence of polymers, the droplets are frequently flocculated, either by a depletion or bridging mechanism. The ultrasonic creaming data enable the degree of flocculation to be determined, and distinguish between the behaviour of emulsions flocculated by the two mechanisms. It thus provides an insight into the properties of complex multiphase systems such as food emulsions.

Acknowledgements

The work described here was carried out by scientists at the Institute of Food Research, in particular David Hibberd, Annette Fillery-Travis and Paul Gunning. We are grateful to the UK Ministry of Agriculture, Fisheries and Food, Shell UK and Système Bioindustries, France for past funding. The Biotechnology and Biological Sciences Research Council under the ROPA scheme currently supports work in this area.

12.8 References

Ball R.C. and Richmond P. (1980) Dynamics of colloidal suspensions. *Phys. Chem. Liq.*, **9**, 99–116.

Batchelor G.K. (1972) Sedimentation in a dilute polydisperse system of interacting spheres, *J. Fluid Mech.*, **52**, 245–68.

Becher P., Editor (1983) *Encyclopedia of Emulsion Technology*, Vol. 1, Marcel Dekker, New York.

Carter C., Hibberd D.J., Howe A.M. Mackie A.R. and Robins M.M. (1988) Non-intrusive determination of particle size distribution in a concentrated dispersion. *Prog. Colloid Poly Sci.*, **76,** 37–41.

Dickinson E. and Stainsby G. (1982) *Colloids in Food*, Applied Science, Essex.

Dickinson E., Ma J. and Povey M.J.W. (1994a) Creaming of concentrated oil-in-water emulsions containing xanthan. *Food Hydrocolloids*, **8**, 481–97.

Dickinson E., Ma J., Pinfield V.J. and Povey M.J.W. (1994b) Ultrasonic studies of the creaming of concentrated emulsions, in *Food Macromolecules and Colloids*, Editors E. Dickinson and D. Lorient, Royal Society of Chemistry, London, pp. 223–34.

Dickinson E., Golding M. and Povey M.J.W. (1997) Creaming and flocculation of oil-in-water emulsions containing sodium caseinate. *J. Coll. Interface Sci.*, **185**, 515–29.

Fillery-Travis A.J., Gunning P.A., Hibberd D.J. and Robins M.M. (1993) *J. Coll. Int. Sci.*, **159**, 189–97.

Hibberd D.J. (1997) Use of ultrasound to characterise polymer-induced flocculation. PhD thesis, University of East Anglia.

Howe A.M., Mackie A.R. and Robins M.M. (1986) Technique to measure emulsion creaming by velocity of ultrasound, *J. Disp. Sci. Technol.*, **7**, 231–43.

Hunter R.J. (1992) *Foundation of Colloid Science*, Oxford University Press, Oxford.

Javanaud C. (1988) Applications of ultrasound to food systems. *Ultrasonics*, **26**, 117–23.

Napper D.H. (1983) *Polymeric Stabilization of Colloidal Dispersions*, Academic Press, London.

Parker A., Gunning P.A., Ng K. and Robins M.M. (1995) How does xanthan stabilise salad dressing? *Food Hydrocolloids*, **9**, 333–42.

Pinfield V., Povey M.J.W. and Dickinson E. (1996) Interpretation of ultrasound velocity creaming profiles. *Ultrasonics*, **34**, 695–8.

Povey M.J.W. (1989) Ultrasonics in food engineering II, *J. Food Engng*, **9**, 1–20.

Povey M.J.W. (1995) Ultrasound studies of shelf-life and crystallization, in *New Physico-chemical Techniques for the Characterization of Complex Food Systems*, Editor E. Dickinson, Blackie, Glasgow, pp. 196–213.

Povey M.J.W. (1997) *Ultrasonic Techniques for Fluids Characterization*, Academic Press, San Diego, CA.

Russel W.B., Saville D.A. and Schowalter W.R. (1989) *Colloidal Dispersions*, Cambridge University Press, Cambridge.

Urick R.J. (1947) A sound velocity method for determining the compressibility of finely divided substances. *J. Appl. Phys.*, **18**, 983–7.

13 Ultrasound and the computer modelling of food

V.J. PINFIELD

13.1 Introduction

Low-power ultrasound is a useful diagnostic tool in the field of food colloids, especially for studying their stability (Fillery-Travis *et al.*, 1993; Dickinson *et al.*, 1993a, 1994, 1995, 1997). The ultrasound signal has no adverse effect on the sample being measured, and is used to determine some physical property of the sample, such as the oil concentration (in the case of an emulsion) or solid fat content (Dickinson *et al.*, 1991, 1993b). This chapter is concerned in particular with the application of ultrasound measurements to the detection of instability by creaming (gravitational instability) in oil-in-water emulsions. The effect on creaming behaviour of the addition of food components such as polymers and proteins, and changes in ionic strength holds information on the physical and chemical interactions within the emulsion. Computer modelling of these food systems can assist in understanding the emulsion behaviour detected using the ultrasound technique, and in identifying potential difficulties and new approaches to the interpretation of the ultrasound measurements. This chapter discusses two computer models of the creaming of oil-in-water emulsions, and how the models contribute to the understanding of ultrasound measurements on real systems.

Many foods are emulsions, i.e. dispersed particles of oil in an aqueous continuous phase, or vice versa. In this chapter, emulsions are hereafter assumed to be oil in water. Particle sizes vary, but are typically in the range of 0.1–10 μm, with an average around 0.5 μm. The system is stabilized against separation into its two phases by an emulsifier or stabilizer which adsorbs at the oil–water interface. One of the principal instability mechanisms in food emulsions is creaming, a gravitationally driven process caused by the density difference between the oil particles and the surrounding aqueous phase. The less-dense oil particles tend to rise through the emulsion, developing the familiar cream layer at the top, and an oil-deficient serum region at the bottom of the emulsion. Many food oils are of very similar density to water, and therefore creaming is relatively slow, perhaps allowing a year's storage before noticeable concentration changes occur.

However, emulsion behaviour can be strongly affected by typical modifications to food components, such as the addition of salt, or the presence of unadsorbed protein in the aqueous phase. Measurement of the creaming properties not only indicates the relative stability of these systems but also incorporates information on the fundamental processes occurring within the emulsion.

Ultrasound has been demonstrated previously to be a very useful technique for studying the creaming behaviour of emulsions. A variety of such measurements is discussed by Margaret Robins in Chapter 12, and the interpretation of ultrasound measurements is detailed by Malcolm Povey in Chapter 3. The experimental system used to make the measurements has been described elsewhere, but a schematic diagram is shown in Figure 13.1. The emulsion sample is placed in a square-section glass cell, which is mounted in a water bath (for temperature control and acoustic coupling). The piezoelectric transducers on either side of the cell act as a transmitter and a receiver of the low-power ultrasound signal. A timer-counter records the time-of-flight for the signal to pass between the transducers. The probes can be moved up and down to allow measurements to be taken at different vertical positions within the sample. The measured time-of-flight is converted into the ultrasound velocity in the emulsion by calibration against distilled water. The principle of the technique is that the ultrasound velocity is related to the volume fraction of oil at the measurement position. Thus, a time-series of concentration profiles as a function of height can be constructed from the measurements.

This technique can be used at a basic level for the prediction of long-term emulsion stability. Changes in ultrasound velocity at the top and bottom of the sample are very sensitive to the small initial changes in oil concentration, which are indicative of the onset of creaming. These changes are too small to

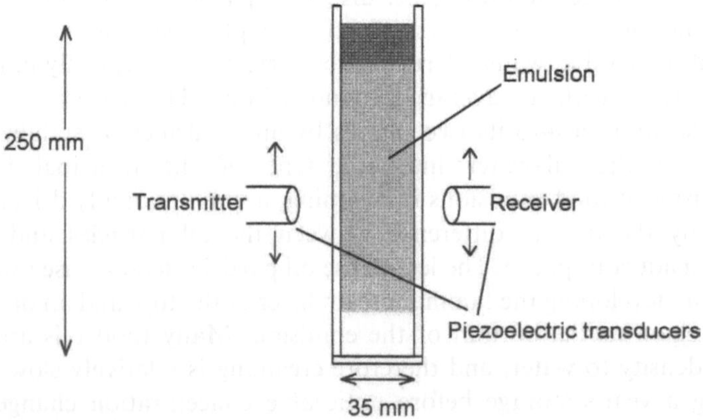

Figure 13.1 The ultrasound profiler used for studying creaming behaviour of emulsions.

be detected by eye or by turbidity measurements. On a more fundamental level, the measured concentration profiles contain a large quantity of information on the physical and chemical processes occurring within the emulsion which manifest themselves in a modification to the 'standard' creaming behaviour. A wide variety of different creaming behaviour has been observed by ultrasound measurements in such systems, demonstrating differing interparticle interactions which affect creaming. Hence, the technique is able to give some insight, by inference, into the particle-level processes in the emulsion, not just creaming itself.

13.2 The use of computer modelling

In order to maximize the useful information obtained from these measurements, a number of elements are essential:

- the interpretation of the ultrasound measurements as oil concentration profiles
- the relationship of these profiles to the fundamental physical processes in the emulsion
- the understanding of these processes in terms of the components in the system.

The first of these has been discussed elsewhere in the literature, as well as in this book; for non-flocculating emulsions in the long-wavelength limit, the theoretical analysis is well established. The second relates to the influence of factors such as the initial particle size distribution, concentration and particle interactions on creaming behaviour. Finally, it is necessary to interpret the manifested effects in terms of the contributions and interactions of the components in the system, e.g. by what mechanism do the particles flocculate and to what degree?

Having satisfactorily converted the ultrasound measurements into concentration profiles, the more difficult analysis is to understand the creaming behaviour that is observed in food systems. In many of these samples, the addition of non-adsorbing polymer leads to rapid creaming, or to stabilization of the emulsion through the induced flocculation (Dickinson *et al.*, 1995, 1997; Fillery-Travis *et al.*, 1993). However, it is not known how the flocculation processes lead to the detailed characteristics seen in the concentration profiles. In particular, at low concentrations of added polymer, depletion flocculation causes the rapid development of a clear serum layer at the bottom of the sample. The speed of this process and the subsequent lack of significant changes in the rest of the sample are not well understood in terms of floc formation and movement.

Computer modelling can assist in the second and third stages of the interpretation process. A computer model of the system behaviour is constructed with carefully defined components and physical and chemical interactions. The results of such a model can be compared with experiments, either to confirm the model (and hence any suggested interaction mechanisms or flocculation characteristics) or to identify any additional effects which were not included. This is the broad aim of computer modelling studies. There are also a number of more specific benefits arising from the use of computer models, and these are discussed in the succeeding paragraphs.

The development of a computer model forces the contributing processes to be quantified. Although it is simple to estimate, for example, the creaming speed and diffusion coefficient for the average particle size in an emulsion, it is far from straightforward to deduce the creaming behaviour of a concentrated polydisperse emulsion of such particles. The interactions between the individual particle motions lead to a non-trivial result. Computer modelling enables the individual processes to be defined, but for the overall resultant behaviour to be determined implicitly in the numerical technique. A computer model allows the microscopic processes occurring at the level of individual particles to be related to the macroscopically observed system behaviour. For example, the macroscopic creaming phenomenon is related to the force balance of the weight, buoyancy forces and viscous drag on a single particle. The 'creaming speed' of this single particle can be used to determine the volume flux of particles crossing a horizontal plane in any time interval, and hence the observable concentration changes. Similarly, the Stokes–Einstein thermal diffusion coefficient of a single particle is used in a macroscopic description of the effect of diffusion on a concentration gradient.

The effects of one component or process can be studied independently of other parameters in a computer model. For example, it is possible in a numerical system to 'switch off' the effects of thermal diffusion in order to assess its contribution to the creaming profiles. In a real emulsion, this could in effect be achieved by choosing larger particles, but this requires the modification of a second parameter (the particle size). Thus it is possible through modelling techniques to build up a more complete picture of how the different component processes contribute to the overall behaviour. A modelled system is fully characterized, e.g. the concentration and particle size distribution are always known. In creaming measurements, although the changes in concentration are measured by the ultrasound technique, changes in particle size distribution are not directly detected. Thus it is possible through computer modelling to examine the behaviour in more detail and also to assess the experimental technique. In the case of ultrasound creaming measurements, this is particularly concerned with the effects of particle size distribution on ultrasound velocity.

The benefits of computer modelling which have been described in the preceding paragraphs are now put into the context of understanding the creaming behaviour of emulsions studied using the ultrasound profiling technique. Two models of creaming are described, one on the scale of the experimental system, and the other an attempt to understand the possible influence of flocculation on creaming by studying the particle-level processes.

13.3 A macroscopic model

The first model, which is described here, represents the macroscopic creaming phenomenon on the scale of the experimental ultrasound profiler. This was intended to allow direct comparison between the model and experimental measurements. It is clear that at this scale, the emulsion behaviour can only be represented by a phenomenological description, i.e. one which describes the resultant macroscopic effects of the particle-level processes. It is not possible to trace the motion of all the particles in the system as in a molecular dynamics simulation (Allen and Tildesley, 1989), or a colloidal Brownian dynamics simulation (Ansell and Dickinson, 1986) or a lattice simulation (van der Knaap *et al.*, 1994). Hence at this scale, the significant contributing processes in the system are modelled by a suitable phenomenological description.

13.3.1 Description of model

The model has previously been described in some detail, although with minor differences (Pinfield *et al.*, 1994; Pinfield, 1996). A schematic diagram of the representation of an emulsion in the model is shown in Figure 13.2. The model is one-dimensional, considering flow only in the vertical

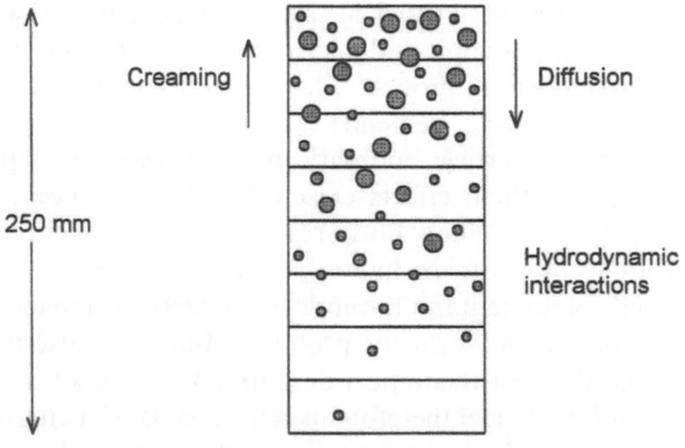

Figure 13.2 Pictorial representation of the macroscopic model of creaming. The sample properties are defined within each horizontal layer by the volume fraction of particles of each size at each time. The processes of creaming and diffusion and the hydrodynamic interactions are included in the model.

direction. Hence, there are effectively no walls to the 'cell', although the hydrodynamic effects of fluid flow within a finite container are included. The modelled emulsion is split into horizontal layers, as shown; 500 layers were used in the calculations, each 0.5 mm thick. The concentration of oil in each layer is defined by the volume fraction of particles of each size in that layer at a particular time. The particle size distribution in a polydisperse sample is divided into discrete size fractions.

Changes in the system are modelled by calculating the volume flux of particles of each size, which crosses between each pair of adjacent layers in a short time interval. This process is repeated many times to determine the developments over a substantial time period. In addition, special attention must be paid to the top and bottom extremes of the concentration profiles for each particle size fraction (Pinfield *et al.*, 1994). The flux across each layer boundary is determined by the physical processes that (in this model) are creaming and thermal diffusion, and by the local conditions (concentration, particle size distribution, etc.). Average hydrodynamic interactions are also included, which modify the effective creaming and diffusion properties of particles, due to the flow of the continuous phase fluid and of the other particles.

The use of the particle flux at the layer boundaries is a phenomenological description of the physical processes of creaming and diffusion. For creaming, this is a simple flux calculation for particles travelling at a defined speed. For diffusion, the rate is determined by the local concentration gradient, which is the familiar macroscopic description of diffusion. However, these macroscopic processes are related to the properties of individual particles. The creaming speed is calculated from the familiar Stokes' law (based on the force balance on a particle between the gravitational forces and the Stokes' viscous drag), and the diffusion rate is controlled by the Stokes–Einstein diffusion coefficient for a single particle with thermal energy. Hence the macroscopic and microscopic properties of the system are connected in the modelling process. The hydrodynamic effects of the continuous phase fluid flow can be defined from ensemble-averaged results for the modification of particle mobility in a dispersion. Unfortunately, analytical results for these effects exist only at low concentrations, and therefore successful modelling must resort to established semi-empirical relationships to account for the hydrodynamic interactions.

An additional component in this model of creaming is the inclusion of the calculation of ultrasound velocity profiles. Multiple scattering theory is considered to be the best theoretical description for sound propagation in such systems, and the model therefore uses the results of scattering theory to simulate the ultrasound velocity profiles that correspond to the model's predictions of creaming behaviour. The calculations include the effects of changes in particle size distribution in the sample as creaming proceeds. In the experimental system, only the volume fraction is deduced from the ultrasound velocity measurements, so changes in particle size distribution are undetected. The details of the scattering theory formulation used in the model are given elsewhere (e.g. McClements and Povey, 1989; Povey, 1997b). It is based on the scattering properties of a single particle (Epstein and Carhart, 1953; Allegra and Hawley, 1972) and the multiple scattering theory formulations for random distributions of scatterers (Waterman and Truell, 1961; Fikioris and Waterman, 1964; Lloyd and Berry, 1967). The reader is referred to Pinfield *et al.* (1994) for further details on all aspects of the model.

The model was used to calculate the expected creaming behaviour of a system with the properties of sunflower oil-in-water emulsion, such as might be studied experimentally. The parameters of the sample modelled are given in Table 13.1. The model does not include direct particle interactions or any effects caused by the presence of the emulsifier. Hence, any flocculation is neglected. The particle size distribution is log-normal and was taken to be similar to those measured experimentally in typical laboratory-produced emulsions.

13.3.2 Results of macroscopic model

The concentration profiles predicted by the macroscopic model for the sunflower oil-in-water emulsion are shown in Figure 13.3. The volume fraction is plotted as a function of height in the emulsion at different times (zero height corresponds to the bottom of the sample). This corresponds to the ultrasound profiler technique in which a series of scans over the height of the emulsion at various time intervals is used to build up a picture of the creaming process. The height of the sample modelled is the same as that of samples used in the experimental system. The general features of the profiles are as expected. The oil is less dense than the water, and so tends to rise to the top of the sample. Thus a more concentrated region forms at the top, and an oil-deficient region (the serum) at the bottom. The time-scale over which creaming occurs in this model is very long. In the absence of particle interactions that lead to instability, an emulsion such as this would be very stable, and the extent of creaming which has been modelled here would be unlikely to be observed in any experiment.

Table 13.1 The parameters of the sunflower oil-in-water emulsion modelled using the macroscopic creaming model

Concentration of sunflower oil	20 vol.%
Modal particle diameter	0.56 μm
98% volume limits	0.2–1.5 μm
Density difference between sunflower and water	83 kg m^{-3}
Sample height	250 mm
Temperature	30°C
Ultrasound velocity in water	1509 m s^{-1}
Ultrasound velocity in sunflower oil	1438 m s^{-1}

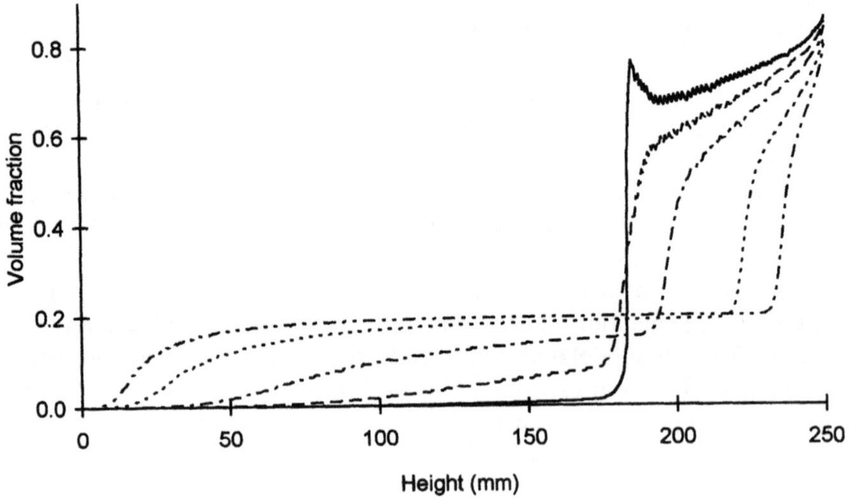

Figure 13.3 Concentration profiles predicted by the macroscopic model for the creaming behaviour of 20% sunflower oil-in-water emulsion. The profiles are after, ----, 50 days; ---, 100 days; ———, 250 days; ————, 500 days; ———, 1000 days.

In a monodisperse sample, the interface between the serum and the region of uniform emulsion is sharply defined (in the absence of thermal diffusion) and moves steadily up the sample (Pinfield, 1996). This is accompanied by an increase in the extent and concentration of the cream. In a polydisperse sample (as shown in Figure 13.3), no clear serum interface exists due to the variation in creaming speed of droplets of different sizes (larger particles have a larger creaming speed). Thus, the concentration profiles show a gradual increase in concentration with increasing height in the sample. The spread of the apparent 'serum interface' increases with time, resulting from the increase in separation of particles of different sizes which were initially at the bottom of the emulsion. The hydrodynamic formulations indicate that the flow of the continuous phase fluid downward is sufficient to carry the smaller particles with it, especially in the early stages of creaming. This accentuates the separation between particles of different sizes in the sample. Thermal diffusion has been shown to have little effect on the concentration profiles in an emulsion like this one (Pinfield, 1996). This may be unexpected for the colloidal particle size range used in the model, but the creaming behaviour is dominated by the larger particles for which diffusion is less significant. Diffusion does, however, have some influence on the particle size distribution in the cream region, but this may not be an observable effect on the macroscopic scale.

The development of the cream is also of interest. The concentration in the cream region increases with height, indicating also that any point within the cream becomes more concentrated as time progresses. The concentration in the cream is rather high compared with the maximum packing fraction for monodisperse particles (around 0.63). The hydrodynamic effects in the model are determined only by the local volume fraction and not by the particle size distribution. Although the creaming speed of all particles does decrease to zero as the concentration increases, the only limit on concentration is at a volume fraction of 1.0. Hence, the predictions of concentration and particle size distribution in the cream are not likely to be accurate but should be qualitatively correct. However, very high volume fractions such as those seen in Figure 13.3 have been observed in experimental measurements, as has the form of the concentration profile in the cream region (Fillery-Travis *et al.*, 1993; Dickinson *et al.*, 1995). The high concentrations are due to the range of particle sizes in the system, which permits more efficient and compact packing than is possible in a monodisperse sample. The peak in volume fraction which occurs after a long time at the bottom of the cream is probably due to the accumulation of relatively fast moving particles which meet the slow-moving particles in the cream layer. A similar feature has been observed experimentally, although it appeared at an earlier stage in the creaming process (Fillery-Travis *et al.*, 1993).

The use of a computer model allows us to assess the influence of the changes in the system on the interpretation of ultrasound measurements. The velocity of sound in sunflower oil and water are given in Table 13.1 and the ultrasound velocity profiles calculated in the model are shown in Figure 13.4. A decrease in ultrasound velocity in the emulsion might be expected where the oil concentration is higher. This is indeed true in most of the

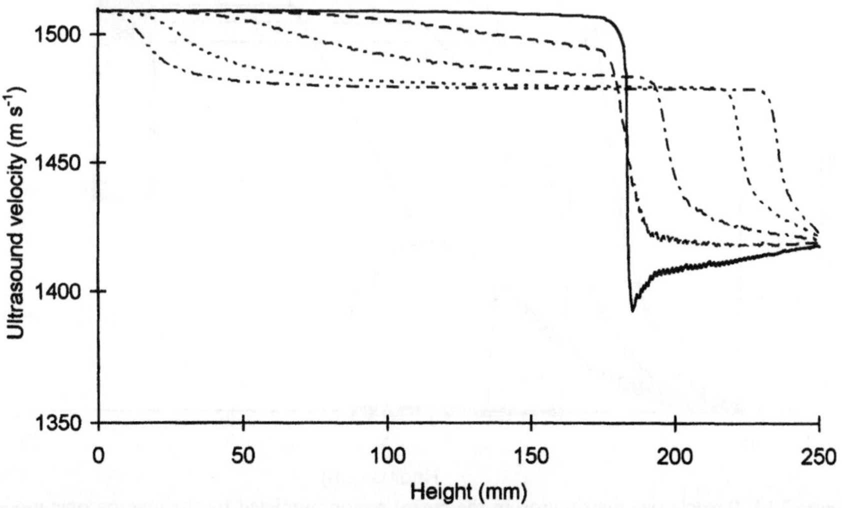

Figure 13.4 Ultrasound velocity profiles predicted by the macroscopic model and using multiple scattering theory, for sunflower oil-in-water emulsion. The legend is the same as for Fig. 13.3.

emulsion, and even in the cream in the early stages of creaming (up to around 250 days). However, in the later stages the variation in ultrasound velocity with height in the cream becomes flatter, and eventually shows an increase towards the top of the sample. The concentration profiles (Figure 13.3) show that the volume fraction of oil always increases towards the top of the sample, except for the small peak at the bottom of the cream. It is clear that ultrasound measurements on this particular emulsion would lead to incorrect results if they were interpreted solely on the basis of changes in oil concentration.

The ultrasound velocity profiles can be understood with reference to the variation in particle size distribution in the cream region (Figure 13.5). The initial size distribution is shown with the distributions at different heights in the cream for comparison. It is clear that a large change in average particle size compared with the initial emulsion can occur due to the relative creaming speeds of different-sized particles. Although the spread of particle size in the initial emulsion has little effect on the sound velocity, the changes seen here in the concentrated region of the cream are very significant. The variation in sound velocity with particle size was sufficient in this case to reverse the expected variation caused by changes in concentration (Figure 13.4). A more detailed discussion of the influence of particle size on ultrasound creaming measurements can be found in the previous work (Pinfield *et al.*, 1994) on the model, and in other papers concerned with the interpretation of ultrasound measurements (Pinfield *et al.*, 1995, 1996).

It should be noted here that these particle size distribution changes are predicted in a model emulsion in which no direct particle interactions or

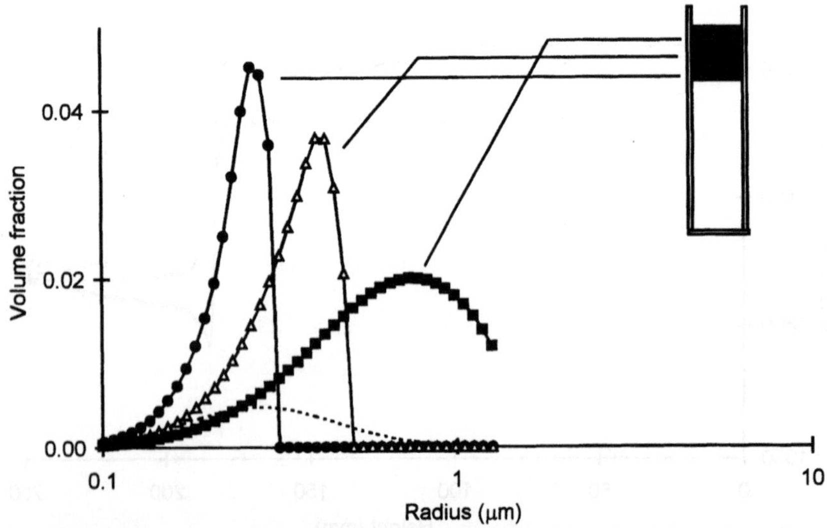

Figure 13.5 Particle size distribution in the cream region predicted by the macroscopic model for sunflower oil-in-water emulsion. The profiles are for a height of •, 200, △, 225, ■, 250 mm from the bottom; ———, is the initial distribution.

flocculation occur. There is evidence that in some circumstances the creaming of flocculated emulsions does not result in significant changes in particle size distribution in this way (Dickinson *et al.*, 1995). In addition, it is often possible to select a particle size range or ultrasound frequency where such size distribution changes will not affect the ultrasound measurements. Hence the effects of particle size changes on the ultrasound interpretation need not always be a problem.

13.4 A real creaming and flocculating sample

The macroscopic model of creaming has demonstrated the effects of a number of the physical processes in an ideal system in which there are no direct particle interactions. Examples of real measurements on a non-ideal system, made using the ultrasound profiling technique, will now be given. The sample was the same as that modelled (Table 13.1), i.e. 20 vol.% sunflower oil in water, with a mode particle diameter of 0.56 μm; however, the experimental system had 0.03 wt% xanthan gum added to the aqueous phase. Ma made the measurements (1995). The ultrasound velocity profiles were converted to volume fraction profiles by calibration against emulsions of known concentration. A selection of the profiles is shown in Figure 13.6.

It is immediately apparent that this system is much less stable than the modelled emulsion. After only half an hour a substantial serum region has developed (which still includes some oil), with a well-defined interface

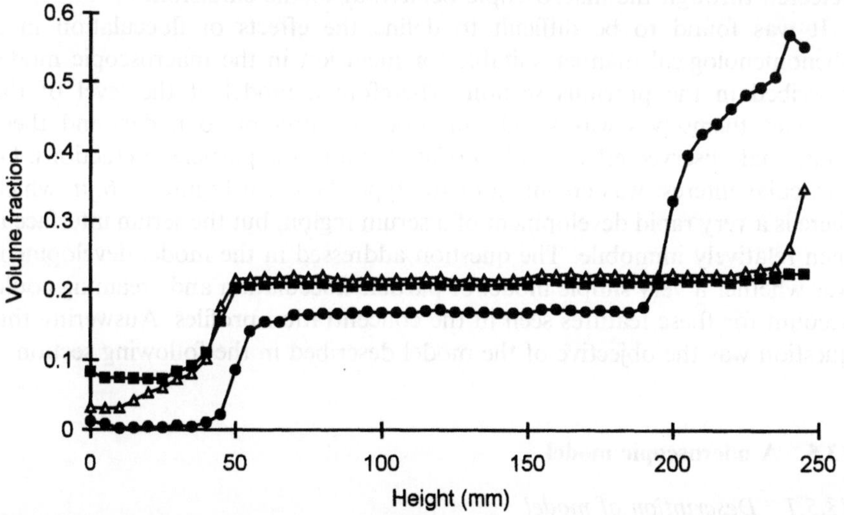

Figure 13.6 Volume fraction profiles measured by the ultrasound profiling technique for sunflower oil-in-water emulsion with xanthan added to the continuous phase. The profiles were measured after ■, 0.5; △, 1.5; h and •, 10 days.

between the serum and a region of uniform concentration. Once this serum has formed, the interface does not move up the sample as expected in an ideal system. Instead, the oil concentration in the serum gradually decreases until it is negligible, but with only slight changes in the serum interface position. The concentration in the middle region of the emulsion decreases in conjunction with the development of a cream layer. The concentration in the cream is lower than that predicted in the model for a non-flocculating sample, but increases towards the top of the sample as seen in the model results. The creaming behaviour observed in this sample is attributed to the formation of flocs which increase the creaming rate in the early stages, but which eventually form a particle network which prevents a significant amount of further creaming (Dickinson *et al.*, 1994, 1995).

The creaming behaviour illustrated in Figure 13.6 is just one example of the variety of characteristics that can be observed using the ultrasound technique. Povey (1997a) gives a wide range of examples of different creaming behaviour. In some cases a clear serum interface forms, in others it is very diffuse. In some samples an extensive cream develops but there is no apparent serum region, whereas other emulsions develop a serum without a noticeable cream. In others the initial creaming rate is very rapid. In each case, the deviation of the concentration profiles from the ideal non-flocculated creaming predictions is a result of the flocculation or particle interactions in the system. Understanding the relationship between the features observed in the creaming profiles and the underlying physical processes can lead to an improved insight into the interactions occurring within the emulsion. These processes are effective at the particle level, but can be detected through the macroscopic behaviour of the emulsion.

It was found to be difficult to define the effects of flocculation in a phenomenological manner suitable for inclusion in the macroscopic model described in the previous section. Therefore a model at the level of the particles themselves was sought in order to attempt to understand these additional observed effects and to relate them to the particle interactions. Of particular interest was creaming of the type shown in Figure 13.6, in which there is a very rapid development of a serum region, but the serum interface is then relatively immobile. The question addressed in the model development was whether a very simple model of particle flocculation and creaming could account for these features seen in the concentration profiles. Answering this question was the objective of the model described in the following section.

13.5 A microscopic model

13.5.1 Description of model

The model of creaming and flocculation at the level of the particles is a very simple one. It is based on a three-dimensional cubic lattice, with the side of each cube equal to one particle diameter (only monodisperse emulsions can be modelled in this way). Thus volume exclusion is satisfied and particles are modelled as hard spheres. Particles are only allowed to sit on the nodes of the lattice, as shown in Figure 13.7, and move about between these nodes according to their creaming speed and diffusion coefficient. The model is described in more detail elsewhere (Pinfield, 1996; Pinfield *et al.*, 1997).

In each of a sequence of time steps, each particle attempts to move upwards (for creaming) and in a random direction (for diffusion). Variable rates of movement are achieved by applying a probability to the attempted movement and comparing this probability each time with a randomly generated number. The movement can also be prevented if it is obstructed by another particle. Flocculation is modelled by allowing particles on neighbouring sites to stick together, with a predefined bonding probability. The bonds can also be allowed to break up to simulate reversible flocculation. As the number of particles in a floc changes, the creaming rate and diffusion coefficient are modified according to fractal scaling rules. As illustrated in Figure 13.7, flocs cream more quickly but diffuse more slowly than single particles. However, they are also more likely to suffer obstruction due to their increased collision cross-section. In this simplified model, no hydrodynamic interactions are included.

Although the model is very simplified, the parameters of the particles were chosen to be as close as possible to a real sample. Again, the properties of sunflower oil-in-water emulsion at 30°C were chosen. The necessary

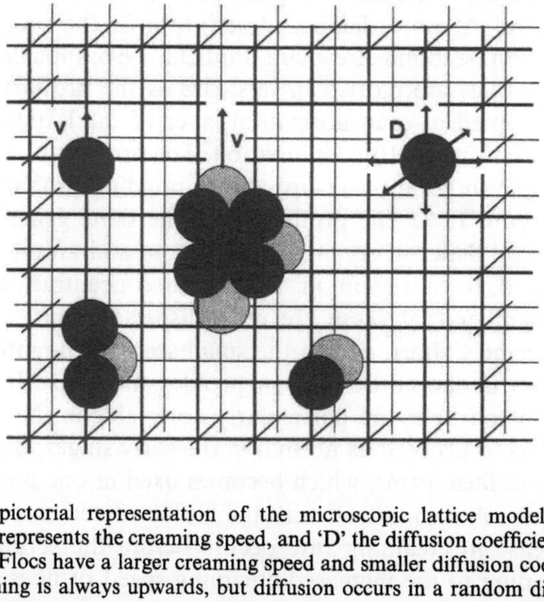

Figure 13.7 A pictorial representation of the microscopic lattice model of creaming and flocculation. 'v' represents the creaming speed, and 'D' the diffusion coefficient for the flocs and single particles. Flocs have a larger creaming speed and smaller diffusion coefficient than single particles. Creaming is always upwards, but diffusion occurs in a random direction.

Table 13.2 The parameters of the sunflower oil-in-water emulsion modelled using the microscopic creaming model

Proportion of lattice sites occupied by particles	10%
Particle diameter	1.2 μm
Density difference between sunflower and water	83 kg m^{-3}
Sample height	200 diameters
Horizontal dimensions	70×70 $diameters^2$
Temperature	30°C

parameters for the model are given in Table 13.2. Initially particles are randomly distributed on the lattice sites, so that 10% of the sites are occupied. For spherical particles this corresponds to a volume fraction of 0.052 (the maximum volume fraction is 0.52). A relatively large particle diameter (1.2 μm) was chosen for this model to avoid spending substantial amounts of computer resources modelling diffusive random walks without observing much creaming. The dimensions of the lattice are 70 × 70 × 200 sites, where the vertical is the largest dimension (equivalent to 98 000 particles at 10% occupation). Periodic boundary conditions were used in the two horizontal dimensions. The limited scale of this model is the penalty for modelling at the particle-level, since only small systems can be represented. The results are unlikely to be directly scaleable to macroscopic dimensions, but the features of the creaming behaviour will be the same.

13.5.2 Results of microscopic model

As with the macroscopic model, only a few results of the microscopic model are presented here. A more detailed description of the model and more extensive results can be found elsewhere (Pinfield, 1996; Pinfield *et al.*, 1997). Concentration profiles are plotted in this case as the proportion of lattice sites that are occupied in each horizontal layer of the lattice. Initially the concentration is uniform at 10% occupation. The profiles for an irreversibly flocculating sample (no floc break-up) with a bonding probability of 0.005 are shown in Figure 13.8. The profiles show the same general features as seen before – an oil-deficient serum at the bottom and an oil-rich cream at the top, separated by a region in which the concentration is virtually unchanged (remember that these are monodisperse emulsions). The time scale is now extremely short, due to the small scale of the model.

The differences in the concentration profiles that arise because of the flocculation of the particles are particularly noticeable in the serum region. Creaming appears to proceed as normal in the early stages, but then a well-defined serum interface forms, which becomes fixed in one position, as seen in the experiments. A comparison with the unflocculated case demonstrates that some increase in creaming rate occurs before the serum interface is fixed, corresponding to the increased creaming speed of flocs. However, as

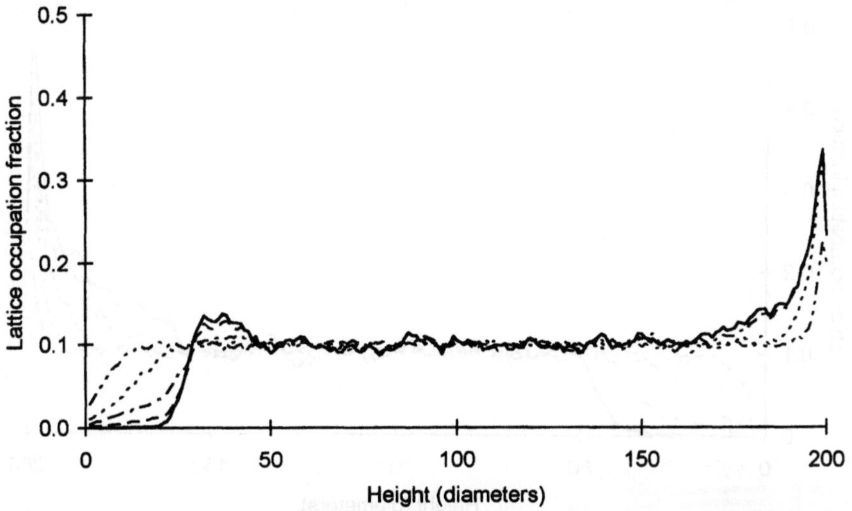

Figure 13.8 Concentration profiles predicted by the microscopic lattice model of creaming and flocculation for an irreversibly flocculated sample, with a bonding probability of 0.005. The profiles are after - - - -, 1 - - -, 2 — - —, 3 — — —, 5 and ——, ≥10 min. (After Pinfield *et al.*, 1996.)

flocculation becomes more extensive the flocs inter-link to form a particle network, and the creaming rate is dramatically reduced. This leads to the formation of the well-defined serum interface, which corresponds to the bottom of the networked region. Freely moving single particles and small flocs continue to move from the more dilute serum region into the network until the serum no longer contains any oil particles. The concentration of the cream is quite low, due to the 'freezing' of the structure by the formation of a particle network before the concentration has increased substantially. The position of the serum interface in this case is around 30 particle diameters from the bottom. Since flocculation is irreversible in this case, no further creaming can occur after all particles are attached to the network.

The effect of floc break-up (reversible flocculation) on the creaming process is illustrated by comparison with Figure 13.9. This is the same bonding probability as before (Figure 13.8), but with a break-up probability of 0.001. The effect of reversibility is to enable the flocs to stay smaller for a longer time, resulting in an extended period of relatively free and rapid creaming. The formation of a particle network is delayed, allowing a more extensive serum region, and a larger and more concentrated cream. In this case, creaming does not stop completely even after a particle network is formed, because particles can break off and move through the system. This is, however, a slow process, especially in comparison with the increased creaming rate in the early stages of flocculation.

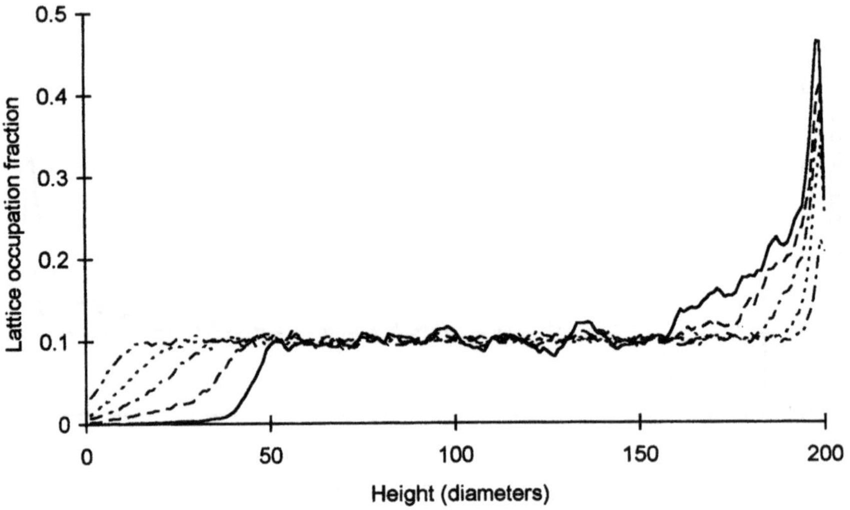

Figure 13.9 Concentration profiles predicted by the microscopic lattice model of creaming and flocculation for a reversibly flocculated sample, with a bonding probability of 0.005, and break-up probability of 0.001. The profiles are after ----—, 1 ---, 2 —-—, 3 ———, 5 and ——, 12 min. (After Pinfield *et al.*, 1996.)

The serum position after 12 min in the reversibly flocculated case has reached around 50 particle diameters from the bottom, compared with 30 diameters in the irreversibly flocculated case. This is an indication that the position of the serum interface observed in the experimental flocculating samples is strongly related to the energies of bonding and break-up. The serum position for four irreversibly flocculating systems modelled using the microscopic lattice model is plotted as a function of bonding probability in Figure 13.10. There is a clear relationship between the two quantities. As already demonstrated, floc break-up will also influence the serum position. Although it is not possible to draw quantitative conclusions from this microscopic model which are applicable to the macroscopic scale, we can deduce that the formation and position of the serum interface seen in experimental systems is indicative of the strength of flocculation. In a real system the flocculation is affected by dynamic, energetic and entropic factors.

Only very brief results have been presented here for the microscopic model. However, these have demonstrated that even this very simplified model of creaming and flocculation can account for some of the features observed in ultrasound measurements of creaming behaviour. In particular, the noticeable feature of an increased creaming rate, followed by the formation of a clearly defined serum interface, which remains fixed in one position, has been reproduced by the model. Unfortunately the dramatic rates of formation of this serum in experimental systems have not been

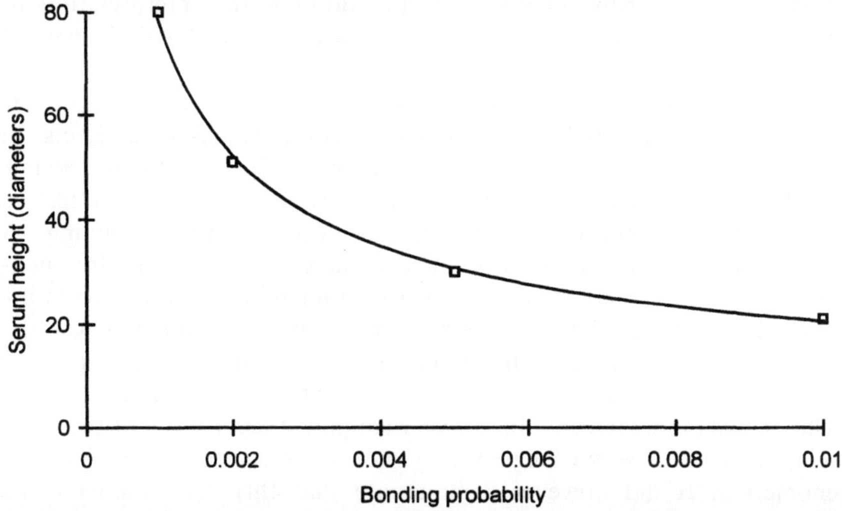

Figure 13.10 Final serum position as a function of bonding probability for irreversibly flocculating samples predicted by the microscopic lattice model of creaming and flocculation. The calculated values are shown by □, and the line is a fitted curve. (After Pinfield *et al.*, 1996.)

reproduced in the model. The degree of increased creaming in the model is significant but is not sufficient to explain the formation of a substantial serum in only half an hour, which has been observed experimentally. It is possible that flocs formed in a polydisperse system are more compact and have a larger creaming speed than that attributed to flocs in the model. Alternatively, the creaming rate in the model may be affected by some unrealistic obstruction in which flocs can be prevented from moving by single particles that lie in their path. Physically, these particles are likely to be swept along with the floc, but this is not a straightforward effect to include in such a simple model. In spite of these considerations, it may be that the model results demonstrate that a simple model of an increased creaming speed for flocs is insufficient to account for the rapid serum development, and a more complicated explanation must be sought.

13.6 Conclusions

This chapter has aimed to show how computer modelling of food systems can help in understanding their behaviour and properties. In particular, this has been related to the creaming properties of emulsions, which are observed using ultrasound profiling techniques. Through computer modelling it is possible to relate features of the measured velocity and concentration profiles to the fundamental processes, e.g. creaming of single particles, or irreversible flocculation. In addition, the interpretation of ultrasound measurements in the experiments was examined with the help of the model.

The models of creaming have attempted to relate the microscopic processes within the emulsion to the macroscopically observed effects. In this case, those processes were creaming, diffusion, hydrodynamic interactions and flocculation. The last of these can have a dramatic effect on the changes of concentration within the sample during creaming. A comparison between the results of the macroscopic model for non-interacting particles and the creaming behaviour of some real flocculating samples demonstrated the influence of flocculation on creaming. A test of a proposed explanation for these effects was achieved through the development of the microscopic model of creaming and flocculation. The results of this model indicate that a very simplified description of the rapid development of a serum region may not be sufficient to explain the phenomenon. It did however demonstrate that this basic model could account for the formation of a well-defined serum interface and some increase in creaming rate. The macroscopic model was also able to show the potential effects of particle size distribution changes on the ultrasound velocity measurements. The interpretation of such measurements in terms of oil concentration is applicable only in the absence of the effects of particle size on ultrasound velocity. This is one effect, which must be remembered when applying the ultrasound technique.

Finally, although ultrasound has been used to study the effects of particle interactions on creaming behaviour, we do not yet understand the influence of those interactions on ultrasound propagation itself. Experimental measurements and past experience have shown that the flocculation of a concentrated emulsion into a particle gel has little effect on ultrasound velocity in many circumstances. Hence, the usual interpretation methods for ultrasound velocity measurements remain valid. However, it may be that in some cases, ultrasound can be used as a probe of the particle interactions themselves, although the theoretical basis for this is currently rather weak (Povey, 1997a; McClements, 1994). This is perhaps an appropriate application for further modelling techniques.

Acknowledgements

This work was carried out with Professor Eric Dickinson and Dr Malcolm Povey and was funded by the BBSRC.

13.7 References

Allegra J.R. and Hawley S.A. (1972) Attenuation of sound in suspensions and emulsions: theory and experiments. *J. Acoust. Soc. Am.*, **51**, 1545–64.

Allen M.P. and Tildesley D.J. (1989) *Computer Simulation of Liquids*, Clarendon, Oxford.

Ansell G.C. and Dickinson E. (1986) Sediment formation by Brownian dynamics simulation: effect of colloidal and hydrodynamic interactions on the sediment structure. *J. Chem. Phys.* **85**, 4079–86.

Dickinson E., McClements D.J. and Povey M.J.W. (1991) Ultrasonic investigation of the particle size dependence of crystallization in *n*-hexadecane-in-water emulsions. *J. Coll. Int. Sci.*, **142**, 103–10.

Dickinson E., Goller M.I. and Wedlock D.J. (1993) Creaming and rheology of emulsions containing polysaccharide and non-ionic or anionic surfactants. *Coll. Surf. A*, **75**, 195–201.

Dickinson E., Kruizenga F.-J., Povey M.J.W. and van der Molen M. (1993) Crystallization in oil-in-water emulsions containing liquid and solid droplets. *Coll. Surf. A* **81**, 273–9.

Dickinson E., Ma J. and Povey M.J.W. (1994) Creaming of concentrated oil-in-water emulsions containing xanthan. *Food Hydrocolloids*, **8**, 481–97.

Dickinson E., Ma J., Pinfield V.J. and Povey M.J.W. (1995) Ultrasonic studies of the creaming of concentrated emulsions, in *Food Macromolecules and Colloids*, Editors E. Dickinson and D. Lorient, Royal Society of Chemistry, Cambridge, pp. 223–34.

Dickinson E., Golding M. and Povey M.J.W. (1997) Creaming and flocculation of oil-in-water emulsions containing sodium caseinate. *J. Coll. Int. Sci.*, **185**, 515–29.

Epstein P.S. and Carhart R.R. (1953) The absorption of sound in suspensions and emulsions I. Water fog in air. *J. Acoust. Soc. Am.*, **25**, 553–65.

Fikioris J.G. and Waterman P.C. (1964) Multiple scattering of waves II 'Hole corrections' in the scalar case. *J. Math. Phys.*, **5**, 1413–20.

Fillery-Travis A.J., Gunning P.A., Hibberd D.J. and Robins M.M. (1993) Coexistent phases in concentrated polydisperse emulsions flocculated by non-adsorbing polymer. *J. Coll. Int. Sci.*, **159**, 189–97.

Lloyd P. and Berry M.V. (1967) Wave propagation through an assembly of spheres IV Relations between different multiple scattering theories. *Proc. Phys. Soc.*, **91**, 678–88.

Ma J. (1995) Creaming and crystallisation in O/W emulsions containing xanthan. PhD thesis, University of Leeds.

McClements D.J. (1994) Ultrasonic determination of depletion flocculation in oil-in-water emulsions containing a non-ionic surfactant. *Coll. Surf. A*, **90**, 25–35.

McClements D.J. and Povey M.J.W. (1989) Scattering of ultrasound by emulsions. *J. Phys. D: Appl. Phys.*, **22**, 38–47.

Pinfield V.J. (1996) Studies of creaming, flocculation and crystallisation in emulsions: computer modelling and analysis of ultrasound propagation. PhD thesis, University of Leeds.

Pinfield V.J., Dickinson E. and Povey M.J.W. (1994) Modeling of concentration profiles and ultrasound velocity profiles in a creaming emulsion: importance of scattering effects. *J. Coll. Int. Sci.*, **166**, 363–74.

Pinfield V.J., Povey M.J.W. and Dickinson E. (1995) The application of modified forms of the Urick equation to the interpretation of ultrasound velocity in scattering systems. *Ultrasonics*, **33**, 243–51.

Pinfield V.J., Dickinson E. and Povey M.J.W. (1997) Modeling of combined creaming and flocculation in emulsions. *J. Coll. Int. Sci.*, **186**, 80–9.

Pinfield V.J., Povey M.J.W. and Dickinson E. (1996) Interpretation of ultrasound velocity creaming profiles. *Ultrasonics*, **34**, 695–8.

Povey M.J.W. (1997) Measuring aggregation in colloids using ultrasound velocity and attenuation, in *Food Colloids – Proteins, Lipids and Polysaccharides*, Editors E. Dickinson and B. Bergenstahl, Royal Society of Chemistry, Cambridge, pp. 150–67.

Povey M.J.W. (1997) *Ultrasound Techniques for Fluids Characterization*, Academic Press, San Diego, CA.

van der Knaap E., Vreeker R. and Hoekstra L.L. (1994) Effect of sedimentation on the formation of particle gels. A computer simulation study. *Coll. Surf. A*, **85**, 265–70.

Waterman P.C. and Truell R. (1961) Multiple scattering of waves. *J. Math. Phys.*, **2**, 512–37.

14 An introduction to ultrasonic food cutting

F.F. RAWSON

14.1 Introduction

Ultrasonic food cutting is yet another example of the very diverse applications of high-power ultrasonics to food production and food processing. Ultrasonic cutting has developed from thermoplastic cut/seal or welding technology. It has been extensively used in the aerospace industry since the 1970s for profile cutting glass and carbon fibre. Ultrasonic guillotining and slitting have gained production status in the food industry during the mid-1990s. This cutting process and equipment are thus well proven within the production requirements of food hygiene and typical factory environment of the industry. Ultrasonic cutting is one of the most important new fully commercial technologies introduced into the food processing industry in the 1990s, and has a far wider application than laser and water (oil) jet cutting technologies introduced in the 1980s.

Ultrasonic cutting uses knife blades or horns (probes) which vibrate at 20 kHz and are essentially 'tuning forks'. The cutting action is a combination of the sharp cutting edge and the mechanical vibration of the blade. The horn/probe equipment is very different from the cleaning bath technology or measurement equipment discussed extensively in other chapters. This new technique can be thought of as enhanced conventional guillotine cutting, with special cutting blades. The blade shapes and sizes are restricted. Ultrasonic cutting blades do contact the product, unlike laser beams which use light energy to vaporize the product. The blades are designed to vibrate in a longitudinal or piston mode, driven by an electromechanical transducer powered by, typically, a 2 kW tuned power supply, very similar to a plastic welding system.

14.2 Food cutting

Commercially, food of virtually every type is cooked or baked in bulk, large sheets, or blocks. To eat or sell the food, it must be reduced in size to portions sutiable for handling or eating. Many diverse terms (see Table 14.1)

Table 14.1 Cutting terms

Cutting	Slicing
Chopping	Dicing
Carving	Splitting
Slitting	Guillotining

are used to describe this essential processing step in food preparation. Many diverse and specialized knives and machines have been developed to make this job easy and effective, especially in large volume automated systems. The paramount requirements are a quick, clean, high-quality, undeformed cut product of visually good appearance with minimal wastage.

Poor cutting results in crushed, crumbled or torn product with significant waste. The multilayered or textured products are smeared and separated. Portions are misshapen, distorted, overweight or underweight.

The factors affecting conventional cutting processes are knife shape and sharpness, cleanliness, knife motion and speed through the product. Applied force is important. All these factors are easily adjusted by the operator and are monitored and controlled in automated cutting. Incorrectly set these variables and the knife jams in the product causing miscutting and a damaged product. All commercial products have to be conditioned or tempered to achieve a compromise between required taste, texture and visual quality and process or cutability. The limits imposed by this cutting process often adds a significant cost to production, or restrict the range of products which can be manufactured (see Table 14.2).

Table 14.2 Factors which affect the cutability or quality of the cut

- Temperature
- Moisture content
- Consistency: brittle, tough, fibrous, sticky

14.3 Ultrasonic food cutting

14.3.1 Benefits of ultrasonic food cutting

Ultrasonic food cutting has many major benefits for the food producer when incorporated into an automated food manufacturing line, or as a free-standing machine. Readily automated, the superb cut quality over a wide range of products is impressive. Table 14.3 lists various aspects of these benefits. These snapshot phrases are self-explanatory and are amplified in the text in other sections.

Table 14.3 Benefits of ultrasonic food cutting

- The quality of the cut face is visually excellent
- The product is virtually undisturbed
- The cutting force required is significantly reduced
- Hard particles can be cut in a soft matrix
- Smearing is reduced
- Multilayered products cut easily
- Blade is self-cleaning
- Ultrasonic cavitation effect kills bacteria
- Ultrasonics cut a wide variety of products
- Crumb and debris is significantly reduced
- Brittle products have less tendency to shatter
- It is a cost effective process with low running costs
- The blade sharpness requirement is less than for conventional blades
- Intervals between sharpening are longer than for conventional blades
- Ultrasonic cutting can be easily automated
- Cutting speeds are usually similar to conventional cutting speeds

14.3.2 Food products suitable for ultrasonic cutting

Most food products can be successfully cut by the ultrasonic cutting blade. The restrictions are primarily product shape and blade shape.

Many difficult-to-cut foods, such as hot bread, sticky confectionery, soft cream cakes, all can be cut successfully, or more successfully than by other methods, using the ultrasonic cutting blade. Table 14.4 lists some generic types of food products successfully being cut commercially. Pure ice and very high water content products are not currently successful. Brittle products cut better than with a conventional blade, but still may require careful preparation, e.g. part cooking.

14.3.3 Costs

It is inappropriate in a book of this type to give actual costs of equipment

Table 14.4 Types of food suitable for ultrasonic cutting

Bakery products	**Frozen products**	**Fresh products**
Bread	Cream cakes	Fish
Pastry	Pies	Meat
Pies	Ice cream	Vegetables
Cakes	Ice cream cakes	Bakery
Swiss rolls	Composite	Confectionery
Rich fruit cakes	Sorbets (not water ices)	Biscuits
Date and nut cakes		Cakes
Cream cakes		Bread
Tarts		
Lemon meringue pie		
Meringue		
Oatmeal biscuits		

because prices vary considerably due to specification, and new manufacturing techniques can change prices overnight. However, a price range can be given relative to other high-technology cutting techniques.

So the capital cost of a fully automated ultrasonic machine falls about halfway between the price of a water jet/laser cutting installation and a conventional cutting system of similar specification.

The operating and running costs will, however, be similar to or much better than the conventional system. No high-pressure system to maintain or laser vacuum tubes to replace, only the cutting blade, which is expensive, but generally has a longer life than the conventional blade. Better plant utilization and less stoppages can significantly reduce costs.

Significant cost savings on wastage and labour can pay back investment in, perhaps, 12–18 months.

14.4 The ultrasonic cutting process

14.4.1 High-power ultrasonic transmission

High-power ultrasonics is mechanical vibrational energy transmitted by metallic horns or probes which are tuned in multiples of half wavelengths. The vibrating horn can transfer the energy to solids, liquids or gases. Applied to solids, the energy can be used to cut, machine, grind or weld, depending on the output configuration. Applied to fluids the cavitation process discussed in other chapters arises. Applied to gases, a variety of processes such as defoaming or particulate clumping can occur (for background see Frederick, 1965).

The basic elements of an ultrasonic cutting system can be seen in Figure 14.1. The metallic high 'Q' horns or probes are specially shaped and frequency tuned as half wavelength longitudinal vibrators (see Figure 14.2). For cutting applications they are wide, thin bar horns with a profiled sharp cutting output tip. Typically the tip movement will be in the range of 50–100 μm peak to peak. The oscillation is sinusoidal. The knife horn or blade is screwed by a high tensile stud to the intermediate booster horn, which has a nodal or non-moving support flange. The booster horn is driven by a piezoelectric-electric multidisc clamped transducer. The transducer converts the input high voltage 20 kHz alternating signal into the mechanical vibration. These complex electromechanical devices are matched to a driving input power supply, which provides up to 2 kW of electrical power to drive the vibrating system.

High-power ultrasonics is widely used in industry and has no major health and safety concerns which are not already addressed in the existing machinery codes. Obviously, the cutting blade needs effective guarding and

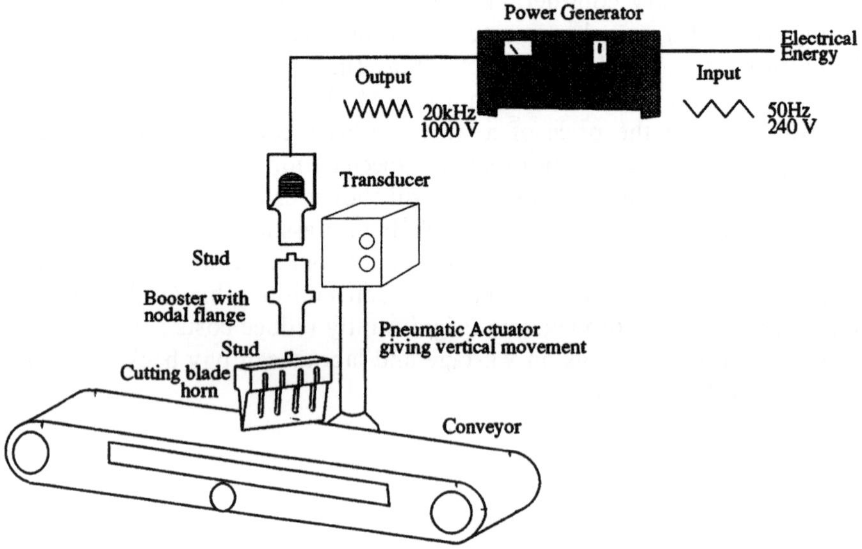

Figure 14.1 Elements of an ultrasonic cutting system.

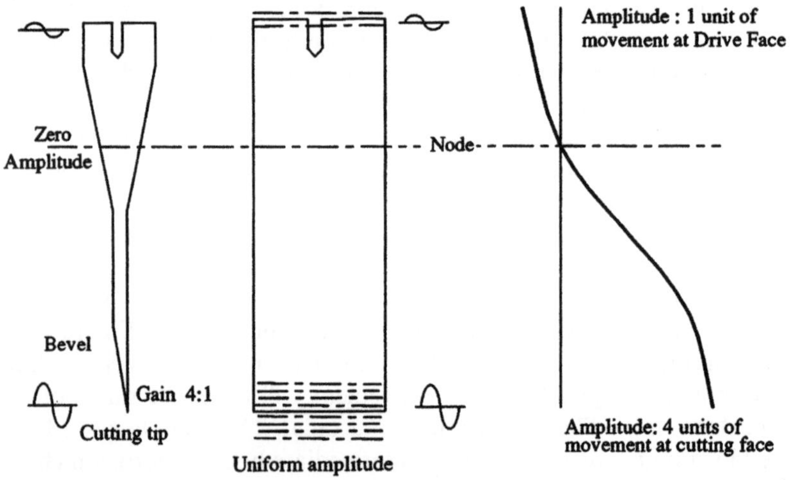

Figure 14.2 Resonant half-wavelength cutting blade horn at 20 kHz showing blade motion/movement with a tip amplitude typically 60 μ and the fall off typical at the nodal point.

the high-voltage electrical cables must be shielded and protected against damage and water.

Noise at 20 kHz is above the level of human hearing, however the sub-harmonics at 10 kHz must be treated as normal audio noise, and can easily be shielded by conventional techniques.

14.4.2 *Ultrasonic cutting*

Ultrasonic cutting, guillotining or slitting, can be thought of as an enhanced conventional cutting process. The blade has a cutting edge and a shank. The ultrasonic vibration of 20 kHz moves the cutting tip rapidly forward and back, superimposed on a forward movement of perhaps 5 m/min. This oscillating movement applies an intermittent cutting force to the crack tip and so controls this crack (cut) propagation or growth minimizing the stress on the bulk material. The repeated application of the cutting tip to the product applies a local fatiguing effect which reduces significantly the overall required force to break the bonds of the bulk material. A useful analogy is the hammer drill or the pneumatic road drill concepts. A second aspect to ultrasonic cutting is the friction reduction of the blade shank as it follows the tip into the opening cut. In conventional cutting the blade has to compress the bulk material to allow a gap the width of the blade to pass through; the compression to the bulk applies a tensile rupturing force at the crack tip. With ultrasonic cutting the whole blade moves or vibrates continuously (see Figure 14.2) as it stretches and contracts. This very high-frequency movement effectively reduces the co-efficiency of friction to a very low level, enabling the blade to slide through the bulk material with considerable reduction to the compression and contact, giving a force reduction and decreased smearing effects.

14.4.3 *Ultrasonic parameters*

Table 14.5 lists the ultrasonic parameters which control the process and are very similar to those which control all high power ultrasonic processes. For further details see texts on plastic welding (e.g. Rawson, 1989, 1994, 1996a).

14.4.4 *Other ultrasonic cutting techniques*

The equipment used to apply high-power ultrasonics is very varied. The techniques outlined below are used in cutting/grinding operations in other industries. They are included, as variations of them will no doubt be adapted for food processing over the next decade.

Table 14.5 Ultrasonic parameters controlling cutting

- Blade tip amplitude
- Blade profile
- Cutting speed
- Power
- Frequency
- Product texture
- Blade shape

14.4.4.1 Medical and microsurgery – ultrasonic scalpel. This is an ultrasonic dissector not a knife. The 20 kHz ultrasonic probe uses ultrasonic cavitation forces generated in fluids to destroy cancerous water-filled tissue. Muscles and healthy cells have little or much less water and so do not absorb the ultrasound at 20 kHz. The technique is thus self-focusing and only damaged tissue is destroyed. It is not cutting. Water is flushed down the outside of the probe and the water and debris is sucked up the high-amplitude tip which is hollow.

14.4.4.2 Ultrasonic machine (USM). This is a plunge impact grinding or erosion technique using a three-dimensional profile on the end of an ultrasonic horn and a slurry of hard abrasive particles in water. Cavitation has a part to play in the process. The process is used on glasses and ceramics.

14.4.4.3 Ultrasonic rotary machine (USRM). USRM is a version of USM (see above). The cutting tool, an ultrasonic cylindrical horn, is coated with a diamond or similar coating and rotated at, typically 3000 rpm, and used as a milling or drilling tool for hard brittle material and glasses and ceramics.

14.4.4.4 Loom slitter – scan welding. Thin material such as fabric, can be cut or slit using a flat horn transmitting the ultrasonics through the thin material on a sharp fixed anvil or wheel. The rim or edge of the anvil is profiled. By adjusting the profile, the cut seal is optimized for a particular fabric thickness and type. A pattern can be engraved on wider anvils and transferred to the fabric. This process is only suitable for thin materials a few millimetres thick.

14.5 Machines and equipment

Successful commercial ultrasonic cutting is achieved by designing a food cutting system round the ultrasonic cutting blade into a cutting head or a package with the food-handling requirements. Machine systems are most successful when built for a specific product or product range with well-defined limits on the size, texture and temperature.

Five typical machine types are discussed below to give an idea of versatility of the ultrasonic approach to cutting. It is not exclusive, and at present most applications tend to be bespoke systems with a significant client specification input.

14.5.1 Guillotine-cutting conveyor machines

The ultrasonic guillotine conveyor machine is the basic module which can be used as either a free-standing machine or be incorporated as a cutting station into a fully automated line. Figures 14.3 and 14.4 show a twin-track, twin-head, with double half wavelength cutting horns giving 180 mm cutting depth. Each blade is 145 mm wide. The three in-line quick-release belts comprise an input station feeding a measuring and cutting section and an output station. A microprocessor is used to control the belts and cutting cycle, with input sizing data from sensors. The construction of food processing equipment must be easy to clean with a minimum of operator adjustments. A microprocessor enables the complex variables to be stored and easily adjusted at the setting up stage, yet gives simple on/off cut control when in production.

The ultrasonic cutting heads are moved vertically by variable stroke and speed pneumatic cylinders with accurate position control. When building the ultrasonic cutting machine, a well-defined product specification is required and it is important to set up and test the product to be cut in the same condition as it will be cut in the factory. Temperature and ageing of food products causes significant textural changes which dramatically alter the cutting characteristics.

Many types of conveyor belt can be used as appropriate to feed and handle the particular food product. Guides, pushers and flights are just a few typical aids used to handle the product.

The aim is always to minimize product damage, yet provide enough feed or cutting force reaction to hold the product securely during the cutting operation.

Round products, cakes and pies, can also be sectioned on this type of machine with a series of extra conveyor modules each with a cutting head set at the required angle, e.g. eight portions require four cutting stations set at 45°.

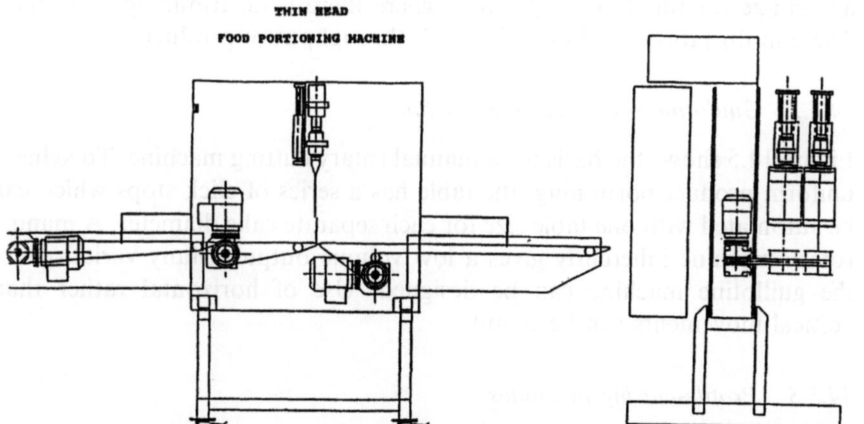

Figure 14.3 Twin-track, twin-head guillotine cutting machine.

Figure 14.4 Twin-track, twin-head guillotine cutting machine type FRGK 9608-2B145-2HWL-4C. (A) cutting blades, (B) twin infeed conveyors, (C) photocell, (D) conveyor tie plates, (E) twin outfeed conveyors. Guards have been removed to show parts.

Positioning the cake or pie to achieve a perfect point on each portion is complex, but easier than on a conventional cutting machine. The high-quality ultrasonic cutting, with little or no debris, has a significant advantage for this type of product where there is no trimming allowance. These multistation machines give high throughput of product.

14.5.2 Guillotine rotary cutting machines

Figure 14.5 shows the basis for a manual rotary cutting machine. To achieve uniform product portioning, the table has a series of click stops which can be automated with one table size for each separate cake diameter. A manual rotary machine inherently gives a low volume output. Many variations of the guillotine machine can be designed. Use of horizontal rather than vertical movements can be useful.

14.5.3 Profile-cutting machines

Three basic machine concepts are available. The moving crosshead *X–Y* machine, the robotic arm and a plunge cutting machine with a two-dimensional shaped 'pastry' cutter. The moving crosshead and robotic arm

Figure 14.5 Manual rotary guillotine cutting machine.

machines are widely used in other industries, including aerospace, automotive and packaging.

The scalpel blade steel or carbide tip (Figure 14.6) is moved by a computer control programme and rotated by a stepper motor to edge cut

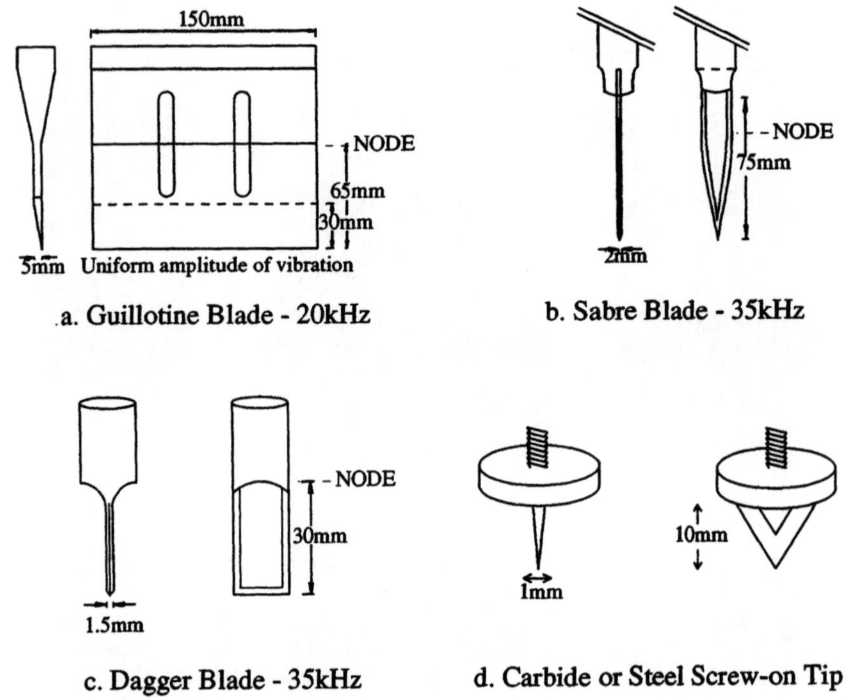

Figure 14.6 Ultrasonic cutting blade types.

the required profile. Depth is limited to 10 mm. The sabre blade, a full wavelength knife, has a greater cutting depth (see Section 14.6.1). Edge following is relatively slow but capable of producing complex shapes, probably of fairly limited application in the food industry but useful for decoration products.

The profile-shaped cutting horn used in a plunge mode like a pastry cutter can be fitted to the guillotining machines or the robotic *X–Y* table giving faster output as the full shape is cut each stroke.

The available shapes and depth cut-able are likely to limit applications initially, but no doubt a series of standard profile horns will be developed if there is a demand.

14.5.4 Slitting and multiple blade systems

Slitting wide products into thin fingers 10–25 mm wide is a common requirement, and can be achieved with the ultrasonic dagger or sabre blades. Product height is limited to perhaps 30 mm, due to the fall off of amplitude at the nodal point on the blade. Cut quality is very dependent on the amplitude requirement to cut a specific food product.

The static blades are each driven by a separate ultrasonic transducer and power supply. The blades can be mounted in-line or in a chosen formation.

To increase the cutting depth, a guillotine blade can be mounted at 45° to give the full cutting amplitude over the full product height (Figure 14.6). The drawback of using guillotine blades is the thickness of the blade, which is 3–5 mm.

Using a separate ultrasonic set for each blade, any number of blades can be mounted across a wide conveyor plant. This gives optimum versatility for setting up and plant maintenance. A worn blade and head can be replaced in a minute or so.

14.5.5 Two-axis portioning machines

Slab products are often up to $1 \times 1.5\,m^2$ in size, and need to be reduced to small portions, typically $80 \times 20\,mm^2$. A combination guillotine and multiple-blade slitting machine can be designed to give the optimum configuration of ultrasonic cutting heads. Often two or three guillotine blades up to 300 mm will be placed side by side. All these machines are built up based on a series of repeat modules.

14.6 Ultrasonic knife blades – longitudinal vibration

In Section 14.4 the longitudinal mode of vibration is described. All the knife blade horns discussed in this section are designed to vibrate in this mode. The amplitude at the tip is a maximum and gives the best cut.

The knife blades are primarily made of titanium alloy 6Al4Va grade. This material will run at 20–40 kHz without self-generation of heat, and so maintain the correct tuned resonance. At the high peak-to-peak amplitude of 50–80 μm no other material has the required ultrasonic fatigue and transmission properties.

14.6.1 Sabre and scalpel blades

Figure 14.2 shows the vibrational mode of the longitudinal blade. It should be noted that the maximum amplitude is at the cutting tip and this blade is ideal for plunge or guillotine cutting. The cutting face can be placed at an angle of typically 45° for use in a slitting action or the 10–20 mm of the blade side edge can be sharpened to give a slitting action. (Figure 14.6b: the sabre blade or similar.)

14.6.2 Guillotining blades

The longitudinal knife blade horn is a thin complex vibrating blade tuned to 20 kHz. This frequency is used to give the maximum cut depth for a single half wavelength, which for titanium alloys at 20 kHz, subject to blade profile, is 140 mm, giving cutting depth of 25–65 mm with cutting section of blade between 3 and 5 mm sharpened to a knife edge (Figure 14.6a). Horns can be up to 300 mm wide with a 25-mm cut depth, reducing to perhaps 150 mm wide for a 65-mm cut depth. Using a full wavelength horn a 145-mm wide cut depth can be increased to 150 mm. These dimensions are all subject to different maximum amplitudes, operating and loading cycles and details are intended to be a guide only. An incorrectly tuned horn will severely damage both the transducer and power generator. It is important to refer to the manufacturer's operating instructions. Under no circumstances can the blade be altered or reshaped without retuning.

The ultrasonic knife is a complex device that gives superb cutting of food products. It is specially designed to vibrate in a single mode and consequently has discrete size ranges. The cutting machines are therefore designed round a specific blade.

14.6.3 Two-dimensional ultrasonic profile cutting

The ultrasonic cutting blades mentioned so far are all for one-dimensional cutting using two-dimensional flat blades for straight cuts. The cutting tip of a pointed sabre or dagger blade can be used for profile cutting on *X–Y* tables controlled by computer (Figure 14.6c, d).

Ultrasonic profile or edge cutting is widely used for glass-fibre material for making aerospace canopies for aircraft and for sampling and short-run cardboard packaging. Figure 14.7 shows a the variety of guillotine and sabre blades that can be produced.

A further type of cutting blade can be made as mentioned previously – the 'pastry cutter' (Figure 14.8). This is used in a plunge mode and can cut a round disc in a single stroke. It is much quicker than edge following. After cutting, a pulse of ultrasound ejects the cut portion. Design of profile cutters is complex, and available shapes are likely to be very restricted because of the vibrational resonance required.

14.7 Future developments

14.7.1 Radial cutting blade

The radial cutting blade offers very substantial benefits and is currently the most significant development to follow the longitudinal cutting blade, as shown in Figures 14.9 and 14.10. The radial blade gives 360° of cutting approach, with uniform amplitude along all the cutting edges and no nodal point; i.e. no point of zero vibration.

The radial mode of vibration is maintained in blades as thin as 1 or 2 mm thick. This offers cutting and slitting with blades similar to conventional

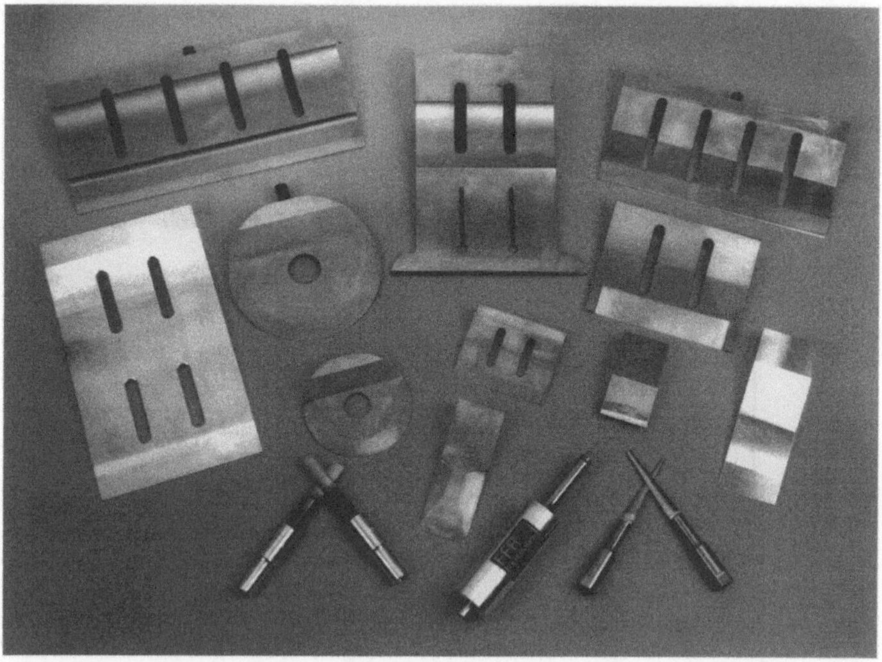

Figure 14.7 Cutting blade horns in commercial use.

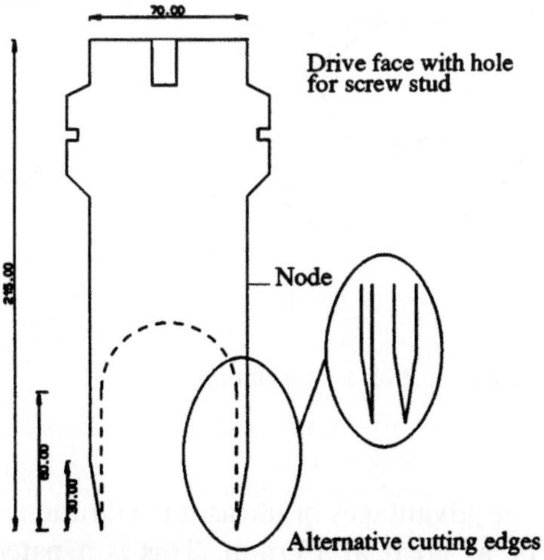

Figure 14.8 The ultrasonic 'pastry cutter'.

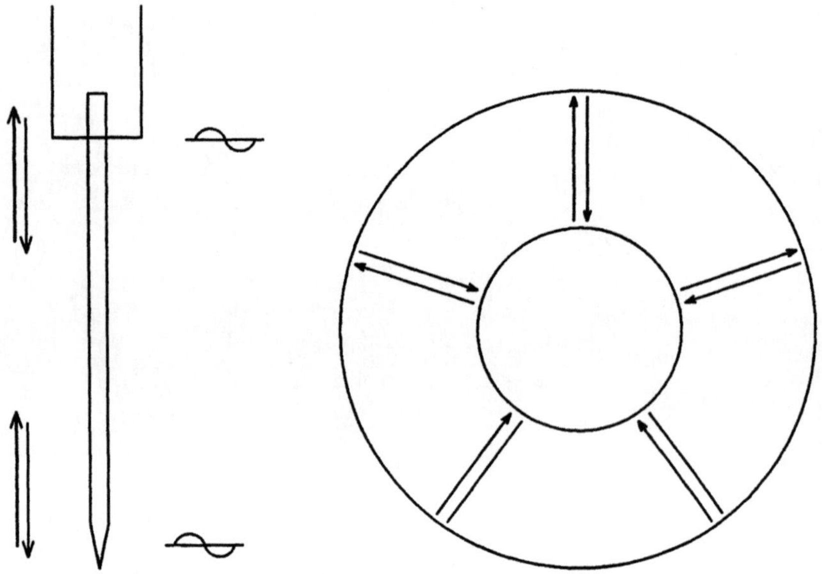

Figure 14.9 Vibrational modes in a radial blade.

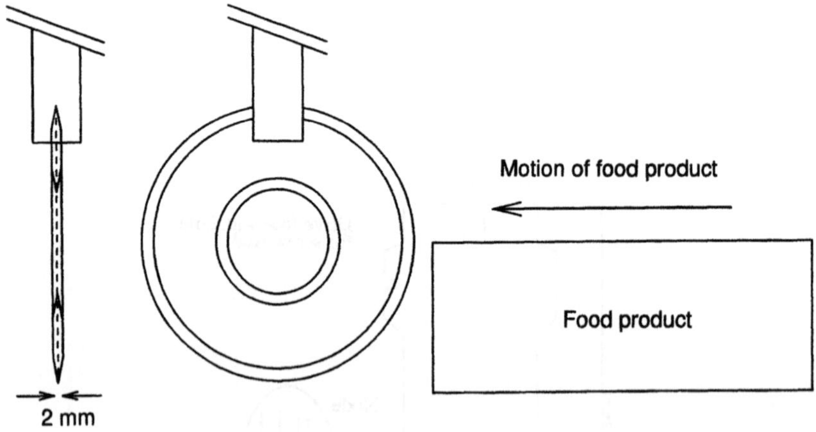

Figure 14.10 The radial cutting blade as a slitting blade.

cutting with all the advantages of ultrasonic vibrations at 20 or 35 kHz. The cut depth can be as great at 150 mm. This is a patented process (British Patent, 1993).

14.7.2 One driving source multivibrator systems

A search of the 1980s and 1990s patent literature shows significant interest in ultrasonic cutting techniques. Multibladed systems feature highly [see British Patent (1988)]. The lack of commercially viable installations shows the complexities of these proposed systems. The interaction of five or 10 resonant blades driven by one source causes unwanted extra vibration due to small variations in construction.

14.7.3 Blade developments

The key to ultrasonic cutting is the knife blade and its interaction with the food product. Extensive use of finite element modelling is being used to develop ever more versatile cutting systems. New fabrication techniques, materials and coating developments will all extend the process over the next decade.

14.8 Conclusions

Ultrasonic cutting, well established in many industries, is now fully commercial in the food industry. A wide range of machines is readily available to produce superb cut quality portioning of food over a wide range of products. It should be noted that much of the technology is subject to UK and other patents. New developments and applications are emerging every few months, and many existing machines can be modified to fit a specific requirement. The scope and potential of ultrasonic cutting is very exciting. There is a significant opportunity for the food industry to develop new products that previously could not be cut (see Rawson, 1996).

Acknowledgements

The author and editors would like to thank FFR Ultrasonics Ltd for permission to publish the photographs and technical information discussed.

14.9 References and bibliography

British Patent (1988) No. 8813089.3.
British Patent (1993) No. PCT/GB93/00159.
Frederick J.R. (1965) *Ultrasonic Engineering*, John Wiley.
Rawson F.F. (1989) *TWI Conference*, Abingdon, Cambridge, September.
Rawson F.F. (1994) *British Plastics and Rubber*, May, 22–5.
Rawson F.F. (1996a) *British Plastics and Rubber*, November, 14.
Rawson F.F. (1996b) *British Baker*, 12 July, 24–5.

Index

MIX
Papier aus verantwortungsvollen Quellen
Paper from responsible sources
FSC® C105338
FSC
www.fsc.org

If you have any concerns about our products,
you can contact us on
ProductSafety@springernature.com

In case Publisher is established outside the EU,
the EU authorized representative is:
Springer Nature Customer Service Center GmbH
Europaplatz 3, 69115 Heidelberg, Germany

Printed by Libri Plureos GmbH
in Hamburg, Germany